AF716946

The Ecology and Management of Invasive *Prosopis* Trees in Eastern Africa

The Ecology and Management of Invasive *Prosopis* Trees in Eastern Africa

Edited by

Urs Schaffner

CABI, Delémont, Switzerland

Brian W. van Wilgen

Centre for Invasion Biology, Department of Botany and Zoology, Stellenbosch University, South Africa

Albrecht Ehrensperger

Centre for Development and Environment (CDE), University of Bern, Switzerland

Ketema Bekele

School of Agricultural Economics, Haramaya University, Dire Dawa, Ethiopia

CABI

CABI is a trading name of CAB International

CABI
Nosworthy Way
Wallingford
Oxfordshire OX10 8DE
UK

Tel: +44 (0)1491 832111
E-mail: info@cabi.org
Website: www.cabi.org

CABI
200 Portland Street
Boston
MA 02114
USA

T: +1 (617)682-9015
E-mail: cabi-nao@cabi.org

©2025 CAB International. [The Ecology and Management of Invasive *Prosopis* Trees in Eastern Africa] is licensed under a Creative Commons Attribution-NonCommercial-NoDerivatives 4.0 International License. All rights, including for text and data mining, AI training, and similar technologies, are reserved.

The views expressed in this publication are those of the author(s) and do not necessarily represent those of, and should not be attributed to, CAB International (CABI). Any images, figures and tables not otherwise attributed are the author(s)' own. References to internet websites (URLs) were accurate at the time of writing.

CAB International and, where different, the copyright owner shall not be liable for technical or other errors or omissions contained herein. The information is supplied without obligation and on the understanding that any person who acts upon it, or otherwise changes their position in reliance thereon, does so entirely at their own risk. Information supplied is neither intended nor implied to be a substitute for professional advice. The reader/user accepts all risks and responsibility for losses, damages, costs and other consequences resulting directly or indirectly from using this information.

CABI's Terms and Conditions, including its full disclaimer, may be found at https://www.cabidigitallibrary.org/terms-and-conditions.

A catalogue record for this book is available from the British Library, London, UK.

ISBN-13: 9781800623620 (hardback)
9781800623637 (OA ePDF)
9781800623644 (OA ePub)

DOI: 10.1079/9781800623644.0000

Commissioning Editor: David Hemming
Editorial Assistant: Helen Elliott
Production Editor: James Bishop

Typeset by Exeter Premedia Services Pvt Ltd, Chennai, India
Printed and bound in the UK by CPI Group (UK) Ltd, Croydon, CR0 4YY

CABI INVASIVES SERIES

Invasive species are plants, animals or microorganisms not native to an ecosystem, whose introduction has threatened biodiversity, food security, health or economic development. Many ecosystems are affected by invasive species and they pose one of the biggest threats to biodiversity worldwide. Globalization through increased trade, transport, travel and tourism will inevitably increase the intentional or accidental introduction of organisms to new environments, and it is widely predicted that climate change will further increase the threat posed by invasive species. To help control and mitigate the effects of invasive species, scientists need access to information that not only provides an overview of and background to the field, but also keeps them up to date with the latest research findings.

This series addresses all topics relating to invasive species, including biosecurity surveillance, mapping and modelling, economics of invasive species and species interactions in plant invasions. Aimed at researchers, upper-level students and policy makers, titles in the series provide international coverage of topics related to invasive species, including both a synthesis of facts and discussions of future research perspectives and possible solutions.

Titles Available

1. Invasive Alien Plants: An Ecological Appraisal for the Indian Subcontinent
   Edited by J. R. Bhatt, J. S. Singh, R. S. Tripathi, S. P. Singh and R. K. Kohli

2. Invasive Plant Ecology and Management: Linking Processes to Practice
   Edited by T. A. Monaco and R. L. Sheley

3. Potential Invasive Pests of Agricultural Crops
   Edited by J. E. Peña

4. Invasive Species and Global Climate Change
   Edited by L. H. Ziska and J. S. Dukes

5. Bioenergy and Biological Invasions: Ecological, Agronomic and Policy Perspectives on Minimizing Risk
   Edited by L. D. Quinn, D. P. Matlaga and J. N. Barney

6. Biosecurity Surveillance: Quantitative Approaches
   Edited by F. Jarrad, S. L. Choy and K. Mengersen

7. Pest Risk Modelling and Mapping for Invasive Alien Species
   Edited by Robert C. Venette

8. Invasive Alien Plants: Impacts on Development and Options for Management
   Edited by Carol Ellison, K. V. Sankaran and Sean Murphy

9. Invasion Biology: Hypotheses and Evidence
   Edited by Jonathan M. Jeschke and Tina Heger

10. Invasive Species and Human Health
    Edited by Giuseppe Mazza and Elena Tricarico

11. Parthenium Weed: Biology, Ecology and Management
Edited by Steve Adkins, Asad Shabbir and Kunjithapatham Dhileepan

12. Biological Control in Latin America and the Caribbean: Its Rich History and Bright Future
Edited by Joop C. van Lenteren, Vanda H.P . Bueno, M. Gabriela Luna and Yelitza C. Colmenarez

13. Plant Invasions: The Role of Biotic Interactions
Edited by Anna Traveset and David M. Richardson

14. Invasive Species and Global Climate Change, 2nd Edition
Edited by L. H. Ziska

15. Parasites and Biological Invasions
Edited by Jamie Bojko, Alison Dunn and April Blakeslee

16. The Ecology and Management of Invasive *Prosopis* Trees in Eastern Africa
Edited by Urs Schaffner, Brian W. van Wilgen, Albrecht Ehrensperger and Ketema Bekele

Contents

Contributors

Beatrice Adoyo – Department of Geography, Population, and Environmental Studies, University of Nairobi, Kenya. Email: beatriceadoyo01@gmail.com. ORCID: 0009-0007-8736-2539.

Tena Alamirew – Water and Land Resource Centre, Addis Ababa University, Ethiopia. Email: tena.a@wlrc-eth.org. ORCID: 0000-0001-7491-4401.

Ketema Bekele – School of Agricultural Economics, Haramaya University, Dire Dawa, Ethiopia. Email: ketmimi@yahoo.com. ORCID: 0000-0002-8107-4807.

María L. Castillo – Institute of Botany of the Czech Academy of Sciences, Průhonice, Czech Republic. Email: mloretocastillo@gmail.com. ORCID: 0000-0003-1228-172X.

Simon K. Choge – Kenya Forestry Research Institute, Marigat, Kenya. Email: skchoge2002@gmail.com.

Sebinasi Dzikiti – Department of Horticultural Science, Stellenbosch University, South Africa. Email: sdzikiti@sun.ac.za. ORCID: 0000-0002-1892-4143.

Sandra Eckert – Centre for Development and Environment, University of Bern, Switzerland; Institute of Geography, University of Bern, Bern, Switzerland. Email: sandra.eckert@unibe.ch. ORCID: 0000-0002-9579-5680.

Albrecht Ehrensperger – Centre for Development and Environment (CDE), University of Bern, Switzerland. Email: albrecht.ehrensperger@unibe.ch. ORCID: 0000-0002-2880-4049.

Sanford D. Eigenbrode – Department of Entomology, Plant Pathology and Nematology, University of Idaho, Moscow, USA; Email: sanforde@uidaho.edu. ORCID: 0000-0003-0054-8511.

René Eschen – CABI, Rue des Grillons 1, Delémont, Switzerland. Email: r.eschen@cabi.org. ORCID: 0000-0002-0464-6680.

Staline Kibet – Department of Resource Management and Agricultural Technology, University of Nairobi, Kenya. Email staline@uonbi.ac.ke. ORCID: 0000-0002-0376-6111.

Charles J. Kilawe – Department of Ecosystems and Conservation, Sokoine University of Agriculture, Morogoro, Tanzania. Email: ckilawe@sua.ac.tz. ORCID: 0000-0002-3504-3014.

Boniface Kiteme – Centre for Training and Integrated Research in ASAL Development (CETRAD), Nanyuki, Kenya. Email: b.kiteme@africaonline.co.ke. ORCID: 0009-0006-7608-4740.

Johannes J. Le Roux – School of Natural Sciences, Macquarie University, Sydney, Australia; Centre for Invasion Biology, Department of Botany and Zoology, Stellenbosch University, South Africa. Email: jaco.leroux@mq.edu.au. ORCID: 0000-0001-7911-9810.

Theo E.W. Linders – Management, Nature and Landscape Department, Central Government Real Estate Agency of the Dutch Ministry of the Interior and Kingdom Relations, The Hague, The Netherlands. Email: theolinders@gmail.com. ORCID: 0000-0002-4930-9004.

Purity R. Mbaabu – Institute for Climate Change and Adaptation, University of Nairobi, Kenya. Email: purityrima@yahoo.com. ORCID: 0000-0003-3334-5278.

John R. Mbwambo – Tanzania Foretsry Research Institute, Morogoro, Tanzania. Email: jorijomb@yahoo.com. ORCID: 0000-0002-9146-530X.

Linus K. Munishi – Landscape and Conservation Ecology, Nelson Mandela African Institute of Science and Technology, Arusha, Tanzania. Email: zt.ca.tsia-mn@ihsinum.sunil.

Mickfanaka Mwihomeke – Department of Land Resource Management and Agricultural Technology, University of Nairobi, Kenya. Email: fanakae@gmail.com.

Wai-Tim Ng – Institute for Surveying, Remote Sensing and Land Information, University of Natural Resources and Life Sciences, Vienna, Austria. Email: wai-tim.ng@vito.be. ORCID: 0000-0002-6926-1546.

Michael O'Rourke – Toolbox Dialogue Initiative Center, Michigan State University, East Lansing, USA; Department of Philosophy, Michigan State University, East Lansing, USA. Email: orourk51@msu.edu. ORCID: 0000-0002-4629-0811.

David M. Richardson – Centre for Invasion Biology, Department of Botany and Zoology, Stellenbosch University, South Africa. Email: rich@sun.ac.za. ORCID: 000-0001-9574-8297.

Urs Schaffner – CABI, Rue des Grillons 1, Delémont, Switzerland. Email: u.schaffner@cabi.org. ORCID: 0000-0002-3504-3014.

Ross T. Shackleton – Swiss Federal Institute for Forest Snow and Landscape Research, Birmensdorf, Switzerland; Centre for Invasion Biology, Department of Botany and Zoology, Stellenbosch University, South Africa. Email: rtshackleton@gmail.com. ORCID: 0000-0001-5628-4506.

Hailu Shiferaw – Water and Land Resource Centre, Addis Ababa University, Ethiopia. Email: hailu2nd@gmail.com. ORCID: 0000-0002-3697-083X.

Brian W. van Wilgen – Centre for Invasion Biology, Department of Botany and Zoology, Stellenbosch University, South Africa. Email: bvanwilgen@sun.ac.za. ORCID: 0000-0002-1536-7521.

Preface and Acknowledgements

Charles S. Elton's *The Ecology of Invasions by Animals and Plants*, published in 1958, is widely regarded as the work that laid the foundation for what we know today as the science of biological invasions. Over the succeeding decades, our understanding of these invasions, and their consequences, has grown substantially. Despite this, large gaps still exist due to taxonomic and geographic biases in the research that has been carried out globally. For example, a review of the field in 2008 established that 2670 research papers had been published in the peer-reviewed literature since 1980, but that only 3.4% of these were from Africa, with three-quarters of those being from a single country, South Africa. This dearth of relevant understanding has dire consequences for the management of a significant environmental problem on the African continent.

The work reported in this book is a modest attempt to address some of that problem. Over the past 10 years, we have been engaged in a large, collaborative research project that examined the ecology and management of alien *Prosopis* trees in eastern Africa. These trees, originally introduced from the Americas to provide a range of benefits, later became invasive and impacted negatively on biodiversity, water resources, rangeland productivity, animal and human health, and the wellbeing of people in the region. The problem is complex because some people benefit while others carry the cost. It is complex because net values change rapidly over time, but people's perceptions are slower to change, and it is complex because there are no clear solutions. If the problem is not effectively managed, the people of the region will carry the significant costs of these invasions.

Much of the information generated over the past decade is scattered in numerous scientific publications in a wide range of journals, many of them not readily accessible. Many of these publications have a relatively narrow focus, and bringing these insights together will provide a more complete and potentially useful treatment of the topic. In this book, we address how the problem came about, what we know about how it works, and how it can potentially be managed at different scales. The book has a focus on eastern Africa but where appropriate, it incorporates research findings from elsewhere in Africa, and globally, where such inclusion will strengthen the degree of confidence that we have when interpreting lessons.

To the best of our knowledge, this is the first regional synthesis in book form of all aspects of the biology, invasion history, impact and management of trees in the genus *Prosopis*, one of the world's most important invasive taxa. It is our hope that the material presented here will provide students, ecologists, farmers, conservationists, aid agencies, managers and policy makers with the information they need to address the problem more effectively. To increase the chances of reaching this wide potential audience, we have chosen to publish the book as an open access resource. The publication of a synthesis of recent work in eastern Africa in an easily accessible and affordable format should allow many readers to gain insights to the full spectrum of issues, and in so doing help them to formulate appropriate policy and management responses.

Many people have helped us to bring this book to fruition. We would like to thank all the contributing authors for their inputs, many of whom had to find time for writing and revisions while simultaneously coping with their demanding day jobs and family commitments. We are also grateful to those experts from around the globe who provided reviews of the draft chapters and in so doing improved the content and clarity of the component chapters. These were: Beatrice Adoyo, Christian Buchmann, Chad Cheney, Corli Coetsee, Liam Cogill, Martin de Wit, Karen Esler, Llewellyn Foxcroft, Jema Haji, Martin Hill, Philip Ivey, Peter Little, Fekadu Gelaw Mersha, Ezekiel Edward Mwakalukwa, Francis Mwaura, Ana Novoa, Tim O'Connor, Olusegun Osunkoya, Alanna Rebelo, Prabhat Rai, Dirk Roux, Ross Shackleton, Izak Smit, Peter Stoett, Helga van der Merwe, Rieks van Klinken, Klaas Vrieling, Russell Wise, and Kris Wyckhuys. We would also like to thank Susan Abraham who redrafted most of the illustrations in this book, and Philip Hulme for writing the Foreword.

Many of the findings and insights reported in this book arise from the research conducted under the Woody Weeds project that ran for 7 years between 2015 and 2021 and was funded by the Swiss National Science Foundation and the Swiss Agency for Development and Cooperation under the Swiss Programme for Research on Global Issues for Development (r4d, Grant Number: 400440_152085). Additional funding was obtained from the Swiss Agency for Development and Cooperation for the Woody Weeds + project, which ran for 4 years from 2021 to 2024 (Project number 7F.10651.01). Brian van Wilgen also acknowledges the South African National Research Foundation for providing additional financial support through Grant numbers 150260 and 151653 that enabled his contributions to the writing and editing of this book.

We express our gratitude to all project partners and friends in the Woody Weeds and Woody Weeds + projects, as well as to the stakeholders who have collaborated with the project partners over the past 10 years. We are particularly grateful to the many PhD and MSc students who collected much of the novel information reported here, and to all those who assisted the students in completing their dissertation and/or thesis works and publishing their research findings. Richard Bardgett, Ray Callaway, John Maron, Heinz Müller-Schärer, Dave Richardson and Rieks van Klinken are thanked for many stimulating discussions on invasive species and/or grassland management.

Last, but of course not least, we thank our families for coping with our frequent absences during our travels to eastern Africa.

Brian W. van Wilgen, Stellenbosch University, South Africa
Urs Schaffner, CABI, Switzerland
Albrecht Ehrensperger, University of Bern, Switzerland
Ketema Bekele, Haramaya University, Ethiopia

The editors at an editing and writing workshop at Hunter's Lodge, Kenya, in March 2024. From left to right: Albrecht Ehrensperger, Brian van Wilgen, Urs Schaffner and Ketema Bekele.

Foreword

Invasive alien plants pose a major threat to the environment as well as to the economies of nations worldwide. Yet, many plant species have been deliberately introduced into new regions for ornament, timber production, pasture enhancement, erosion control or as biofuel feedstocks. As a result, the management of invasive alien plants can be fraught with conflict as stakeholders argue over the costs and benefits of species introductions. There are few better examples of such conflicts than the case of the widespread promotion of *Prosopis* species as multipurpose trees by international aid agencies. Several *Prosopis* species were heralded as solutions to poverty in arid and semi-arid regions in Africa, Asia and South America since they could establish in poor soils, combat erosion, and provide shade and fodder.

This volume, entitled *The Ecology and Management of Invasive* Prosopis *Trees in Eastern Africa*, presents an outstanding overview of the problems arising from misguided policies that favoured short-term results over longer-term environmental and social consequences. Distilled from over a decade's worth of interdisciplinary research in eastern Africa, the volume brings together multiple perspectives on the problems posed by *Prosopis* trees that are relevant to many arid regions in the world where these species have been planted.

The scale of the topics covered in the volume is impressive, bringing together perspectives on the problems and their possible solutions from Ethiopia, Kenya and Tanzania, and ensuring that the findings are relevant to a wide range of situations. By building on the historical narrative of the species introductions, it brings to light the wide impacts these species have caused not only to biodiversity and ecosystem hydrology but also to rural livelihoods and human health. This robust evidence base that quantifies negative impacts sends a clear warning to those individuals, institutions and organizations that continue to consider *Prosopis* species as valuable multipurpose trees. The prolific growth rate and high dispersal efficiency of *Prosopis* trees mean that, despite the best intentions of those seeking to use it as a multipurpose tree, the ultimate outcome has been widespread invasion and a subsequent worsening of the environment as well as the quality of human life.

A unique aspect of this volume is that, rather than simply presenting a negative perspective, the authors also set out guidance on best practice management as well as outlining the key role that improved governance must play in any future solutions. My own experience of the problems arising from widespread planting of *Prosopis* trees stems from research in the semi-arid caatinga ecosystems of Brazil, where plantations of *Prosopis* trees provide pods that are used both as animal feed and as flour for human consumption. Environmental concerns have been paid scant attention despite the evidence of establishment in native vegetation and spread along ephemeral water bodies and rivers. Despite the contrasting geography, the lessons and examples from eastern Africa brought together in this excellent volume are equally pertinent to managing the problem in Brazil. By ensuring that the key information on the biology and management of *Prosopis* trees has been brought together in a single comprehensive volume, the editors are making accessible an incredible resource that has the potential to result in a step change in the management of *Prosopis* trees worldwide.

Philip Hulme is Distinguished Professor of Plant Biosecurity in the Department of Pest-Management and Conservation, Lincoln University, New Zealand

Glossary

Adaptive management – An intentional approach to making decisions and adjustments in response to new information and changes in context.

Aggregate economic welfare loss – A measure of the degree to which people are worse off because of certain circumstances (e.g. the presence of invasive *Prosopis* trees) compared to where they would have been had the situation not existed.

Agroforestry – A land management system that combines the cultivation of trees or shrubs with crops or livestock production on the same land.

Agropastoralism – Livelihoods that integrate crop production and livestock production (see also **pastoralism**).

Alien species (syn. exotic, non-native, foreign) – A species that is present in a region outside its natural range due to human actions (intentional or accidental) that have enabled it to overcome biogeographic barriers.

Allelopathy – The chemical inhibition of one plant by another, due to the release into the environment of substances acting as germination or growth inhibitors.

Barazas – A place where public meetings are held. By extension, also the public meeting itself.

Benefit – The positive outcomes associated with an action, event or the presence of a species (see also **impact**).

Bimodal rainfall – A rainfall pattern with two distinct wet periods separated by a dry period.

Biodiversity – The variety of life on Earth at all its levels, from genes to ecosystems.

Biological control – The utilization of natural enemies, such as host-specific insects or pathogens, to reduce the density or spread of an invasive alien species to an acceptable level (usually defined in terms of perceived impact).

Biological invasions – The phenomenon of and suite of processes that are involved in determining the transport of organisms to areas outside their natural range by human activities and the fate of the organisms in their new ranges.

Carrying capacity – The number of grazing animals a piece of land can support in the long term while maintaining or improving the rangeland resources (vegetation, soils and water).

Chemical control – Any technique that involves the application of a chemical (herbicide) to weeds (i.e. undesirable alien plant species) or soil to control the germination or growth of the weed species.

Community Forest Associations – Associations established under the Kenyan Forest Act of 2005, which provides for the establishment, development and sustainable management of forests, including conservation and rational utilization of forest resources for the socio-economic development of the country.

Community-based resource management – An approach to natural resource management that involves the full participation of indigenous peoples, local communities and resource users in decision-making activities, and the incorporation of local institutions, customary practices and knowledge systems in management, regulatory and enforcement processes.

Competition – The interaction between organisms or species in which both require a resource that is in limited supply (such as food, water or territory).

Conflict species – Invasive alien species that simultaneously provide benefits and cause harm.

Containment – The goal of preventing spread or reducing or reversing the rate of spread of invasive species.

Control by utilization – An attempt to reduce the range or abundance of an invasive species by deliberately harvesting it and producing products for use or sale.

County Assembly – A body established in terms of Kenya's Constitution, in which the legislative authority of a county is vested.

County Implementation Group – The County Implementation Group is composed of stakeholder groups covering all major sectors affected by *Prosopis juliflora* and covering actors from the county to the community level.

Cross-disciplinary – A generic term used to talk about integration involving multiple disciplines in its many manifestations, including multi-, inter- and transdisciplinarity (see also **interdisciplinary**, **multidisciplinary** and **transdisciplinary**).

Customary natural resource management institutions – Bodies that regulate land use on communal or tribal land through traditional laws.

Cytotype – The cellular constitution of an organism; a group of individuals sharing the same caryological characteristics (e.g. the same ploidy level).

Deciduous (of plants and shrubs) – Shedding foliage at the end of the growing season (see also **drought deciduous**).

Deliberative multi-criteria decision support – An approach that combines multi-criteria decision analysis with the benefits of stakeholder participation (see also **multi-criteria decision making**).

Drought deciduous – refers to plants that shed their leaves during periods of drought or in the dry season (see also **deciduous**).

Early detection and rapid response – A coordinated set of actions to find and eradicate potential invasive species in a specific location before they spread and cause harm.

Eastern Africa – In the context of this book, includes Djibouti, Eritrea, Ethiopia, Kenya, Somalia, South Sudan, Tanzania and Uganda.

Economic value – The worth of something in terms of money or market exchange.

Ecophysiology – The branch of physiology that deals with the physiological processes of organisms with respect to their environment.

Ecosystem services – The many and varied benefits that humans gain from natural environments and ecosystems.

Enemy release hypothesis – The notion that alien species have a better chance of establishing and becoming dominant when released from the negative effects of natural enemies that, in their native range, lead to high mortality rates and reduced fecundity and/or productivity.

Eradication – The complete removal of all individuals and propagules of a population of an alien species from a particular area to which there is a negligible likelihood of reinvasion. The probability of reinvasion must have been explicitly assessed, and if it is negligible it can result in a reallocation of management resources (i.e. ongoing control and monitoring is no longer required).

Expansion (syn. spread) – The unaided movement of alien organisms within a defined area.

Extirpation – The result of a control operation whereby all individuals in a population are removed. Other populations might be close by or pathways or introduction and dispersal are still operating such that the probability of reinvasion is probable or not known.

Fractional cover – The ratio of the vertical projection area of above-ground vegetation (or individual species) organs on the ground to the total area in which the vegetation or species occurs.

Gross Domestic Product – The total monetary value of goods and services produced within a country over a specific period of time.

Host specificity – A term used to describe the range of species that an introduced biological control agent will attack and/or develop on. Biological control agents are usually selected when they exhibit a high level of host specificity, i.e. they will attack only the target invasive species, or one or more closely related invasive species, and pose no risk to native or crop species.

Hybrid – The offspring of two plants or animals of different species or varieties.

Hydraulic redistribution – The passive movement of water from moist to dry soil through plant roots, including the lifting of water from deeper to shallower soil layers (hydraulic lift) and the movement from shallow to deep soil layers (downward hydraulic redistribution).

Impact – The description or quantification of how an alien species affects the physical, chemical and biological environment, in either negative or positive ways.

Indigenous species – see **Native species**

Interdisciplinary – Research involving multiple disciplines, where the disciplinary perspectives are integrated to produce research output that is more than the sum of the research inputs (see also **cross-disciplinary**, **multidisciplinary** and **transdisciplinary**).

Invasive alien species – Alien species that sustain self-replacing populations over several life cycles, produce reproductive offspring, often in very large numbers at considerable distances from the parent and/or site of introduction, and have the potential to spread over long distances.

Livelihoods – The capabilities and assets (including both material and social resources) needed for people or households to support their way of life.

Local Implementation Group – Groups of people formed in the Woody Weeds project to connect researchers and external stakeholders with local stakeholders in order to coselect and test practices for effectively managing *Prosopis* tree invasions.

Mainstreaming – The process of bringing something into the main culture or practice.

Mechanical control – A physical activity (manual or using equipment or machinery) that inhibits unwanted plant growth.

Multi-criteria decision making – A discipline that explicitly evaluates multiple conflicting criteria in decision making.

Multidisciplinary – Research involving multiple disciplines that make contributions which are loosely integrated without the disciplinary perspectives themselves being integrated (see also **cross-disciplinary**, **interdisciplinary** and **transdisciplinary**).

National Environment Management Authority – Established under the Kenyan Environmental Management and Co-ordination Act No. 8 of 1999 as the principal instrument of the Kenyan Government for the implementation of all policies relating to the environment.

National Land Commission – A constitutional commission established under Article 67 (1) of the Constitution of Kenya 2010. It is mandated, *inter alia*, to manage land, recommend policy and encourage traditional dispute resolution in instances of conflict.

Native species (syn. Indigenous species) – Species that are found within their natural range where they have evolved without human intervention. Also includes species that have expanded their range as a result of human modification of the environment that does not directly impact dispersal (e.g. species are regarded as native if they increase their range as a result of watered gardens).

Net economic value – The overall economic value of something after subtracting costs from benefits.

Net present value – The value of all future cash flows (positive and negative) over the entire life of an investment discounted to the present.

Non-cognitive abilities – Abilities and skills that are related to motivation, integrity and interpersonal interaction ('soft skills'). Cognitive skills involve conscious intellectual effort, such as thinking, reasoning or remembering.

Non-linear impact – An impact that is not based on a simple proportional relationship between cause and effect. Therefore, such changes are often abrupt, unexpected and difficult to predict.

One Health – An integrated, unifying approach that aims to sustainably balance and optimize the health of plants, animals, the environment and people. It recognizes that the health of humans, domestic and wild animals, plants and the wider environment (including ecosystems) are closely linked and interdependent.

Pastoralism – A mode of subsistence that involves raising domestic animals in rangelands using herd and household mobility.

Pathways (of spread) – A broadly defined term that refers to the combination of processes and opportunities that result in the movement of alien species from one place to another.

Phenology – A branch of science dealing with the relations between climate and periodic biological phenomena (e.g. seasonal flowering or loss of leaves in plants).

Phylogeny – The representation of the evolutionary history and relationships between groups of organisms.

Radiometric resolution – The precision with which a sensor can distinguish different levels of light intensity, relevant for remote sensing.

Rangeland – Any extensive area of land that is occupied by native herbaceous or shrubby vegetation which is grazed by domestic or wild herbivores.

Rehabilitation (see also **restoration**) – Reinstating degrees of ecosystem functionality on degraded sites where restoration is not the aspiration, to permit ongoing provision of ecosystem goods and services.

Restoration (see also **rehabilitation**) – The process of assisting the full recovery (to original pristine state) of an ecosystem that has been degraded, damaged or destroyed.

Sensu lato – In the broad sense.

Social-ecological system – An ecological system intricately linked with and affected by one or more social systems.

Spread – see **Expansion**.

Stakeholder – A person, group or organization with a vested interest, or stake, in the decision making and activities of a business, activity, organization or project.

Sustainable land management – The use of land resources, including soils, water, animals and plants, for the production of goods to meet changing human needs, while simultaneously ensuring the long-term productive potential of these resources and maintenance of their environmental functions.

Total economic value – The value derived by people from a natural resource, compared to not having it.

Transdisciplinary – Research that transcends disciplinary boundaries by involving inputs from multiple disciplines, where the disciplinary perspectives are integrated to produce research output that is more than the sum of the research inputs and the disciplinary inputs are themselves changed in the process (see also **cross-disciplinary**, **interdisciplinary** and **multidisciplinary**).

Transhumance – The practice of moving livestock from one grazing area to another in a seasonal cycle.

Upscaling – The conversion of data collected at smaller scales (for example, on experimental plots) to larger scales (for example, landscapes or countries).

Water Resources Authority – a state corporation established under Section 11 of the Kenyan Water Act 2016, mandated to safeguard the right to clean water by ensuring regulation of the management and use of water resources.

Water resources users associations – Associations of water users, riparian landowners and other **stakeholders** who have formally and voluntarily associated for the purposes of cooperatively sharing, managing and conserving common water resources. These associations are registered by the Kenyan **Water Resources Authority**.

Wicked problem – A problem that is difficult or impossible to solve because of incomplete, contradictory and changing requirements that are often difficult to recognize.

Willingness to pay – The maximum price at or below which a person will definitely pay for, or contribute towards, an action.

Abbreviations

CABI	Centre for Agriculture and Biosciences International
CBNRM	Community-Based Natural Resource Management
CEC	County Executive Committee
CFA	Community Forest Association
CGIAR	Consultative Group on International Agricultural Research
CHP	Compensation Heat Pulse Method
CIC	County Implementation Committee
CIDP	County Integrated Development Plan
CIG	County Implementation Group
COVID	Coronavirus Disease
CPA	Charcoal Producer Association
CSG	County Steering Group
DNA	Deoxyribonucleic Acid
DMCD	Deliberative Multi-criteria Decision
EDRR	Early Detection and Rapid Response
EICAT	Environmental Impact Classification for Alien Taxa
ENMAP	Environmental Mapping and Analysis Program
EO	Earth Observation
ES	Ecosystem Services
ESV	Ecosystem Service Value
ETB	Ethiopian Birr
EUR	Euros
FAO	Food and Agriculture Organization
FLEX	Fluorescence Explorer
FSC	Forest Stewardship Council
GBIF	Global Biodiversity Information Facility
GDP	Gross Domestic Product
GIS	Geographic Information System
HRM	Heat Ratio Method
IAS	Invasive Alien Species
ICRAF	International Centre for Research in Agroforestry
IPBES	Intergovernmental Panel on Biodiversity and Ecosystem Services

IUCN	International Union for Conservation of Nature
KEFRI	Kenya Forest Research Institute
KFS	Kenya Forest Service
KSH	Kenyan Shillings
KWS	Kenya Wildlife Service
LIG	Local Implementation Group
MCA	Member of County Assembly
MSP	Multi-stakeholder Platform
NBSAP	National Biodiversity Strategy and Action Plan
NDVI	Normalized Difference Vegetation Index
NEMA	National Environmental Management Authority (Kenya)
NGO	Non-governmental Organization
NLC	National Land Commission (Kenya)
NPS	National *Prosopis* Strategy
NRM	Natural Resource Management
OH	One Health
SES	Social-ecological System
SDC	Swiss Agency for Development and Cooperation
SDM	Species Distribution Model
SEICAT	Socio-economic Impact Classification for Alien Taxa
SLM	Sustainable Land Management
SMART	Specific, Measurable, Achievable, Realistic and Time-bounded
SNP	Single Nucleotide Polymorphism
SNSF	Swiss National Science Foundation
TDI	Toolbox Dialogue Initiative
TEV	Total Economic Value
ToR	Terms of Reference
TZS	Tanzanian Shillings
UAV	Unmanned Aerial Vehicle
USD	United States Dollar
WOCAT	World Overview of Conservation Approaches and Technologies
WRA	Water Resources Authority (Kenya)
WRUA	Water Resource User Association

1 Addressing the Ecology and Management of *Prosopis* Species in Eastern Africa

Brian W. van Wilgen[1]*, Urs Schaffner[2], Albrecht Ehrensperger[3] and Ketema Bekele[4]

[1]Centre for Invasion Biology, Department of Botany and Zoology, Stellenbosch University, South Africa; [2]CABI, Delémont, Switzerland; [3]Centre for Development and Environment (CDE), University of Bern, Switzerland; [4]School of Agricultural Economics, Haramaya University, Dire Dawa, Ethiopia

Abstract

Prosopis trees have intentionally been introduced to eastern Africa to provide a suite of benefits. The trees have subsequently become invasive, resulting in detrimental impacts on ecosystems and people's livelihoods. Their management is therefore complex, with no apparent simple solutions being available. Addressing this will require integrated interdisciplinary research approaches involving ecologists, economists and social scientists, as well as close collaboration with the potential end-users of research findings. This book describes how these problems were addressed in a 8-year research initiative, the Woody Weeds project, and a three-year implementation project, the Woody Weeds + project, which sought to integrate findings from different disciplines to scale these up to appropriate spatial resolutions, and to involve potential end-users at all stages. This chapter outlines the intention of the book to draw together scattered studies into a single, open-access volume that should be useful to researchers and managers alike.

The book is in four parts. The first deals with the history, evolution and spread of *Prosopis* trees in eastern Africa – how the trees arrived and adapted to the new environment, the rate at which they spread, and their current and potential future distribution. The second examines what is known about the ecological, social and economic consequences of *Prosopis* trees in the region. The third deals with how *Prosopis* trees can best be managed by outlining both the management techniques themselves and how they could be effectively implemented. The final part addresses how to bridge silos in practice and research by encouraging the broad involvement of all parties and integrating research across disparate disciplines. It is hoped that these accounts will allow readers to gain insights into the complexity of the issue, and in so doing assist them in the formulation of appropriate policy and management responses.

1.1 Introduction

In the arid and semi-arid parts of eastern Africa (broadly defined here as including Djibouti, Eritrea, Ethiopia, Kenya, Somalia, South Sudan, Tanzania and Uganda; www.britannica.com/place/eastern-Africa), the problem of invasions by alien species has manifested very prominently in the form of trees in the genus *Prosopis*. These trees were introduced from the Americas to

*Corresponding author: bvanwilgen@sun.ac.za

© CAB International 2025. *The Ecology and Management of Invasive Prosopis Trees in Eastern Africa* (eds U. Schaffner *et al.*)
DOI: 10.1079/9781800623644.0001

Africa deliberately, to provide fodder and shade for livestock, a source of fuelwood and a means to counter erosion and desertification (Box 1.1). However, the trees soon became invasive, spreading across large swathes of arid and semi-arid land across the region.

Prosopis trees are difficult to manage. This is not only because of their ecological features (such as the ability to produce copious amounts of seed and to outcompete native species for resources, see Box 1.2) but also because they have beneficial uses as well as negative impacts. This makes them 'conflict' species, leading to situations where people cannot agree on the goals of management and how to achieve them (Woodford *et al.*, 2016; Zengeya *et al.*, 2017).

The available literature on the ecology and management of invasive *Prosopis* trees in Africa and elsewhere is dominated by studies that have been conducted within specific disciplines. For example, ecological studies have been conducted to determine the impact of *Prosopis* trees on biodiversity (e.g. Steenkamp and Chown, 1996; Dean *et al.*, 2002; Shackleton *et al.*, 2015), water resources (e.g. Dzikiti *et al.*, 2013), rangeland carrying capacity (e.g. Ndhlovu *et al.*, 2011) and interspecific interactions (e.g. Schachtschneider and February, 2013); economic studies have sought to quantify their impacts and benefits in monetary terms (e.g. Wise *et al.*, 2012); social studies have used questionnaire surveys to gauge people's perceptions (e.g. Shackleton *et al.*, 2017b); studies on the distribution of invasions have used remote sensing to map invasions (e.g. Van den Berg, 2010); and management studies have sought to identify control options, for example biological control (Zimmermann, 1991; Zachariades, 2021). Generally, these studies have given more weight to ecological problems than socio-economic problems of the invasion (Schüttler *et al.*, 2011). Further, they have been conducted at varying spatial scales, at different localities, and at different times, which significantly complicates the task of integrating available information and formulating practical and robust policy interventions.

Against this backdrop, researchers have more recently endeavoured to collect, analyse and interpret pertinent information that can be used to develop sustainable approaches for addressing the problem of *Prosopis* invasions in Africa. The fact that the potential costs and benefits of *Prosopis* trees create conflict presents a significant challenge to their sustainable management. Addressing this will require interdisciplinary research approaches involving ecologists, economists and social scientists, as well as close collaboration with the potential end-users of research findings. These end-users should ideally be involved throughout the duration of research projects that seek to formulate relevant solutions.

This book, authored by experts and postgraduate students who have gained extensive experience through research conducted between 2015 and 2023 in eastern Africa, offers a comprehensive overview of the knowledge that has been acquired on the ecology and management of invasive alien *Prosopis* trees in the region. This introductory chapter outlines what the book intends to do and how it is structured.

1.2 What This Book Intends to Do

This book arises primarily from research work conducted under the Woody Weeds project

Box 1.1. The global redistribution of trees in the genus *Prosopis*.

Since 1882, *Prosopis* trees have been widely planted outside their native ranges. There have been several reasons for their introduction, including for soil stabilization, fencing, rehabilitation of disturbed sites, fodder for livestock, shade and firewood. Most introductions were intentional, although in some cases the trees have entered countries by crossing the border from neighbouring countries to which they had been intentionally introduced (for example, they have been introduced unintentionally to Botswana, Nigeria, Tanzania and Yemen through livestock trading with neighbouring countries). Currently, *Prosopis* trees occur in 129 countries worldwide. Of these countries, 26 have only native species, 64 have only introduced Prosopis species, and 39 have both native and introduced species (Shackleton *et al.*, 2014).

Continued

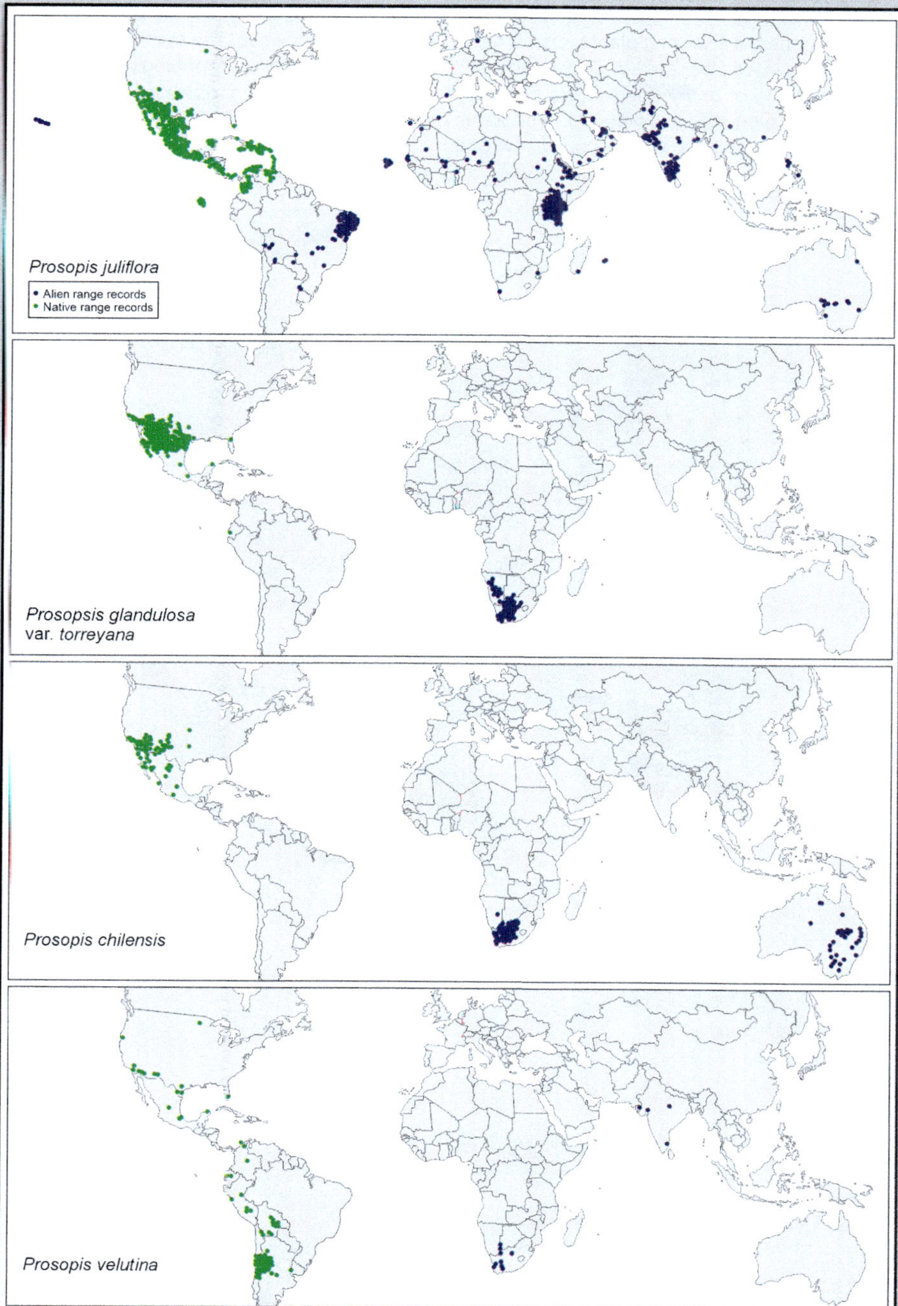

Box Fig. 1.1. Examples of *Prosopis* species that have been introduced outside their native ranges. Green circles indicate the native ranges and blue circles the alien ranges of these species. Records are from the Global Biodiversity Information Facility (www.gbif.org), downloaded 18 February 2023.

Box 1.2. Features of trees in the genus *Prosopis*.

Prosopis is a genus of flowering plants, mainly trees, in the family Fabaceae. The taxonomy of *Prosopis* species is problematic because they are difficult to separate based on morphological features, the native distributions of many species are not clearly understood, and hybridization occurs between some species (Chapter 3, this volume). The genus comprises 44 species from the Americas, south-west Asia and North Africa, which are mostly found in arid and semi-arid regions. Only one species (*Prosopis africana*) is native to Africa (see Box 1.3 for recent taxonomic revisions).

Continued

Box Fig. 1.2. (A) The pods of *Prosopis* trees can provide nutrition to livestock and wildlife if not consumed in large quantities. (B) The branches of *Prosopis* trees carry numerous thorns which hamper access and cause injury to people and animals. (C) Many tree species in the genus *Prosopis* become invasive, forming dense and impenetrable thorny thickets. Photos courtesy of B.W. van Wilgen (A, B) and K. Bekele (C).

Prosopis trees spread by means of seeds. The trees produce one crop of pods annually and each seed pod holds 5–20 seeds, with potentially hundreds of thousands of seeds produced per mature plant. The major dispersal agents are livestock and wildlife that consume the pods and spread them in dung.

Several features of *Prosopis* trees contribute to their ability to invade and dominate in new areas. They have deep taproots that give them access to groundwater, they fix nitrogen, they have allelopathic effects on native plants and are able to produce copious amounts of seeds in the absence of natural enemies such as seed-feeding insects in their native ranges.

As a result of these competitive advantages, *Prosopis* trees have negative impacts on biodiversity, water resources and the ability of rangelands to support livestock and wildlife. They can access water at deeper levels, thus depriving native trees of water during dry spells. Grasses are shaded out and this severely reduces the capacity of invaded rangelands to support both wildlife and domestic livestock. Although the pods can be nutritious, *Prosopis* thickets form dense, impenetrable stands that prevent their utilization by domestic and wild ungulates. The thorns are also responsible for injuries to animals and humans. Habitat changes brought about by the replacement of natural vegetation with a monoculture of alien trees have also resulted in the decline or local disappearance of native mammals, birds and invertebrates (Chapter 5, this volume).

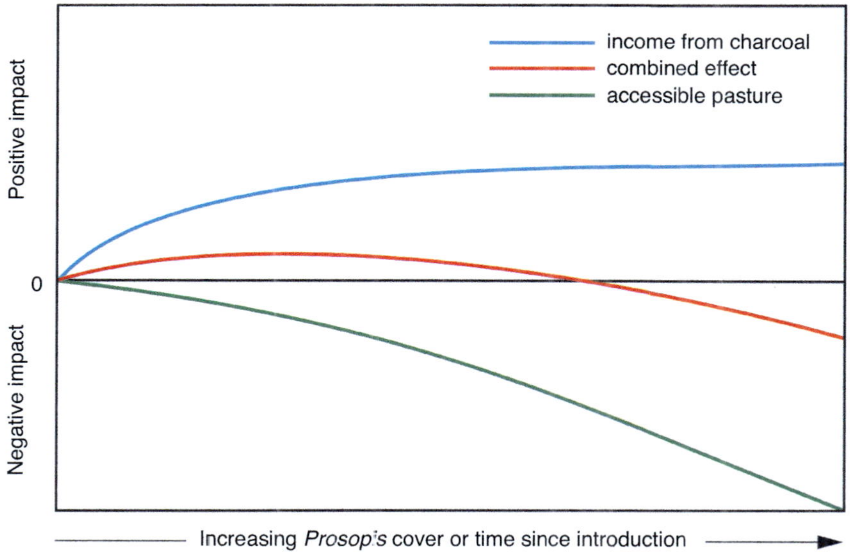

Fig. 1.1. The perceived or actual impacts of *Prosopis* trees on individual elements of a socio-ecological system may be positive or negative, and their magnitude may depend on the abundance of the trees in a location or region. For example, income derived from wood may increase if *Prosopis* trees become abundant, while access to and quality of pasture may decrease simultaneously. The combined net effect may, however, respond differently, with positive impacts at low densities that turn negative once the species abundance crosses a certain threshold.

(www.woodyweeds.org), which ran from 2015 to 2021, with a follow-up project (Woody Weeds +) that ran from 2021 to 2024. Woody Weeds was funded by the Swiss National Science Foundation and Woody Weeds + by the Swiss Agency for Development and Cooperation. These two projects involved six universities and eight institutes in Ethiopia, Kenya, Tanzania, South Africa, Switzerland and the USA. It spanned at least five research disciplines, trained graduate students and engaged stakeholders from local to national levels. The project addressed a complex problem with no clear solutions. Many people and communities experience both positive benefits and negative impacts of *Prosopis* trees, with some gaining from the positive uses while others carry the costs of negative impacts. This picture also changes over time. When the trees become overabundant, the net effects become negative and apparent. Successful management of *Prosopis* trees therefore requires early implementation, before the negative effects become apparent to all (Fig. 1.1), and codesign with multiple stakeholders. Few projects have addressed the complexity of invasive woody plants comprehensively and this has impeded development of ecologically and socio-economically sound and effective responses.

This book provides the first attempt to bring together, in a single, open-access volume, the experience of 26 authors from six countries over the past 7 years directed towards addressing a complex problem in a deliberately integrated way. It is not just simply a review of research outputs but rather an integration of all the findings to recognize the links between research and implementation across spatial and temporal scales. The integration of knowledge about the ecology, history of introduction and spread creates a common basis for potential actions. It integrates known impacts across sectors and scales, addresses conflicts of interest and presents these impacts in monetary terms that allow for the direct comparison of impacts in a common currency. It also outlines how future research projects can build on the lessons we have learnt about how to structure research aimed at understanding the full impacts of invasive alien species and how they can effectively be addressed.

One might ask why we have compiled this synthesis. Primarily, it is an attempt to collate all the available studies and to place them into a broader context. The findings of individual studies presented in this book have been published in a wide range of journals but these are not always easily accessible to stakeholders, particularly in developing countries. In addition, there remained much scope for a broader synthesis of the lessons that have been learnt. The publication of a synthesis and the distillation of recommendations for policy and management in a single, open-access book will be a useful source of information for students, researchers, managers and policy makers. It should allow readers to gain insights into the full spectrum of issues, and in doing so help them to formulate appropriate policy and management responses.

1.3 How This Book Is Structured

1.3.1 A focus on four aspects

This book is in four parts. The first part deals with history, evolution and spread, where we outline how *Prosopis* trees arrived in eastern Africa, how they adapted to the new environment, how far they have spread and how far they still may go. The second part deals with impacts and examines the ecological, social and economic consequences of *Prosopis* trees in the region. The third part deals with management, outlining how *Prosopis* trees can best be managed and how the available interventions could be effectively implemented by all stakeholders involved. The final part deals with effective integration in both research and management, by moving beyond the concerns of individual disciplines and the attempts to propose simple one-dimensional solutions to the problem.

1.3.2 Part 1: History, evolution and spread

The first part of this book contains three chapters that set the scene for parts that follow. It broadly addresses the questions of how and why *Prosopis* trees were introduced, the degree to which they evolved and became invasive in the new environment, and where and how they spread across the landscape.

The first chapter reviews the rationale for introducing and propagating *Prosopis* trees in eastern Africa, and the growing awareness of harm that followed the onset of invasion (Chapter 2). There is a description of how attitudes have changed over time, leading to sometimes conflicting proposals on how to deal with the problem. Although an improved understanding of the costs and benefits of *Prosopis* trees has recently influenced the development of national strategies for dealing with *Prosopis* tree invasions, contradictory initiatives (for example, a global drive to plant more trees; Bond *et al.*, 2019) continue to add further complexity to the problem.

Research into the ecology and management of *Prosopis* trees has been complicated by the lack of a clear understanding of their taxonomy and systematics. Uncertain biogeography of some species in the genus, coupled with high intraspecific variation and extensive hybridisation has partly contributed to this situation. As a result, researchers face various problems when trying to investigate the biology and management of invasive *Prosopis* species and their hybrids. These issues are addressed in Chapter 3, where the systematics, taxonomy and evolutionary ecology of the genus are reviewed, with a special emphasis on *Prosopis juliflora*, the main invasive *Prosopis* species in eastern Africa and the primary subject of this book. Although a recent taxonomic revision has placed most of the *Prosopis* species into the genus *Neltuma* (Box 1.3), we have continued to refer to *Prosopis* species in this volume to avoid confusion that might arise from adopting this recent name change.

Chapter 4 examines the current and potential future distribution of *Prosopis* trees in eastern Africa. The rate at which introduced trees spread over time has been assessed using satellite remote sensing at two important locations: the Afar Region in Ethiopia and the Baringo County in Kenya. The development of a clear picture of where the invasions are now, where they potentially may go in future and the rate at which they spread is vital for scaling up assessments of impact at finer scales, and for developing robust estimates of the extent and consequences of invasion at landscape scales.

1.3.3 Part 2: The impacts of invasive *Prosopis* species

It is necessary to obtain a balanced, integrated assessment of the positive benefits and negative consequences of invasive alien tree species that have been deliberately introduced by governments and NGOs. There is thus a need for well-designed studies that attempt to quantify how these positive and negative effects change with increases in the extent and densities of invasion. Such impact studies should be conducted in the region where existing policies are under review, and at scales relevant for policy makers. Also, rather than making statements about single positive or negative impacts, policy and management decisions should be based on a more holistic approach where the multiple impacts and benefits are considered in an integrated way.

The second part of this book therefore examines what is known about the ecological, social and economic consequences of *Prosopis* trees. While there has been a growing recognition of the detrimental effects of invading *Prosopis* trees for some time now, much of the evidence in this regard has been anecdotal. It is only relatively recently that quantitative studies have been carried out to document actual measured impacts in the field. Part 2 of the book reviews this evidence in Chapters 5–9, based on case studies in eastern Africa but including information from other parts of Africa (notably South Africa) that corroborate the local evidence and increase the degree of confidence in generalized patterns and trends of impact. *Prosopis* trees have both positive and negative impacts. As a result, not all stakeholders perceive these impacts in the same way, which can lead to conflict. It is thus important to be able to describe these impacts in a comparable way, so that the net overall outcome (value of benefits minus cost of impacts in monetary terms) can be reduced to a common currency, and a picture of who gains from the benefits and who suffers from the negative impacts can be determined and compared.

Invasive *Prosopis* trees are superior competitors for resources and this allows them to spread into and dominate in natural ecosystems, displacing native species and impacting negatively on biodiversity and ecosystem functioning. This

Box 1.3. Recent changes in the naming of *Prosopis* species.

Hughes *et al.* (2022) published a disintegration of the mimosoid clade of the Caesalpinioideae (which includes the genus *Prosopis*), proposing name changes to almost all the species that are the topic of this book. They resurrected three segregate genera (*Anonychium*, *Neltuma* and *Strombocarpa*) and proposed 57 new name combinations where necessary. This resulted in the genus *Prosopis* itself being reduced to just three species, with the native African species *Prosopis africana* being retained. The remainder of the *Prosopis* species of interest in this book were renamed in the genus *Neltuma*. An editorial decision was made to refer to *Prosopis* species in the title and throughout this book, because all the available literature and policy documents refer to *Prosopis* and not *Neltuma*. It could therefore be confusing to use these recent name changes in a book intended to raise awareness and disseminate information to a wider audience. The most important name changes are listed below.

Nomenclature in this volume	Proposed new name (Hughes *et al.*, 2022)	Vernacular name	Notes
Prosopis africana (Guill. & Perr.) Taub.	*Prosopis africana* (Guill. & Perr.) Taub.	African mesquite, iron tree	The only Prosopis species indigenous to Africa, occurring in west tropical Africa to Uganda, and on the Arabian Peninsula
Prosopis chilensis (Molina) Stuntz	*Neltuma chilensis* (Molina) C.E. Hughes & G.P. Lewis	Chilean mesquite	Hybridizes with other Prosopis species in South Africa, making identification difficult
Prosopis glandulosa var. *torreyana* (L.D. Benson) M.C. Johnst.	*Neltuma odorata* (Torr. & Frém.) C.E. Hughes & G.P. Lewis	Honey mesquite	Hybridizes with other Prosopis species in South Africa, making identification difficult
Prosopis juliflora (Sw.) DC.	*Neltuma juliflora* (Sw.) Raf.	Mesquite; Mathenge, Promi, Biskuti mjinga (Kenya); Mrashia, Mgunga taveta, Mpopote (Tanzania); Woyane Zaf, Dergi Hara, Biskut Zaf (Ethiopia)	The major invasive species in eastern Africa
Prosopis pallida (Humb. & Bonpl. ex Willd.) Kunth	*Neltuma pallida* (Humb. & Bonpl. ex Willd.) C.E. Hughes & G.P. Lewis	Algaroba	Introduced to eastern Africa but not invasive
Prosopis velutina Wooton	*Neltuma velutina* (Wooton) Britton & Rose	Velvet mesquite	Invasive in South Africa

issue is addressed in Chapter 5, where the direct impacts of invasion on the species richness and composition of native natural ecosystems, as well as the impacts on ecosystem functioning that depend on biodiversity, are reviewed.

Water is a vital resource that supports ecosystems and the people who live in them. This is particularly important in the arid and semi-arid lands into which *Prosopis* trees have been introduced in eastern Africa and elsewhere, and where they can substantially reduce water resources. These impacts are described in Chapter 6. It reviews the physiological features of *Prosopis* trees that allow them to be superior competitors for water and the impacts that this has on groundwater reserves. The chapter also describes how these impacts are scaled up to provide estimates of water use at a landscape scale, and what the value of these impacts are in monetary terms.

Pastoralism is an important way of life in eastern Africa, and pastoralists are increasingly faced with reductions in the extent and livestock carrying capacity of rangelands due to *Prosopis* tree invasions. These impacts are covered in Chapter 7, which describes how there have been significant declines in livestock and wildlife populations in invaded areas, leading in turn to substantial negative impacts on the quality of life of nomadic pastoralists and the sustainability of their lifestyles.

Chapter 8 examines the impacts of *P. juliflora* on local livelihoods, perceptions of the impacts within local communities and willingness to pay for managing invasions. All these aspects are important for informing the development of effective management strategies. The chapter also describes how the net benefits change over time, provides an account of how these costs will escalate as the invasions grow in future and reports on how much communities would be willing to pay to bring the invasions under control. The current and potential future total impacts of the *Prosopis* trees for the region are provided in monetary terms.

The final account in this part of the book (Chapter 9) examines the holistic impacts of *Prosopis* invasions using the One Health approach, which collectively considers current impacts in four dimensions, namely human, animal, plant and environmental health. The analysis clearly shows that some stakeholders (such as charcoal producers and 'developmental' NGOs) can benefit from the trees but that most people are severely negatively affected. Holistic assessments such as these can raise awareness and help to secure the funding that will be needed to contain and hopefully reverse the problems associated with invasion by *Prosopis* trees.

1.3.4 Part 3: Options for managing *Prosopis* trees

The third part of the book addresses the available options for managing *Prosopis* trees in eastern Africa. Originally, tree planting was promoted to provide wood, fodder and shade, and to prevent erosion, and more recently tree planting has become very popular in the context of climate change mitigation and adaptation (Chapter 2, this volume). Because some alien tree species possess special traits (e.g. high growth rates, ease of propagation and resistance to climate change, pathogens and pests), they can thrive in novel regions, and they were and still are favoured over native species in tree planting programmes. However, the traits that help them to survive, grow and reproduce successfully in novel environments also increase the likelihood that they will become invasive, which has frequently been the case for many introduced species (Rejmánek and Richardson, 2013).

Over recent decades, invasion science has developed into a rapidly expanding discipline (e.g. Richardson, 2011) but the sustainable management of widely established invasive alien plants remains a challenge (Shackleton *et al.*, 2020; Sun *et al.*, 2022). Conflicts of interest between different stakeholders regarding different management objectives can hinder implementation of targeted and coordinated management practices. This challenge is not restricted to invasive alien tree management; it is a topic that cuts across attempts to manage limited natural resources in general. It is therefore reasonable to explore the extent to which approaches developed in other areas of sustainable land management can also be adopted in the management of invasive non-native trees species.

Grice *et al.* (2011) and Shackleton *et al.* (2017a) provided the first examples of how geographic differentiation of management objectives can be used as a key element of a strategy to manage invasive alien plants at the landscape scale. Using the example of *P. juliflora* in eastern Africa, this book further elaborates this idea and proposes a management approach that is based on a spatially explicit management plan. Following the concept of land use planning (e.g. Metternicht, 2017), spatially explicit management approaches offer the possibility to integrate the different and sometimes conflicting needs of stakeholders spatially. They can help to plan local interventions in a way that they also contribute to invasive species management at a broader scale, and to emphasize that, in essence, the management of invasive alien tree species such as *P. julilfora* is a component of land use planning and should thus be treated as such.

Prosopis species can be difficult to manage, and Chapter 10 examines how the various available techniques can be optimally combined into a set of 'best practices' to achieve the goals of management. The importance of tailoring appropriate management goals in relation to the abundance of *Prosopis* trees in a particular location is stressed. In addition, emphasis is placed on the need for adequate planning, regular review and possible adaptation of management goals throughout the management cycle. The chapter also outlines tested management methods to prevent introduction, control invasions and restore land.

Chapter 11 addresses the need for spatially implicit approaches to management. The appropriate form of management to adopt will depend on the stage of invasion in a particular part of the distribution range of an invasive species. This chapter considers four broad management categories in this regard. When invasive *Prosopis* trees are not yet present in an area, then awareness creation and prevention of introduction are called for. Early incursions to an area would rely on early detection and rapid response, while in areas where the species are well established and widespread, control to reduce their impacts is needed. Chapter 11 therefore outlines key factors that influence management options and proposes a decision support tool to aid the selection of appropriate management objectives for specific spatial management units. New technologies that can assist in selecting management options, including unmanned aerial vehicles (drones) and spatial modelling in a geographic information system, are also described.

Chapter 12 addresses the importance of stakeholder engagement when managing *Prosopis* trees. The chapter explores the shortcomings of existing governance and institutional structures in Kenya (the only country in which we tested novel institutional approaches at different scales in selected counties). It then elaborates the rationale for creating 'new' institutions (Local Implementation Groups and County Implementation Groups), which will have the aim of improving stakeholder engagement. The chapter describes the role of participative decision-making processes and supporting tools used in these group processes, and their importance for stakeholder engagement, co-ownership and capacity building.

Although the long-term effectiveness of these interventions cannot yet be assessed, indications are that social learning among the stakeholders involved has helped to mainstream invasive species management into planning and budgetary processes at the local, subnational and national level.

1.3.5 Part 4: Bridging silos in practice and research

Part 4 of the book addresses how effective solutions need to be based on information from a wide range of stakeholders, and from a range of scientific and social disciplines. The Woody Weeds project (see 1.2 above) was designed as a 'real-life laboratory' that sought to gain a better understanding of how to integrate research and management through transdisciplinary approaches to research as well as to involve all affected stakeholders, including policy makers.

Chapter 13 explores the operational environment within which *Prosopis* species must be managed in an eastern African context. The researchers involved in conducting studies on the management of *Prosopis* species have found that multiple stakeholders have very heterogeneous levels of knowledge, skills and awareness in relation to the management of *Prosopis* trees, as well as very different interests and levels of authority and/or capacity to implement their decisions. Chapter 13 therefore proposes a generic 'stakeholder mobilization model' based on a synthesis of the experiences of researchers in finding practical and sustainable solutions to complex management issues.

Chapter 14 outlines the deliberate interventions that were implemented throughout the duration of the Woody Weeds and Woody Weeds + research projects. It shows how cross-disciplinary and interinstitutional research collaborations can be designed to deliver solutions to complex challenges to regional sustainability. Many lessons were learnt in this process, and the chapter addresses those that worked well, those that could be improved based on our experience and where the challenges to this type of work will lie in future. As such, it provides a framework for the design and implementation of future research projects that seek to have real impact.

1.4 Concluding Remarks

Chapter 15 summarizes the key findings of recent studies and examines the implications of these findings for policy and practice. The invasion of eastern Africa by *P. juliflora* is relatively young, and many areas suitable for invasion have not yet been colonized. Monitoring systems based on modern remote sensing techniques should be developed so that early action can be taken to direct interventions to where they can prevent the invasion of new areas. The impacts of these invasions are substantial and they considerably outweigh any benefits that the trees may provide. However, the fact that some benefit can still be extracted from invasive *Prosopis* trees leads to conflicting ideas on how they should be managed. Like all established invasive alien species, *Prosopis* trees cannot be eradicated, and ways to sustainably contain them need to be found to prevent the growth of substantial impacts. In this respect, the introduction of effective biological control options will be essential in future. Our research has confirmed that managing the *Prosopis* tree invasions is a complex socio-ecological problem, and that all stakeholders will need to be identified and actively engaged for control to be achieved.

This book synthesises information relevant to an important environmental problem – the invasion of natural ecosystems in Africa by alien *Prosopis* trees and how they should be managed. Much of this information has until now been scattered in numerous scientific publications in a wide range of journals. Many of these publications have a relatively narrow focus and bringing these insights together provides a more complete and potentially useful treatment of the topic. In this book, we address how the problem came about, what we know about how it works and how it can potentially be managed at different scales. The publication of a summary of recent work in eastern Africa and a synthesis of this work in a single, open-access book will, we hope, provide a relevant and useful overview of the problem. It essentially should allow readers to gain insights to the full spectrum of issues that are embedded in a complex problem, and in so doing assist them in the formulation of appropriate policy and management responses.

Close communication with stakeholders in eastern Africa across sectors and scales has started to bear fruit; in Kenya, the Cabinet has approved a National *Prosopis* Strategy which was influenced by the findings of the Woody Weeds and Woody Weeds + projects. First counties have integrated *Prosopis* management into their County Integrated Management Plans using the County *Prosopis* Management Plans, which were codeveloped with stakeholders in the Woody Weeds + project, as management guidance documents. In Tanzania, the lessons learnt in the Woody Weeds projects fed into the National Invasive Species Strategy and Action Plan (NISSAP) which came into effect in 2019.

The synthesis presented in this book is intended to provide an accurate and up-to-date account that can guide stakeholders as they deal with this significant environmental problem in the future. Each of our chapters concludes with a short section on the relevant implications of the chapter's contents and makes suggestions for future management options. Much remains to be done and addressing the issue effectively will require urgency. It is our hope that future decisions on the management of *Prosopis* tree invasions will be based on the best available scientific evidence, and that this book will provide much of that evidence.

Acknowledgements

This work was supported by the Swiss Programme for Research on Global Issues for Development (r4d), funded by the Swiss National Science Foundation and the Swiss Agency for Development and Cooperation, through the project 'Woody invasive alien species in East Africa: Assessing and mitigating their negative impact on ecosystem services and rural livelihood' (Grant Number: 400440_152085) and by the Swiss Agency for Development and Cooperation through the project 'Woody Weeds +' (Project number 7F.10651.01). We thank Ashlyn Padayachee for preparing distribution maps. BWvW thanks the National Research Foundation of South Africa for funding (Grant Numbers 150260 and 151653). US was supported by CABI with core financial support from its member countries (see http://www.cabi.org/about-cabi/who-we-work-with/ key-donors/ for full details).

References

Bond, W.J., Stevens, N., Midgley, G.F. and Lehmann, C.E.R. (2019) The trouble with trees: Afforestation plans for Africa. *Trends in Ecology and Evolution* 34, 963–965.

Dean, W.R.J., Anderson, M.D., Milton, S.J. and Anderson, T.A. (2002) Avian assemblages in native *Acacia* and alien *Prosopis* drainage line woodland in the Kalahari, South Africa. *Journal of Arid Environments* 51, 1–19.

Dzikiti, S., Schachtschneider, K., Naiken, V., Gush, M., Moses, G. *et al.* (2013) Water relations and the effects of clearing invasive *Prosopis* trees on groundwater in an arid environment in the Northern Cape, South Africa. *Journal of Arid Environments* 90, 103–113.

Grice, A.C., Clarkson, J.R. and Calver, M. (2011) Geographic differentiation of management objectives for invasive species: A case study of *Hymenachne amplexicaulis* in Australia. *Environmental Science and Policy* 14, 986–997.

Hughes, C.E., Ringelberg, J.J., Lewis, G.P. and Catalano, S.A. (2022) Disintegration of the genus *Prosopis* L. (Leguminosae, Caesalpinioideae, Mimosoid clade). *PhytoKeys* 205, 147–189..

Metternicht, G. (2017) Land use planning. Global Land Outlook. Available at: https://www.unccd.int/resources/publications/land-use-planning (accessed 23 July 2024).

Ndhlovu, T., Milton-Dean, S.J. and Esler, K.J. (2011) Impact of *Prosopis* (mesquite) invasion and clearing on the grazing capacity of semiarid nama karoo rangeland, South Africa. *African Journal of Range and Forage Science* 28, 129–137.

Rejmánek, M. and Richardson, D.M. (2013) Trees and shrubs as invasive alien species – 2013 update of the global database. *Diversity and Distributions* 19, 1093–1094.

Richardson, D.M. (2011) Invasion science: The roads travelled and the roads ahead. In: Richardson, D.M. (ed.) *Fifty Years of Invasion Ecology: The Legacy of Charles Elton*. Oxford, Blackwell Publishing, pp. 397–407.

Schachtschneider, K. and February, E.C. (2013) Impact of *Prosopis* invasion on a keystone tree species in the Kalahari desert. *Plant Ecology* 214, 597–605.

Schüttler, E., Rozzi, R. and Jax, K. (2011) Towards a societal discourse on invasive species management: A case study of public perceptions of mink and beavers in Cape Horn. *Journal for Nature Conservation* 19, 175–184.

Shackleton, R.T., Le Maitre, D.C., Pasiecznik, N.M. and Richardson, D.M. (2014) Prosopis: A global assessment of the biogeography, benefits, impacts and management of one of the world's worst woody invasive plant taxa. *AoB PLANTS* 6, lu027.

Shackleton, R.T., Le Maitre, D.C., van Wilgen, B.W. and Richardson, D.M. (2015) The impact of invasive alien *Prosopis* species (mesquite) on native plants in different environments in South Africa. *South African Journal of Botany* 97, 25–31.

Shackleton, R.T., Le Maitre, D.C., van Wilgen, B.W. and Richardson, D.M. (2017a) Towards a national strategy to optimise the management of a widespread invasive tree (*Prosopis* species; mesquite) in South Africa. *Ecosystem Services* 27, 242–252.

Shackleton, R.T., Witt, A.B.R., Pirons, F.M. and van Wilgen, B.W. (2017b) Distribution and socio-ecological impacts of the invasive alien cactus *Opuntia stricta* in Eastern Africa. *Biological Invasions* 19, 2427–2441.

Shackleton, R.T., Foxcroft, L.C., Pyšek, P., Wood, L.E. and Richardson, D.M. (2020) Assessing biological invasions in protected areas after 30 years: Revisiting nature reserves targeted by the 1980s SCOPE programme. *Biological Conservation* 243, 108424.

Steenkamp, H.E. and Chown, S.L. (1996) Influence of dense stands of an exotic tree, *Prosopis glandulosa* benson, on a savanna dung beetle (Coleoptera: Scarabaeinae) assemblage in Southern Africa. *Biological Conservation* 78, 305–311.

Sun, Y., Müller-Schärer, H. and Schaffner, U. (2022) Fighting neobiota with neobiota: Consider it more often and do it more rigorously. *Biological Conservation* 268, 109506.

van den Berg, E. (2010) Detection, quantification and monitoring *Prosopis* spp. in the Northern Cape province of South Africa using remote sensing and GIS. MSc thesis, North-West University, Potchefstroom, South Africa.

Wise, R.M., van Wilgen, B.W. and Le Maitre, D.C. (2012) Costs, benefits and management options for an invasive alien tree species: The case of mesquite in the Northern Cape, South Africa. *Journal of Arid Environments* 84, 80–90.

Woodford, D.J., Richardson, D.M., MacIsaac, H.J., Mandrak, N.E., van Wilgen, B.W. *et al.* (2016) Confronting the wicked problem of managing biological invasions. *NeoBiota* 31, 63–86.

Zachariades, C. (2021) A catalogue of natural enemies of invasive alien plants in South Africa: Classical biological control agents considered, released and established, exotic natural enemies present in the field, and bioherbicides. *African Entomology* 29, 1077–1142.

Zengeya, T., Ivey, P., Woodford, D.J., Weyl, O., Novoa, A. *et al.* (2017) Managing conflict-generating invasive species in South Africa: Challenges and trade-offs. *Bothalia* 47(2), a2160.

Zimmermann, H.G. (1991) Biological control of mesquite, *Prosopis* spp. (Fabaceae), in South Africa. *Agriculture, Ecosystems and Environment* 37, 175–186.

Part 1: History, Evolution and Spread

This Part includes three chapters that provide an overview of how, why and when *Prosopis* trees were introduced to eastern Africa, how they became invasive, and what the current and potential future extent of their distribution is. In Chapter 2, van Wilgen *et al.* review the reasons for introducing *Prosopis* trees over 100 years ago, and how these introductions continued until the 1990s when it became increasingly apparent that they represented a growing negative problem. This follows similar patterns of changing perceptions regarding the relative value of these species in other parts of the world, where it is now recognized that new approaches will be needed to manage them.

In Chapter 3, Le Roux *et al.* examine the taxonomy and systematics of the notoriously problematic genus *Prosopis*. In doing so, they recognize that species do not only become problematic because they possess traits that predispose them to becoming invasive, but also that they have been shown to evolve in response to conditions in the new environment. A number of factors, including aspects of the genetic diversity in the introduced population, propagule pressure, rapid post-introduction evolution and phenotypic plasticity have all contributed to *Prosopis juliflora* becoming invasive in eastern Africa, while the congeneric *P. pallida* did not. These findings will be useful for understanding why some species become invasive, and how closely related species could be managed and regulated.

In Chapter 4, Eckert *et al.* review recent studies that have sought to quantify the rate at which *Prosopis juliflora* has spread across eastern Africa, and what we know about its past and potential future spread. Currently, millions of hectares are already invaded in the region, and local case studies in Kenya and Ethiopia have estimated that average annualized rates of spread have been around 12% over the past 30 years. These estimations have been made possible by recent advances in remote sensing tools, which are also described in Chapter 4. Many as yet uninvaded parts of eastern Africa are at risk from future invasion, and the extent of these suitable areas has been modelled using climatic, topographical, hydrological and anthropogenic variables. These findings underline the importance of taking steps to prevent the further spread of *Prosopis julilfora* as soon as possible, and where to prioritize interventions.

2 A Brief History of the Introduction and Management of *Prosopis* Trees in Eastern Africa

Brian W. van Wilgen[1]*, Purity R. Mbaabu[2,3] and Simon K. Choge[3]
[1]Centre for Invasion Biology, Department of Botany and Zoology, Stellenbosch University, South Africa; [2]Institute for Climate Change and Adaptation, University of Nairobi, Kenya; [3]Kenya Forestry Research Institute, Marigat, Kenya

Abstract
This chapter reviews the history of introduction of *Prosopis* trees to Africa, with a focus on eastern Africa, how their perceived and actual value has changed over time, and how management policies have accordingly been adapted. People were initially encouraged to use *Prosopis* trees to combat desertification, and as a source of firewood and fodder for livestock. Some *Prosopis* species, notably *P. juliflora* in eastern Africa, subsequently became invasive, causing substantial negative impacts. The benefits provided by these trees were initially welcomed but this changed as the negative impacts grew. *P. juliflora* had thus become a conflict species, in which some people reaped the benefits of the trees while others suffered the negative consequences. This led, in one notable case, to rural communities seeking redress from the Kenyan government in court in 2006. While the realization that these trees were more harmful than useful was slow to develop, several recent studies have clearly demonstrated that the net economic value has become negative, with any value derived from the plants being offset by greater losses in the value of ecosystem services. Management has had to adapt to these new realities. The first policies promoted the active distribution of *Prosopis* trees. When the trees became invasive and negative impacts became apparent, management switched to policies of control through utilization, mainly in the form of subsidised charcoal production projects. The failure of this approach is now apparent and new initiatives have been proposed. These have been informed by recent transdisciplinary research projects and include spatially explicit management approaches, the use of best management practices (including biological control) and raising awareness. Key challenges remain but the chances of successful outcomes would be increased by embracing biological control opportunities and setting clear and achievable management goals, both of which have been absent to date.

2.1 Introduction

A large proportion of eastern Africa's landscapes are arid or semi-arid. These areas are characterized by environmental degradation brought about by frequent drought, combined with overgrazing and overharvesting of wood as a source of energy, and leading to shortages of grazing and fuelwood, as well as to erosion. The introduction of fast-growing, drought-resistant alien trees was seen as a potential way of alleviating these problems. The repeated introduction of trees to regions beyond their native range, mainly to provide timber and fodder for

*Corresponding author: bvanwilgen@sun.ac.za

© CAB International 2025. *The Ecology and Management of Invasive Prosopis Trees in Eastern Africa* (eds U. Schaffner *et al.*)
DOI: 10.1079/9781800623644.0002

livestock, or to help with the rehabilitation of degraded areas, has been a global feature of landscape management over the past century (Richardson, 1998; Richardson and Rejmánek, 2011).

Over the years, many of these introduced woody plant species (e.g. 178 species in southern Africa; Rejmánek and Richardson, 2013) have escaped cultivation and invaded grasslands, croplands, wetlands and other ecosystems. This has had serious consequences for nature and human wellbeing, and several of these invasive alien trees are considered to be among the worst invasive alien species worldwide (Lowe *et al.*, 2000). In particular, trees in the genus *Prosopis* that were introduced from central America to eastern Africa to provide a range of benefits have now become problematic invaders.

The introduction of alien trees to new environments across the world has almost entirely been based on the benefits that they could provide. These positive perceptions about trees often shape the attitudes of people in general (van Wilgen, 2012), and several organizations have been founded to protect trees and/or to promote their use across the world, including, for example, World Agroforestry, sections of the Food and Agricultural Organization of the United Nations and the Forestry Stewardship Council. When some of these trees become invasive and start to generate negative impacts, people's perceptions and opinions can become divided, setting the stage for conflict (Dickie *et al.*, 2014; van Wilgen and Richardson, 2014). The introduction of trees in the genus *Prosopis* to several countries in eastern Africa provides a good example of this phenomenon.

In this chapter, we review the rationale behind the introduction and management of *Prosopis* trees in eastern Africa, and the growing realization that their harmful characteristics may outweigh their positive attributes. We examine how attitudes have changed over time, leading to sometimes conflicting proposals on how to deal with the problem. These attitudes have also been shaped by international commitments that countries in eastern Africa have made, and that are directly relevant to the management of *Prosopis* trees. We conclude with an examination of the challenges for the future in terms of maximizing benefits and minimizing harm.

2.2 The History of *Prosopis* Introduction in Africa

The first recorded introduction of *Prosopis* trees to Africa was in 1822, when the then French Governor of Senegal introduced *Prosopis pallida* to the estuary of the River Senegal (Diagne, 1992; Choge *et al.*, 2022; Table 2.1). It was not until almost 60 years later that further introductions of *Prosopis* trees took place in South Africa in 1880, and later in Namibia in 1912 (Poynton, 2009). The planting and propagation of these trees were actively encouraged by the South African government for over half a century, and by the time it was realized that they had become invasive, the populations had established over very large areas, impeding effective control (Zimmermann, 1991).

The earliest introductions of *Prosopis* trees to eastern Africa have been traced to Khartoum, Sudan, in 1917 and were made by the government botanist R.E. Massey. The seeds originated from South Africa and Egypt (Broun and Massey, 1929; Magid *et al.*, 2014). These plantings were aimed at stabilizing sand dunes, soil conservation and socio-economic benefits such as fuel wood, timber and fodder supply (Magid *et al.*, 2014). Encouraging results from initial plantations of *Prosopis* trees in Khartoum spurred widespread planting programmes in the rest of Sudan. In 1938, a nation-wide planting programme was launched after field trials (with several tree species) in 1928 showed that only *Prosopis* species were able to survive on sandy soils in areas receiving less than 150 mm of rainfall annually (Magid *et al.*, 2014). This culminated in a massive airborne seeding campaign in Sudan's dry areas and riparian lands along the Nile River (Babiker, n.d.). Subsequently, Sudan's Forestry Service adopted *Prosopis* species for sand fixation, desertification control and rehabilitation of degraded lands (Magid *et al.*, 2014). Over the decades, these trees have invaded natural and managed landscapes, particularly in the eastern parts of Sudan. Presently, *Prosopis* trees cover more than 1.5 million ha across the country (Gafaar, 2011).

News of the successful *Prosopis* plantings in Sudan provided the impetus for their introduction into neighbouring countries such as Kenya and Ethiopia. In Kenya, *Prosopis* trees

Table 2.1. Timeline of events relating to the introduction, spread and management of alien *Prosopis* trees in Africa.

Date	Event	Source
1822	*Prosopis pallida* was introduced to Senegal	Diagne (1992)
Late 19th century	Several species of *Prosopis* introduced to South Africa (*P. glandulosa* in 1880, *P. chilensis* in 1894 and *P. juliflora* in 1897)	Poynton (2009)
1912	*Prosopis chilensis* introduced to Namibia	Poynton (2009)
1917	*Prosopis* species introduced to Sudan	Broun and Massey (1929)
1938	Nation-wide *Prosopis* tree planting plan initiated in Sudan, including aerial sowing of seeds along dry river beds	Magid *et al*. (2014); Babiker (n.d.)
1948	*Prosopis juliflora* and *P. pallida* introduced to Kenya	Choge *et al*. (2010)
1973	Plantations of *Prosopis juliflora* and *P. pallida* established in Mombasa, Kenya	Johansson (1990)
Late 1970s	*Prosopis juliflora* introduced to Ethiopia	Sertse and Pasiecznik (2005)
Early 1980s	*Prosopis juliflora* and *P. pallida* planted in Baringo County, Kenya	Johansson (1990)
1987	*Algarobius bottimeri* (a seed-feeding biological control agent) released in South Africa	Zachariades (2021)
1988	*Prosopis* trees spread across the border from Kenya to Tanzania	Kilawe *et al*. (2017)
1993	*Neltumius arizonensis* (a seed-feeding biological control agent) released in South Africa	Zachariades (2021)
1995	Working for Water programme established in South Africa in a nation-wide attempt to assist landowners with the control of invasive alien plants, including *Prosopis* species	van Wilgen and Wannenburgh (2016)
1999	A workshop in Ethiopia concludes that *Prosopis* trees should be eradicated because their costs outweighed the perceived benefits	Desissa and Binggeli (2003)
2002	*Prosopis* trees noted as having become invasive in Kenya	Choge *et al*. (2002)
2006	Residents of Baringo petition the Kenyan government for compensation because of impacts of introduced *Prosopis* trees	Petition 466 of 2006
2007	Study of *Prosopis* trees in the Northern Cape Province, South Africa, provides estimates of annual spread rates of between 7.4 and 27.5%	van den Berg (2010)
2012	Study on the economic costs and benefits of invasive *Prosopis* trees published	Wise *et al*. (2012)
2015	Inaugural meeting of the Woody Weeds project, Nairobi	
2016	Presence of *Prosopis chilensis* and *P. cf. glandulosa* reported from Angola; date of introduction unknown	Rejmánek *et al*. (2017)
2017	Strategy for the management of *Prosopis* in South Africa published; Ethiopia publishes national strategy on *Prosopis juliflora* management	Shackleton *et al*. (2017); Anonymous (2017)

Continued

Table 2.1. Continued

Date	Event	Source
2019	Tanzania publishes National Invasive Species Strategy and Action Plan	Anonymous (2019a)
2020	Kenya publishes national strategy for the management and control of invasive *Prosopis* species	Ndufa *et al*. (2020)
2021	2021 *Evippe* species (a leaf tier, miner and feeder) released as a biological control agent in South Africa	Zachariades (2021)
2022	Final meeting of the Woody Weeds project, Nairobi	

were first introduced in 1948 in Taita Taveta (previously known as Mrashe Township) by colonial veterinary workers, using seeds obtained from South Africa (Choge *et al.*, 2022). Between 1950 and 1960, *Prosopis* species (mainly *P. pallida* and *P. juliflora*) were further planted as shade and ornamental trees in Kenya's hinterland by Christian missionaries and non-governmental organizations. In attempts to rehabilitate derelict land, *P. pallida* seeds sourced from Brazil and Hawaii were planted in quarries near Bamburi in Mombasa in 1973. From this plantation, seedlings were distributed to other degraded lands such as Tharaka, Nakuru and Baringo for various perceived benefits including environmental rehabilitation and forage provision. Currently, the species is found in about 20 out of 47 counties in Kenya including Mandera, Kwale, Kajiado, West Pokot, Marsabit and Samburu. Counties with the largest invasions include Turkana, Garissa, Baringo and Tana River (see Chapter 4, this volume). The years 1996–1999 were characterized by above-average rainfall, which triggered extensive seed dispersal and subsequent colonization of previously uninvaded grasslands, making the aggressively invasive nature of these trees clearly visible for the first time in Kenya (Choge *et al.*, 2022). Estimates indicate that over 1.5 million ha are currently covered by *Prosopis* tree invasions in Kenya, which includes invasion of croplands, grazing lands, protected areas, communal lands, riparian zones and road reserves (Ministry of Environment and Forestry, 2022).

In Ethiopia, historical accounts of the introduction of *Prosopis* trees differ with respect to the exact place, time and actors involved in their introduction (Tessema, 2012). Some sources suggest that the trees may have been introduced in the late 1970s at Goro Nursery, Dire-Dawa, with seeds sourced from India (Etana, 2013), while others indicate that *P. juliflora* was introduced from Sudan to Ethiopia during the establishment of an irrigation project at Middle Awash (Shiferaw *et al.*, 2004; Etana, 2013). It is also thought that the Afar region was the putative starting point of *P. juliflora* spread. This matches with information that *Prosopis* trees may have been introduced from Kenya or Sudan by foreign workers during the construction of the Middle Awash irrigation scheme in the 1970s and 1980s. Tilahun *et al.* (2017) also indicated that these introductions first occurred in a very small number of agricultural research stations for use as barrier plants or for water and soil conservation. The only reported *Prosopis* species in all these introduction accounts is *P. juliflora*. This identification has been confirmed through research by the Ethiopian Agricultural Research Organisation and the Henry Doubleday Research Association (Etana, 2013).

Prosopis juliflora spread from Kenya to Tanzania in 1988, probably by being unintentionally carried across the border by traders and their donkeys travelling between Taita Taveta in Kenya and Mwanga District of Tanzania. By 1992, the species had become established in trading centres in the Mwanga region, and was further spread by livestock that consumed the pods (Kilawe *et al.*, 2017). The tree was also planted near schools to provide shade. The current area invaded by *Prosopis* trees in Tanzania has not yet been accurately quantified.

2.3 Benefits, Costs and Changes Over Time

Prosopis species, like many other invasive alien tree taxa (e.g. Australian *Acacia* species, van Wilgen *et al*., 2024; or *Pinus* species from the northern hemisphere, Richardson, 1998), provide both benefits and costs to society and the environment. This mix of benefits and costs often generates disagreement between stakeholders (Kueffer, 2013; Zengeya *et al*., 2017) and consequently leads to conflicts around how to manage them (van Wilgen and Richardson, 2014; Zengeya *et al*., 2017). Several studies now exist that have identified and quantified positive and negative outcomes of the introduction of *Prosopis* species in eastern Africa and elsewhere. The benefits centred largely on the sale of charcoal and of pods (Choge *et al*., 2022), while the costs include negative impacts on biodiversity, water resources, rangeland carrying capacity and the health of people, livestock and ecosystems (see Chapters 5–9, this volume, where each of these impacts is discussed in more detail).

The picture that emerges is a recipe for conflict between people who benefit from utilization on the one hand and those who bear the costs of negative impacts on the other. For example, in Kenya, *Prosopis* trees have been widely used and promoted for charcoal production as a short- and medium-term strategy for meeting the energy demand deficits in the country (Choge *et al*., 2022). To the rural communities in the Baringo region in Kenya, charcoal production from *Prosopis* trees through legalized charcoal producer associations, which were established in 2009, earned individual members a monthly income of up to USD 200 (Choge *et al*., 2022). In Afar, Ethiopia, the aggregated household benefit from *P. juliflora* was estimated at USD 303 per month (Ilukor *et al*., 2016; see also Chapter 8, this volume). However, many local people did not benefit from the sale of charcoal and bore the consequences of invasions of the land on which their livelihoods depended (Linders *et al*., 2021). In 2006, matters came to a head when local communities from the Baringo region sought compensation from the Kenyan government as a result of the substantial net negative effects on their livelihoods and wellbeing (Box 2.1).

If all costs and benefits are considered together, the net value of an invasive species will vary in space and time depending on the magnitude and value of component costs and benefits, and these will in turn depend on the socio-economic circumstances of communities (Shackleton *et al*., 2007; Linders *et al*., 2021). Based on their relative degree of harm or benefit, van Wilgen and Richardson (2014) categorized invasive alien species into four types. The first type includes alien species with low negative impact and benefit (inconsequential species; Fig. 2.1). The second type includes species with minimal negative impact but with significant benefit (beneficial species). The third type comprises alien species that provide few benefits but cause substantial negative impacts (destructive species). Unlike species in the first three types, conflict-generating species are those that are both useful and harmful, and they tend to generate controversy among different stakeholders and contexts. It is in this category that trees in the genus *Prosopis* reside (see also Chapter 8, this volume).

The introduction of *Prosopis* trees to eastern Africa was initially appreciated because of their ability to proliferate rapidly, their capacity to withstand harsh climatic conditions and because of their high survival rates even in severely degraded lands. However, attitudes began to change as impacts grew and came to exceed benefits (Fig. 2.2; see also Fig. 15.1, Chapter 15, this volume). Seid *et al*. (2020) reported that in the Afar Region, Ethiopia, some 78% of household heads were in favour of *Prosopis* trees at the time they were introduced to the region. Today, less than 1% still favour the trees, while ~90% are not in favour of them. Similarly, Little (2019) cited a senior Il Chamus leader who, while resting under a *Prosopis* tree, initially praised the cool shade and the breezes that the tree provided. A few years later, the same person was one of the main advocates for completely removing *Prosopis* trees and suing the government for introducing them to Baringo County (see Box 2.1).

Quantitative estimates of this dynamic dichotomy have been documented in South

Box 2.1. Seeking redress for impacts: the case of *Prosopis* before the courts in Kenya.

The planting of *Prosopis juliflora* trees in Kenya's Baringo County was actively encouraged by the United Nations Food and Agricultural Organization in conjunction with the Kenyan government in the mid-1970s. Two decades later, during which time the trees had become invasive and spread to cover a large proportion of the county, members of the local community started to complain about the situation to the government. However, although the government initiated long-term research on the impacts of the species on environment, livelihoods and land use systems, they retained the view that *Prosopis* trees were bringing benefits to the area, including reversing desertification and soil erosion, and as a source of fuelwood (Odhiambo, 2015). This left the community with little other option than to seek redress in a court of law (Little, 2019; see http://kenyalaw.org/caselaw/cases/view/41547/ for a full account of the case).

In their submission to the court, lodged on 23rd August 2006, the Il Chamus community members cited the loss of valuable pastures, loss of livestock due to injury by *Prosopis* thorns, impeding the movement of people, livestock and machinery, blockage of rivers causing flooding, displacement from homes, loss of playing fields, perpetual anxiety and apprehension due to adverse effects, and the destruction of the community's economic base, leading to poverty. The list also included harm inflicted on people by thorns, in some cases leading to the amputation of limbs. In an unusual move, the communities brought a toothless goat into the court as evidence of the effects of the species on livestock (Wabusya, 2013).

After considering the evidence, a panel of three judges found in favour of the community, and that the government of Kenya was liable for the damage and loss brought about by the invasion. On 7th December 2007, they ordered that a commission comprising technical and local experts be appointed by the government to assess:

- the impacts on the environment and the residents of Baringo County
- the loss resulting from the introduction and non-action by the Government, and
- the injuries suffered by people.

The panel was to report on its findings within 60 days of its appointment, in which they should also recommend commensurate monetary compensation for losses. However, the court order was never implemented following the postelection violence of 2008, a case that was reported in the local press as 'justice denied' 2 years later (https://nation.africa/kenya/news/justice-denied-as-court-orders-to-save-the-illchamus-delayed--562392). Little (2019) noted that 'the government has rarely implemented judicial rulings in favour of indigenous communities in Kenya'.

The struggle of the local community at Baringo has also highlighted the plight of minority groups with little political influence, and that the general policies of African governments are unsympathetic to rural pastoralism (Little, 2019), both of which have been highlighted by the case involving the introduction and subsequent invasion by *Prosopis* trees in Kenya.

Africa by Wise *et al.* (2012). Their study found that the net economic value of *Prosopis* trees was positive in the short term. However, the monetary value of benefits would be exceeded by the monetary value of negative impacts within 4 years (if spread rates were 30%) or within 22 years (if spread rates were 15%). This is because *Prosopis* trees do not only expand their range but also increase in density within that range, increasing the magnitude of impacts and decreasing the magnitude of benefits respectively. Similar projections of the future trends in benefits and costs of *P. juliflora* trees were reported by Ilukor *et al.* (2016), who estimated that the net present value of costs and benefits over the next 30 years in the Afar region of Ethiopia would be negative if the invasion was not halted. A net negative value is apparently already the case in both the Afar region of Ethiopia and Baringo County in Kenya (Bekele *et al.*, 2018). Shiferaw *et al.* (2019) estimated that the average loss of ecosystem service values was USD 602 million over the last 31 years, noting that this had serious consequences for the livelihoods of rural people in the area.

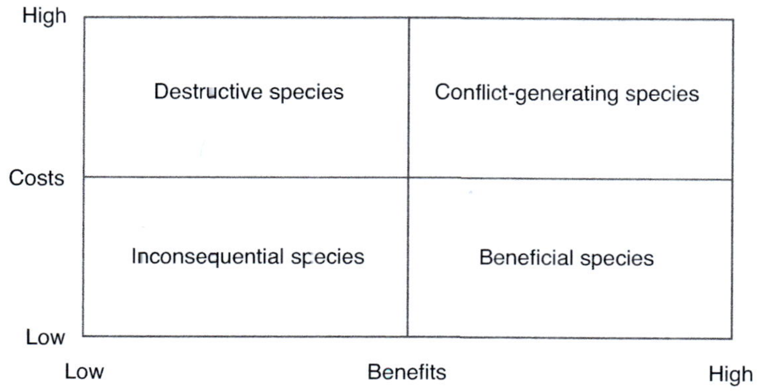

Fig. 2.1. Categorization of alien species according to their degree of harm and benefit. Adapted from van Wilgen and Richardson (2014).

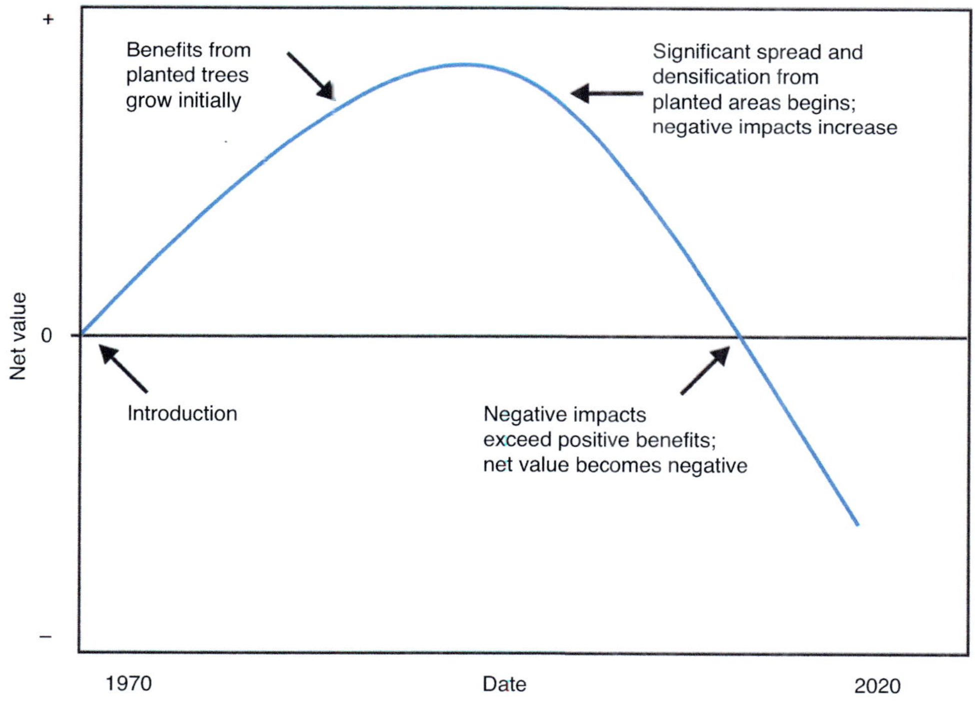

Fig. 2.2. Conceptual illustration showing changing net values (sum of benefits minus sum of impacts) over time associated with *Prosopis* trees in eastern Africa. Adapted from van Wilgen and Richardson (2014).

2.4 Strategic Approaches to Dealing with *Prosopis* Trees

2.4.1 Initial approaches to management

Prosopis trees were initially promoted as 'wonder plants', a phase of management that continued well into the late 20th century. By the time their invasive potential had been realized, they had already spread from the initial plantings and reached alarming proportions (Zimmermann, 1991). For example, the organized planting of *P. juliflora* was carried out in numerous refugee camps in Somaliland, with the active support of many non-governmental organizations, until as late as 1988 (Awale and Sugule, 2006).

In South Africa, a biological control programme was initiated in 1987, but the agents were restricted to seed-feeding insects only, so as not to damage the plant itself, which was still regarded as useful (Kleinjan *et al.*, 2021). In eastern Africa, an appreciation of the negative impacts of *Prosopis* trees began to emerge in the late 1990s. For example, in Ethiopia, *Prosopis* trees were recognized at this time as aggressive and problematic invaders, especially in the Afar region (Desissa and Binggeli, 2003; Shiferaw *et al.*, 2004).

In 1999, during a workshop in Ethiopia on agricultural weeds, it was concluded (after some debate) that the trees should be eradicated because their costs outweighed the perceived benefits (Desissa and Binggeli, 2003). About 10 years later, the local community and government in Afar declared invasion by *Prosopis* trees a top-priority issue and consequently requested support from development agencies to mitigate further expansion and restore invaded lands (Haregeweyn *et al.*, 2013).

2.4.2 Attempts at control through utilization

The Kenyan government implemented a pilot programme in Baringo County between 2005 and 2008 and subsequently formally adopted a policy of control through utilization in 2009, in which it intended to harvest the substantial biomass available from invasive *Prosopis* trees with a view to reducing their range and density (Choge *et al.*, 2022). The policy was based on the availability of a largely unexploited resource that could help to alleviate a substantial wood fuel shortage in the short to medium term, as well as to reduce negative impacts. This approach has been (and in many cases still is) promoted as the optimum solution that would minimize harm and maximize benefits (Pasiecznik *et al.*, 2001; Tewari *et al.*, 2011; Ilukor *et al.*, 2016). The approach was, however, not successful in the case of Baringo County, and invasions by *Prosopis* trees continued at an exponential rate despite the establishment of numerous government-subsidised charcoal-producing enterprises (Ng *et al.*, 2017; see also Chapter 4, this volume).

2.4.3 Developing new approaches to management

The growing awareness of the substantial impacts of *Prosopis* tree invasions, together with the realization that control by utilization was not working, led to a quest for alternative approaches to management (Choge *et al.*, 2022).

International cooperative research projects were initiated in 2015, with a focus on developing a sound understanding of the ecological and socio-economic dynamics around *Prosopis* tree invasions. This work was intended to provide the basis for the development of sustainable land use practices (Choge *et al.*, 2022). Further motivation for an expanded research programme and a change of management approach was provided by the Convention on Biological Diversity, to which all countries in eastern Africa are party. Article 8(h) of this convention requires participating countries to prevent the introduction of, control or eradicate those alien species which threaten ecosystems, habitats or species (which quite clearly includes invasive *Prosopis* species).

In 2020, the International Union for the Conservation of Nature (IUCN) produced guidelines for the prevention of biodiversity loss caused by alien invasive species (IUCN, 2000). These guidelines were intended to assist managers to give effect to Article 8(h) of the Convention on Biodiversity, and to contribute to the development of strategies. The guidelines are not binding but nevertheless useful in developing and adopting policies for invasive species management.

Recent research has explored options for the sustainable management of invasive *Prosopis* trees in eastern Africa. A wide range of stakeholders have been identified, and ways in which they can be meaningfully involved in management have been proposed (Chapters 12 and 13, this volume). Spatially explicit management, in which a range of specific management objectives relevant to the stage and extent of invasion have been identified and mapped, has been developed to guide project planning and management (Chapter 11, this volume). Best management practices, which include guidelines for setting goals, planning, monitoring and adaptive management, are outlined in Chapter 10. Chapter 10 also addresses the integration of appropriate combinations of mechanical, chemical and biological control that can be used to achieve the goals of management.

The messages from international agreements or conventions are, however, sometimes confusing or at least do not address the issue of how to deal with invasive species. For example, the United Nation's binding Convention to Combat Desertification has been silent on the issue of using potentially invasive species (Nawata, 2012) and the IUCN's Bonn Challenge to increase tree cover also significantly influenced thinking on the use of trees, and also prompted the continuous planting of invasive alien tree species (Box 2.2).

There are now three countries in eastern Africa that have developed explicit strategies to manage *Prosopis* invasions, based to a large degree on new information generated by ongoing research. These strategies have all been developed relatively recently, in response to changing perceptions and new evidence. The strategies for Ethiopia, Tanzania (including other high-priority invasive species) and Kenya were published in 2017, 2020 and 2022, respectively (Anonymous, 2017; Anonymous, 2019a; Ndufa *et al.*, 2020), and thus are all less than 5 years old at the time of writing (late 2022). While a great deal of effort has gone into their development, none has yet been formally implemented.

There are several aspects that are common to the three strategies, and these indicate broad agreement on some important aspects. All of them recognize the need for collaboration and coordination involving multiple government departments or entities and other stakeholders. All call for the identification of research priorities and increases in research effort, which indicates the universal need for better understanding of the problem. Efforts to increase the levels of awareness of the problem, and to build and maintain capacity to manage it, are also included. Further strengths are evident in the explicit recognition of the international nature of the problem in both the Ethiopian and Tanzanian strategies. More important, perhaps, is that all recognize the inadequacy of current levels of funding, and this represents a significant challenge to the implementation of the strategies.

There are, however, several weaknesses apparent in the strategies. The first area of weakness is around how goals are set. In the Kenyan strategy, the goals are framed in terms of inputs (e.g. to carry out control, to build capacity or to establish committees) rather than outputs (reductions in the extent of invasions or the restoration of functioning ecosystems). The Ethiopian strategy aims to prevent expansion and reclaim and restore functional ecosystems, but the extent of desired restoration is not specified. The Tanzanian strategy, on the other hand, is quite explicit about the desired outcomes but the targets that are set are considerably over-ambitious. For example, there are targets for eradicating 95% of emerging invasive species, securing 70% of the necessary funding at a national level, and reducing the abundance and spread of all national priority invasive species by at least 50%, all due to be achieved by 2029 (which is almost certainly unattainable). All the strategies lack indicators by which the progress towards these goals can be monitored (see Wilson *et al.*, 2018 for examples of country-level indicators). Other weaknesses include ambivalence about utilization as part of control operations (Ethiopia), no clear statement on the use of biological control (Ethiopia and Tanzania), and confusion about how to resolve conflicting goals of removing *Prosopis* trees while meeting a target to increase tree cover (Kenya; see Box 2.2).

Box 2.2. *Prosopis* and the global drive to increase tree cover.

In 2011, the International Union for the Conservation of Nature (IUCN) and the government of Germany issued the Bonn Challenge (www.bonnchallenge.org). The goal of this initiative was to bring 150 million ha of 'degraded and deforested landscapes' into restoration by 2020 and 350 million ha by 2030. The challenge's website notes that it had received pledges for 150 million ha by 2017. There were 31 pledges made by African countries, including 15 million ha in Ethiopia, 5.1 million ha in Kenya, 14.6 million ha in Sudan, 5.2 million ha in Tanzania and 2.5 million ha in Uganda. In addition, Kenya adopted a revised national constitution in 2010 (Anonymous, 2010), in which it was proclaimed [Section 69(1)(b)] that 'The State shall work to achieve and maintain a tree cover of at least ten per cent of the land area of Kenya'.

The basis for the Bonn Challenge has been disputed, however. For example, Bond *et al.* (2019) pointed out that pristine and species-rich grasslands and savannas in Africa were often flagged as degraded and deforested landscapes due to their low tree cover, thus qualifying them for planting. These areas included pristine and iconic conservation areas such as the Serengeti National Park in Tanzania and the Kruger National Park in South Africa. However, converting these and other areas by planting trees may not only lead to ambiguous results regarding ecosystem carbon balance (Jackson *et al.*, 2002), it would also impact negatively on their significant biodiversity and on the ecosystem services that they deliver to millions of people. Planting trees for carbon sequestration would require trade-offs with other ecosystem services, such as water, because trees use more water than the vegetation they replace (Jackson *et al.*, 2005). Furthermore, if areas invaded by alien *Prosopis* trees are cleared and restored to grassland, this would be at least as effective in replenishing soil organic carbon levels in degraded rangelands as encroachment by the invasive *Prosopis* trees themselves (Mbaabu *et al.*, 2020). Smith *et al.* (2016) concluded that 'negative emission technologies', including tree planting schemes, are unlikely to have much impact and are merely distractions from attending to the real issue of reducing fossil fuel consumption.

In the process of drafting a national strategy to deal with invasive trees in Kenya, it was initially proposed that invasions by *Prosopis* trees could assist the country in achieving their constitutional goal of 10% tree cover in the country. An early draft of the proposal noted the constitutional requirement to increase tree cover and stated that 'this applies to all woody vegetation types and *Prosopis* is one of the tree species that can be managed to contribute to the achievement of this objective'. This was changed in later versions as a result of inputs from invasion ecologists, who pointed to the dangers of such an approach.

The final version (Ndufa *et al.*, 2020) now states that 'the long-term ecological, environmental and socio-economic losses due to *Prosopis* invasions are higher than the benefits derived from this invasive tree species', and it should therefore be excluded from any afforestation projects. The (over-ambitious) strategy for achieving and maintaining over 10% tree cover by 2022 (Anonymous, 2019b) currently only lists the non-invasive *Prosopis pallida* as one of the candidate species for planting.

The use of alien trees that are known to be invasive remains a threat, though. This was illustrated by the recent granting of certification for the use of such trees, including neem (*Azadirachta indica*) and jacaranda (*Jacaranda mimosifolia*) by the Forestry Stewardship Council (https://wwf.panda.org/wwf_news/?19414/Kenya-carves-its-place-in-forests-history). Both of these species are known to be invasive (Witt and Luke, 2017). These conflicts and trade-offs will remain one of the challenges when managing invasive trees.

2.5 Implications for Management

Only a few *Prosopis* trees (primarily *P. juliflora* and *P. pallida*) were introduced into eastern Africa to provide benefits to rural communities. One of these (*P. pallida*) did not become invasive, or only at a local level, but *P. juliflora* proved to be an aggressive invader that caused many significant impacts, as outlined in other parts of this book. The focus of management has had to shift from establishing the trees across wide areas to preventing their spread and reducing their impacts. In future, managers will have to agree on how to develop policies and approaches that can be used to maximize benefits and minimize harm.

There are two issues that are currently being debated. The first relates to whether ongoing utilization can provide at least some benefit from invasive stands while at the same time allowing for effective control. One way to address such conflicts may be to codevelop a spatially explicit management approach with the stakeholders, which primarily focuses on preventing the invasion of new land and on clearing high-value land or land with low densities of *P. juliflora* and continuously removing new seedlings. In such a scenario, utilization of *P. juliflora* may still be possible in the highly invaded areas, but only if it does not interfere with the overall management objective of stopping and reversing the invasion elsewhere (see Chapter 11, this volume, for a more detailed description of spatially explicit approaches). The second issue relates to the use of biological control agents, which offer the only real hope of bringing invasions under control over larger areas (Henderson and Wilson, 2017; Sun *et al.*, 2022). To date, countries in eastern Africa have only made limited use of biological control, and a general aversion to the use of this option needs to be overcome (Chapter 12, this volume).

Acknowledgements

This work was supported by the Swiss Programme for Research on Global Issues for Development (r4d), funded by the Swiss National Science Foundation and the Swiss Agency for Development and Cooperation, through the project 'Woody invasive alien species in East Africa: Assessing and mitigating their negative impact on ecosystem services and rural livelihood' (Grant Number: 400440_152085). This work was also supported in part by the National Research Foundation of South Africa (Grant Numbers 150260 and 151653 to BWvW).

References

Anonymous (2010) The constitution of Kenya, 2010. National Council for Law Reporting with the Authority of the Attorney General, Nairobi, Kenya.

Anonymous (2017) National strategy on *Prosopis juliflora* management. Federal Democratic Republic of Ethiopia, Ministry of Livestock and Fisheries, Addis Ababa.

Anonymous (2019a) National invasive species strategy and action plan (NISSAP) (2019–2029). Office of the Vice President, Dodoma, Tanzania.

Anonymous (2019b) National strategy for achieving and maintaining over 10% tree cover by 2022. Ministry of Environment and Forestry, Nairobi, Kenya.

Awale, A.I. and Sugule, A.J. (2006) Proliferatio nof Honey Mesquite (*Prosopis juliflora*) in Somaliland. Case study on opportunities and challenges. Candlelight for Health, Education and Environment, Abu Dhabi, United Arab Emirates.

Babiker, A.G.T. (n.d.) Mesquite (*Prosopis* spp.) in Sudan: History, distribution and control. Available at: https://vonheimendahl.wordpress.com/mesquite-prosopis-spp-in-sudan-history-distribution-and-control/ (accessed 27 January 2018).

Bekele, K., Haji, J., Legesse, B., Shiferaw, H. and Schaffner, U. (2018) Impacts of woody invasive alien plant species on rural livelihood: Generalised propensity score evidence from *Prosopis* spp. invasion in Afar region in Ethiopia. *Pastoralism: Research, Policy and Practice* 8, 1–17.

Bond, W.J., Stevens, N., Midgley, G.F. and Lehmann, C.E. (2019) The trouble with trees: Afforestation plans for Africa. *Trends in Ecology and Evolution* 34, 963–965.

Broun, A.F. and Massey, R.E (1929) *Flora of the Sudan*. Thomas Murby and Company, London, UK.

Choge, S., Mbaabu, P.R. and Muturi, G.M. (2022) Management and control of the invasive *Prosopis juliflora* tree species in Africa with a focus on Kenya. In: Puppo, M.C. and Felker, P. (eds) *Prosopis as a Heat Tolerant Nitrogen Fixing Desert Food Legume*. Academic Press, London, pp. 67–81.

Choge, S.K., Ngunjiri, F.D., Kuria, M.N., Busaka, E.A. and Muthondeki, J.K. (2002) Status and impact of *Prosopis* in Kenya. Unpublished Technical Report. Kenyan Forestry Research Institute, Nairobi, Kenya.

Choge, S.K., Wahome, R., Kyuma, R. and Tendwa, J. (2010) Management and utilization of *Prosopis* species: Experiences from Peru and Argentina in South America. Technical Trip Report. Kenyan Forestry Research Institute, Nairobi, Kenya.

Desissa, D. and Binggeli, P. (2003) *Prosopis* sp. Available at: www.mikepalmer.co.uk/woodyplantecology/ethiopia/invasives/prosopis.html (accessed 30 April 2023).

Diagne, O. (1992) Current developments on *Prosopis* species in Senegal including work on nitrogen fixation. In: Dutton, W. (ed.) Prosopis *Species:Aspects of Their Value, Research and Development*. University of Durham, Durham, UK, pp. 47–60.

Dickie, I.A., Bennett, B.M., Burrows, L.E., Nuñez, M.A., Peltzer, D.A. *et al*. (2014) Conflicting values: Ecosystem services and invasive tree management. *Biological Invasions* 16, 705–719.

Etana, B. (2013) Distribution and challenges of an invasive exotic species, *Prosopis juliflora* (sw.) DC. (Fabaceae) in Ethiopia, East Africa. *International Journal of Green and Herbal Chemistry* 2, 110–120.

Gafaar, A. (2011) *Plantations Forestières et Ilots Boisés au Soudan*, Working Paper Series, Vol. 1. African Forest Forum, 113 pp.

Haregeweyn, N., Tsunekawa, A., Tsubo, M., Meshesha, D. and Melkie, A. (2013) Analysis of the invasion rate, impacts and control measures of *Prosopis juliflora*: A case study of Amibara District, Eastern Ethiopia. *Environmental Monitoring and Assessment* 185, 7527–7542.

Henderson, L. and Wilson, J.R. (2017) Changes in the composition and distribution of alien plants in South Africa: An update from the Southern African plant invaders atlas. *Bothalia* 47, 1–26.

Ilukor, J., Rettberg, S., Treydte, A. and Birner, R. (2016) To eradicate or not to eradicate? Recommendations on *Prosopis juliflora* management in Afar, Ethiopia, from an interdisciplinary perspective. *Pastoralism* 6, 1–8.

IUCN (2000) IUCN Guidelines for the prevention of biodiversity loss caused by alien invasive species. Approved by the 51st meeting of the IUCN Council, Gland Switzerland, February 2000. Available at: https://portals.iucn.org/library/efiles/documents/Rep-2000-052.pdf (accessed 12 May 2023).

Jackson, R.B., Banner, J.L. and Jobbágy, E.G. (2002) Ecosystem carbon loss with woody plant invasion of grasslands. *Nature* 418(6898), 623–626.

Jackson, R.B., Jobbágy, E.G. and Avissar, R. (2005) Trading water for carbon with biological carbon sequestration. *Science* 310(5756), 1944–1947.

Johansson, S. (1990) *Controlling and containing the spreading of* Prosopis *spp. at Bura. An outline of options and required actions. Research component in Bura fuelwood project*. Technical Report. Kenyan Forestry Research Institute Nairobi, Kenya.

Kilawe, C.J., Mbwambo, J.R., Kajembe, G.C., Mwakalukwa, E.E., Amri, A.M. *et al*. (2017) Mrashia: *Prosopis* invading pastures and agricultural lands in Tanzania. Available at: https://www.researchgate.net/publication/317064207_Mrashia_Prosopis_has_started_invading_pastures_and_agricultural_lands_in_Tanzania (accessed 30 April 2023).

Kleinjan, C.A., Hoffmann, J.H., Heystek, F., Ivey, P. and Kistensamy, Y. (2021) Developments and prospects for biological control of *Prosopis* (Leguminosae) in South Africa. *African Entomology* 29, 859–874.

Kueffer, C. (2013) Integrating natural and social sciences for understanding and managing plant invasions. In: Larrue, S. (ed.) *Biodiversity and Societies in the Pacific Islands*. Presses Universitaires de Provence and ANU ePress, Canberra, Australia, pp. 71–96.

Linders, T.E., Schaffner, U., Alamirew, T., Allan, E., Choge, S.K. *et al*. (2021) Stakeholder priorities determine the impact of an alien tree invasion on ecosystem multifunctionality. *People and Nature* 3, 658–672.

Little, P.D. (2019) When "green" equals thorny and mean: the politics and costs of an environmental experiment in East Africa. *African Studies Review* 62, 132–163.

Lowe, S., Browne, M., Boudjelas, S. and Poorter, M. (2000) 100 of the world's worst invasive alien species: A selection from the global invasive species database. UN Invasive Species Specialist Group. Available at: https://portals.iucn.org/library/sites/library/files/documents/2000-126.pdf (accessed 15 October 2024).

Magid, T.D.A., Siddig, E.N.A.E., Ahmed, A. and Ahmed, M.T.E.S. (2014) Mesquite in Sudan: A boon or bane for dry lands? Its socio-economic and management aspects in Kassala State, Sudan. *Journal of Forest Products and Industries* 3, 182–190.

Mbaabu, P.R., Olago, D., Gichaba, M., Eckert, S., Eschen, R. *et al*. (2020) Restoration of degraded grasslands, but not invasion by *Prosopis juliflora*, avoids trade-offs between climate change mitigation and other ecosystem services. *Scientific Reports* 10, 1–13.

Ministry of Environment and Forestry (2022) *National strategy and action plan for the management and control of the invasive* Prosopis juliflora *tree species. 2022–2030*. Ministry of Environment and Forestry, Nairobi, Kenya.

Nawata, H. (2012) To combat a negative heritage of combating desertification: Developing comprehensive measures to control the alien invasive species mesquite (*Propopis juliflora*) in Sudan. *Journal of Arid Land Studies* 22, 9–12.

Ndufa, J.K., Choge, S. and Cheboiwo, J. (2020) *National strategy for the management and control of the invasive* Prosopis *species*. Ministry of Environment and Forestry, Nairobi, Kenya.

Ng, W.T., Rima, P., Einzmann, K., Immitzer, M., Atzberger, C. *et al.* (2017) Assessing the potential of Sentinel-2 and Pléiades data for the detection of *Prosopis* and *Vachellia* spp. in Kenya. *Remote Sensing* 9, 74.

Odhiambo, M.R. (2015) Il chamus versus the state: Vulnerability, litigation and resilience building in the Baringo lowlands of Kenya. PhD thesis, University of Cologne, Germany.

Pasiecznik, N.M., Felker, P., Harris, P.J., Harsh, L., Cruz, G. *et al.* (2001) *The* Prosopis juliflora–Prosopis pallida *Complex: A Monograph*. HDRA, Coventry, UK, p. 172.

Poynton, R.J. (2009) In: *Tree Planting in Southern Africa. Volume 3. Other genera*, Vol. 3. Department of Agriculture, Forestry and Fisheries, Pretoria, South Africa.

Rejmánek, M. and Richardson, D.M. (2013) Trees and shrubs as invasive alien species – 2013 update of the global database. *Diversity and Distributions* 19, 1093–1094.

Rejmánek, M., Huntley, B.J., Le Roux, J.J. and Richardson, D.M. (2017) A rapid survey of the invasive plant species in western Angola. *African Journal of Ecology* 55, 56–69.

Richardson, D.M. (1998) Forestry trees as invasive aliens. *Conservation Biology* 12, 18–26.

Richardson, D.M. and Rejmánek, M. (2011) Trees and shrubs as invasive alien species – a global review. *Diversity and Distributions* 17, 788–809.

Seid, O., Haji, J. and Legesse, B. (2020) Rural households' perception on the effects of *Prosopis juliflora* invasion: The case of Amibara District of Afar National Regional State, Ethiopia. *Pastoralism* 10, 1–9.

Sertse, D. and Pasiecznik, N.M. (2005) *Controlling the spread of Prosopis in Ethiopia by Its utilisation*. Henry Doubleday Research Association, Coventry, UK.

Shackleton, C.M., McGarry, D., Fourie, S., Gambiza, J. and Shackleton, S.E. (2007) Assessing the effects of invasive alien species on rural livelihoods: Case examples and a framework from South Africa. *Human Ecology* 35, 113–127.

Shackleton, R.T., Le Maitre, D.C., van Wilgen, B.W. and Richardson, D.M. (2017) Towards a national strategy to optimise the management of a widespread invasive tree (*Prosopis* species; mesquite) in South Africa. *Ecosystem Services* 27, 242–252.

Shiferaw, H., Teketay, D., Nemomissa, S. and Assefa, F. (2004) Some biological characteristics that foster the invasion of *Prosopis juliflora* (Sw.) DC. at Middle Awash Rift Valley Area, north-eastern Ethiopia. *Journal of Arid Environments* 58, 135–154.

Shiferaw, H., Bewket, W., Alamirew, T., Zeleke, G., Teketay, D. *et al.* (2019) Implications of land use/land cover dynamics and *Prosopis* invasion on ecosystem service values in Afar Region, Ethiopia. *Science of the Total Environment* 675, 354–366.

Smith, P., Davis, S.J., Creutzig, F., Fuss, S., Minx, J. *et al.* (2016) Biophysical and economic limits to negative CO_2 emissions. *Nature Climate Change* 6, 42–50.

Sun, Y., Müller-Schärer, H. and Schaffner, U. (2022) Fighting neobiota with neobiota: Consider it more often and do it more rigorously. *Biological Conservation* 268, 109506.

Tessema, Y.A. (2012) Ecological and economic dimensions of the paradoxical invasive species *Prosopis juliflora* and policy challenges in Ethiopia. *Journal of Economics and Sustainable Development* 3, 1–8.

Tewari, J., Krishnan, P., Harsha, S.L. and Bohra, H.C. (2011) *Prosopis juliflora: Past, Present and Future*. Desert Environmental Conservation Association, Jodhpur and Central Arid Zone Research Institute, Jodhpur, India.

Tilahun, M., Birner, R. and Ilukor, J. (2017) Household-level preferences for mitigation of *Prosopis juliflora* invasion in the Afar region of Ethiopia: A contingent valuation. *Journal of Environmental Planning and Management* 60, 282–308.

van den Berg, E. (2010) Detection, quantification and monitoring *Prosopis* spp. in the Northern Cape Province of South Africa using remote sensing and GIS. MSc thesis, North-West University, Potchefstroom.

van Wilgen, B.W. (2012) Evidence, perceptions, and trade-offs associated with invasive alien plant control in the Table Mountain National Park, South Africa. *Ecology and Society* 17, 23.

van Wilgen, B.W. and Wannenburgh, A. (2016) Co-facilitating invasive species control, water conservation and poverty relief: Achievements and challenges in South Africa's Working for Water programme. *Current Opinion in Environmental Sustainability* 19, 7–17.

van Wilgen, B.W. and Richardson, D.M. (2014) Challenges and trade-offs in the management of invasive alien trees. *Biological Invasions* 16, 721–734.

van Wilgen, B.W., Holmes, P.M., Wannenburgh, A. and Wilson, J.R. (2024) Controlling invasive wattles: The South African story. In: Richardson, D.M., Le Roux, J.J. and Marchante, E. (eds) *Wattles – Australian Acacia Species Around the World*. CABI Publishers, Wallingford, UK, pp. 416–437.

Wabusya, E.N. (2013) Policy outcomes and community perceptions of a dryland invasive species: A case study of *Prosopis juliflora* in Baringo County, Kenya. Master's thesis, International Institute of Social Studies, The Hague, The Netherlands.

Wilson, J.R., Faulkner, K.T., Rahlao, S.J., Richardson, D.M., Zengeya, T.A. *et al.* (2018) Indicators for monitoring biological invasions at a national level. *Journal of Applied Ecology* 55, 2612–2620.

Wise, R.M., van Wilgen, B.W. and Le Maitre, D.C. (2012) Costs, benefits and management options for an invasive alien tree species: The case of mesquite in the Northern Cape. *Journal of Arid Environments* 84, 80–90.

Witt, A. and Luke, Q. (2017) *Guide to the Naturalised and Invasive Plants of Eastern Africa*. CABI, Wallingford, UK.

Zachariades, C. (2021) A catalogue of natural enemies of invasive alien plants in South Africa: Classical biological control agents considered, released and established, exotic natural enemies present in the field, and bioherbicides. *African Entomology* 29, 1077–1142.

Zengeya, T., Ivey, P., Woodford, D.J., Weyl, O., Novoa, A. *et al.* (2017) Managing conflict-generating invasive species in South Africa: Challenges and trade-offs. *Bothalia* 47, 1–11.

Zimmermann, H.G. (ed.) (1991) Biological control of mesquite, *Prosopis* spp (Fabaceae), in South Africa. *Agriculture, Ecosystems and Environment* 37, 175–186.

3 The Biogeography, Taxonomy and Molecular Ecology of *Prosopis* Invasions, with an Emphasis on *P. Juliflora*

Johannes J. Le Roux[1,2]* and María L. Castillo[2,3]

[1]School of Natural Sciences, Macquarie University, Sydney, Australia; [2]Centre for Invasion Biology, Department of Botany and Zoology, Stellenbosch University, South Africa; [3]Institute of Botany of the Czech Academy of Sciences, Průhonice, Czech Republic

Abstract

The taxonomy and systematics of species in the genus *Prosopis* are notoriously problematic. Uncertain biogeography of some species in the genus, coupled with high intraspecific variation and extensive hybridization, has partly contributed to this dilemma. This uncertainty has caused various problems for researchers studying the biology and management of invasive *Prosopis* species and their hybrids. We discuss the complex taxonomy, native-range biogeography and introduction history of *Prosopis*, with an emphasis on *P. juliflora*, a major invader in many parts of the world. We then review studies on the molecular ecology and genetics of the genus. Molecular analyses have helped to resolve some of the biogeographic and taxonomic issues of the genus, including systematics, classification and evolutionary ecology. For example, recent and comprehensive phylogenetic analysis led to the reclassification of *Prosopis* into four genera and population genetic studies have helped delimit the native-range areas and species boundaries in some taxa. Genetic and cytogenetic studies confirmed the occurrence of extensive hybridization between closely related diploid *Prosopis* species (in both their native and non-native ranges) and identified *P. juliflora* as the only polyploid (tetraploid) species in the genus. Genetic improvement studies have also created superior germplasm for forestry and agroforestry purposes. We discuss all these findings in the light of the evolutionary ecology of *P. juliflora*, with a focus on its invasion in eastern Africa.

3.1 Introduction

Reliable taxonomic identification of invasive alien species can be complicated because of high intraspecific variation (Le Roux and Wieczorek, 2009), the cointroduction of closely related taxa such as cryptic species or subspecies (e.g. Thompson *et al.*, 2015) and hybridization between congeners (Ellstrand and Schierenbeck, 2000), amongst others. Correct taxonomic identification is important for effective management of invasive species (Pysek *et al.*, 2013). For example, while species distribution models (SDMs) are valuable to assess risk, predict impact and monitor the spread of invasive species, they require accurate taxonomic identification to provide information useful for management (Franklin, 2009). Another example of how uncertain taxonomy may impact management is the lowered efficiency of biological control

*Corresponding author: jaco.leroux@mq.edu.au

© CAB International 2025. *The Ecology and Management of Invasive Prosopis Trees in Eastern Africa* (eds U. Schaffner *et al.*)
DOI: 10.1079/9781800623644.0003

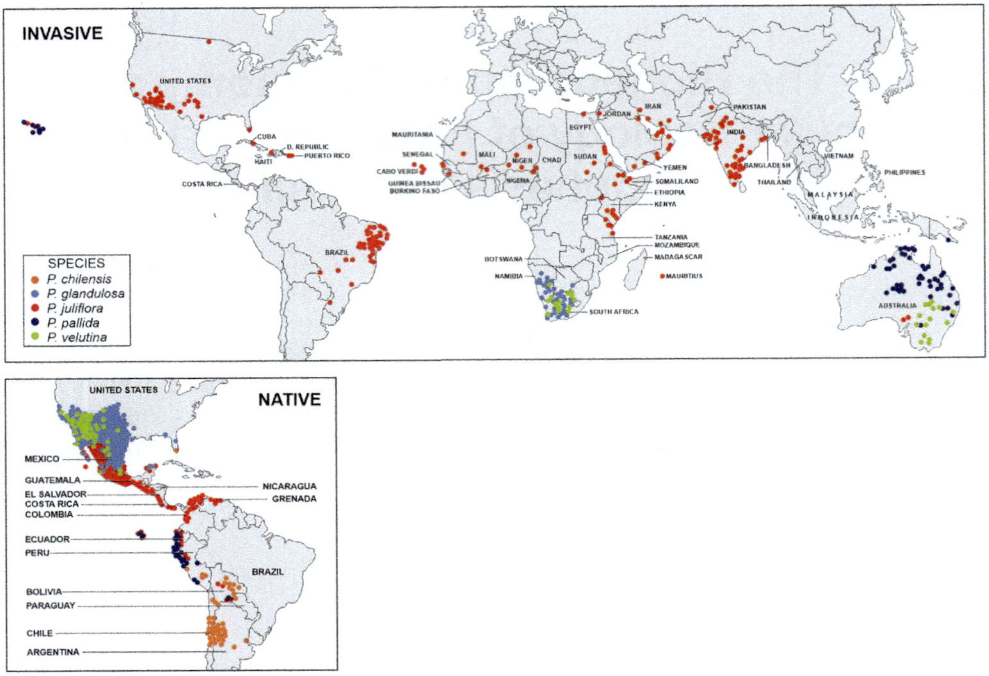

Fig. 3.1. Maps showing invasive and native records of selected *Prosopis* species. Maps were drawn from data reported in Shackleton *et al.* (2014) and the Global Biodiversity Information Facility database. For a full list of countries/regions that form part of the non-native ranges of these *Prosopis* species, see Shackleton *et al.* (2014).

when target hosts are misidentified (Le Roux, 2021).

Species in the genus *Prosopis* (*sensu lato*), commonly known as mesquite, provide a good example of how uncertain taxonomy can complicate the study and management of invasive populations. Several species in the genus (in particular *P. chilensis*, *P. glandulosa*, *P. juliflora*, *P. pallida* and *P. velutina*) have been introduced to over 100 countries outside their native ranges in South, Central and southern North America (Shackleton *et al.*, 2014; Fig. 3.1). Most of these introductions were intentional, for agroforestry, 'environmental services' such as soil improvement, animal fodder and for fuelwood (Shackleton *et al.*, 2014). Some *Prosopis* introductions have resulted in widespread invasions. Several species are also invasive in their native ranges (Shackleton *et al.*, 2014).

The identification of *Prosopis* species is notoriously difficult because of limited morphological variation among some closely related taxa (Burkart, 1976; Díaz Celis, 1995), widespread introgressive hybridization among species (Hunziker *et al.*, 1986) and considerable intraspecific variation (Burkart, 1976; Solbrig *et al.*, 1977). Burkart (1976) originally described 44 *Prosopis* species, divided into five sections, primarily based on armature (see Appendix 3.1). A recent comprehensive phylogenetic analysis based on analyses of 997 nuclear genes challenged this classification (Hughes *et al.*, 2022), supporting previous work suggesting that the genus is polyphyletic and in need of taxonomic revision (Bessega *et al.*, 2006; Burghardt and Espert, 2007). Based on this phylogenetic information, a revised taxonomy was proposed that splits *Prosopis* (*sensu lato*) into four genera (*Anonychium*, *Prosopis*, *Strombocarpa* and *Neltuma*), with most species residing in the resurrected genus *Neltuma*, while only three species remain in the genus *Prosopis* (Hughes *et al.*, 2022; see Appendix

3.1). As discussed in Chapter 1 of this volume, Burkart's (1976) classification of *Prosopis* is adopted here.

The problematic taxonomy of *Prosopis* has complicated an understanding of the ecology and management of invasive species in the genus in several ways. For example, in Australia four species, *P. glandulosa*, *P. juliflora*, *P. pallida*, *P. velutina* and their hybrids, are responsible for invasions over vast areas in the arid and semi-arid parts of the country (Van Klinken and Campbell, 2009). Invasive hybrids in Australia almost certainly never involve *P. juliflora*. This species is the only tetraploid member of the genus and, thus, any hybrids arising from cross-fertilization between it and other (diploid) *Prosopis* species will result in sterile triploid hybrids. While the parentage of hybrids in Australia is poorly understood, morphological characters suggest that all four species are hybridizing (Van Klinken and Campbell, 2009). Moreover, it is likely that additional undetected *Prosopis* species are present in the country but that excessive hybridization makes their identification near impossible. In Australia, some species, such as *P. glandulosa*, *P. velutina*, and their hybrids, also appear to be more fire-tolerant than other species (e.g. *P. pallida*) (Van Klinken and Campbell, 2009), reducing the utility of fire as a control method in many invaded areas (Shackleton *et al.*, 2014). The distributions of different *Prosopis* taxa also differ in Australia, with *P. pallida* commonly found from the east coast of Queensland through to the Northern Territory to the west coast of Western Australia (Van Klinken and Campbell, 2009; Conn, 2023). However, this species is absent in the cooler southern states of the country, where *P. velutina* and *P. velutina* × *P. glandulosa* hybrids dominate (Van Klinken and Campbell, 2009; Conn, 2023). While these distributions may simply reflect a legacy of different introduction histories of different taxa, they may also indicate variation in climate preferences of different species and their hybrids (Castillo *et al.*, 2021a).

Similar to *Prosopis* in Australia, the taxonomy of invasive populations in South Africa is confusing. For example, Ross (1975) initially identified three species in the country: *P. glandulosa*, *P. pubescens* and *P. velutina*, with two subspecies of the former having been introduced. Subsequently, *P. alba*, *P. chilensis*, *P. juliflora* and *P. laevigata* (and hybrids) have been added to the list (Shackleton *et al.*, 2014 and references therein). Of all *Prosopis* species in South Africa, *P. glandulosa* var. *torreyana*, *P. velutina* and their hybrids dominate invasions (Zimmermann, 1991). Interestingly, biological control agents released in Australia and South Africa appear to do well on most 'pure' *Prosopis* species and their hybrids (Zimmermann, 1991; Zachariades *et al.*, 2011), suggesting that these agents are not highly coevolved biotypes of specific *Prosopis* species.

Molecular analyses have been valuable in resolving some of the taxonomic issues associated with various invasive plants (Le Roux and Wieczorek, 2009). In some instances, these studies have also provided valuable insights into the dynamics of invasive populations (e.g. inferences of introduction histories; Hirsch *et al.*, 2021) and post-introduction processes that underlie invasiveness, such as rapid evolution (e.g. Marchini *et al.*, 2018).

In the remainder of this chapter, we discuss examples of molecular studies on *Prosopis*, with a focus on *P. juliflora*, a major invader in many parts of the world and the subject of this book. We start by discussing biogeographic and taxonomic uncertainties of *P. julifora* and then review the literature on molecular studies investigating different aspects of the genus. We discuss the findings of these studies in the light of *Prosopis* invasions, with an emphasis on invasive *P. juliflora* in eastern Africa.

3.2 Biogeography, Introduction History and Taxonomy of *Prosopis juliflora*

3.2.1 Native range biogeography

The native range of *P. julifora* includes northern South America (Venezuela, Colombia) and Central America to Mexico. As one of two tropical *Prosopis* species (the other being *P. pallida*), *P. juliflora* has been extensively introduced into areas around the world with suitable climates (into at least 90 countries; Shackleton *et al.*, 2014; Fig. 3.2). Despite being the only tetraploid member of the genus, the taxonomic

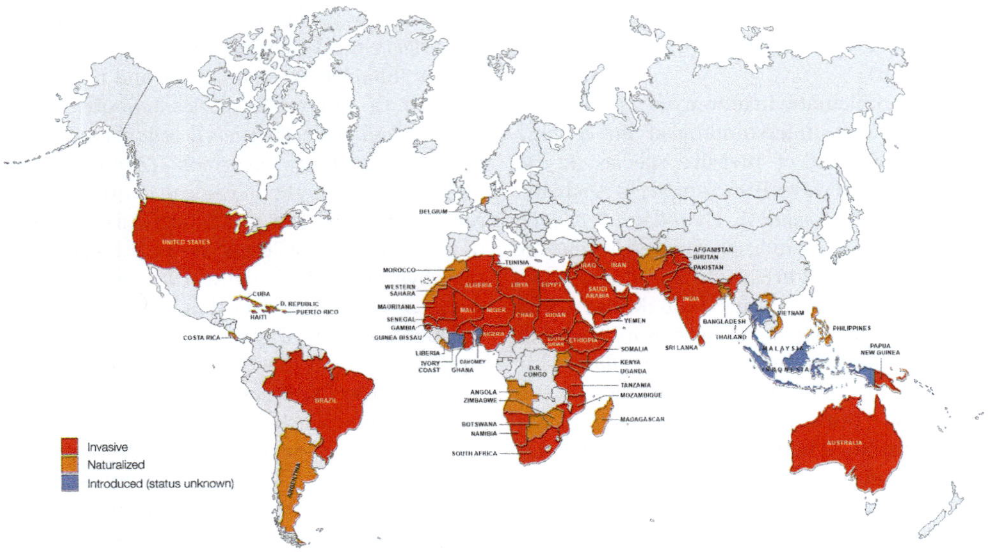

Fig. 3.2. Map showing the status of introduced *Prosopis juliflora* in different parts of the world. Countries/regions where the species is considered invasive are indicated in red and those where it is considered naturalized are indicated in orange. Countries/regions where *P. juliflora* is known to have been introduced but without knowledge of its status as either naturalized or invasive are indicated in blue. Map drawn from data reported in Shackleton *et al.* (2014). Please note that some small island nations where *P. juliflora* have been introduced are not shown on the map. For a full list of countries/regions that form part of the species' non-native range, see Shackleton *et al.* (2014).

and biogeographic delimitations of *P. juliflora* are notoriously problematic, illustrated by the array of synonymy attributed to the species since it was described in 1788 (originally as *Mimosa juliflora*; Pasiecznik *et al.*, 2001). The taxonomic confusion surrounding *P. juliflora* is further exacerbated by old literature often assigning it to species that are now accepted as other taxa (Pasiecznik *et al.*, 2001). As discussed below, there is also evidence that global introductions of some *Prosopis* species have been misidentified as *P. juliflora* (Pasiecznik *et al.*, 2001; Palacios *et al.*, 2012). Given the morphological similarities between *P. juliflora* and many diploid *Prosopis* taxa, flow cytometry analysis or chromosome counts are needed for accurate identification.

Many of the problems in delimiting the species boundaries of *P. juliflora* stem from historical treatments of the genus' biogeographic distribution and taxonomy. In 1875, Bentham grouped all *Prosopis* taxa known at the time from the section Algarobia under *P. juliflora*. He also described the new taxon, *P. limensis*, as the only species from the section Algarobia present in Peru. Burkart (1940) revised the geographic distribution of *P. juliflora* to include Colombia, Cuba, Dominican Republic, El Salvador, Jamaica, Mexico, Nicaragua and Venezuela. The type specimen of *P. juliflora* is from Jamaica, but apparently this material represents an introduced population (Burkart, 1976). In one of the most detailed taxonomic analyses ever undertaken on *Prosopis*, Johnston (1962) observed that *P. juliflora* previously described from the Pacific coast of Central America was morphologically distinct from those in the Caribbean. This was so conspicuous that he suggested the possible reversion of Pacific coastal material to a former name, *P. vidaliana* (also see Pasiecznik *et al.*, 2001), which was subsequently reinstated (Palacios, 2006).

Burkart (1976) revised his earlier (1940) distribution of *P. juliflora* to include coastal areas in Ecuador and Peru where he identified two varieties, var. *inermis* and var. *horrida* respectively, extending the distribution of the species further south. He also synonymized *P. limensis* from Peru with *P. pallida* and included Ecuador and Colombia as part of the latter's distribution along '... drier parts and along the Pacific coast'. *Prosopis juliflora* and *P. pallida* harbour various intraspecific morphological forms and there have been numerous misidentifications in both the native and non-native areas where these two species co-occur (Pasiecznik *et al.*, 2001). To deal with this taxonomic uncertainty, Pasiecznik *et al.* (2001) proposed the classification of the '*P. pallida–P. juliflora* complex'. However, subsequent morphological and genetic studies found that these two species are distinct (e.g. Harris *et al.*, 2003; Landeras *et al.*, 2006; Sherry *et al.*, 2011). Moreover, molecular analyses including *Prosopis* from Peru, Ecuador and Colombia confirmed the separation of *P. juliflora*, *P. limensis* and *P. pallida* (Palacios *et al.*, 2012). Palacios *et al.*'s (2012) analyses also no longer considered *P. juliflora* as being native to Peru or Ecuador, in contrast to the work of Burkart (1976) and others that followed (e.g. Díaz Celis, 1995; Pasiecznik *et al.*, 2001). Thus, all references to *P. juliflora* and its varieties identified by Burkart (1976) in this region should be absorbed into synonymity with *P. pallida*.

Prosopis pallida is also often confused for other *Prosopis* species in the section Algarobia (Pasiecznik *et al.*, 2001). For instance, in the Hawaiian Islands naturalized *Prosopis* populations were initially thought to be either *P. chilensis* or *P. juliflora*, but later identified as *P. pallida* (even though hybrids may also be present on the islands; Johnston, 1962).

The taxonomic and biogeographic issues highlighted above present unique challenges to studies on *Prosopis* species in their non-native regions, in particular for *P. juliflora* and *P. pallida*. In the following section, we briefly discuss one such aspect: the worldwide introduction history of *P. juliflora*.

3.2.2 Introduction history

Prosopis species have been introduced to many parts of the world for several centuries, dating back to the early 1820s (Shackleton *et al.*, 2014). More recently, between the 1970s and 1990s, there was an upsurge in *Prosopis* introductions for reforestation programmes, particularly into Africa and Asia (Pasiecznik *et al.*, 2001; Shackleton *et al.*, 2014). As one of only two tropical species in the genus, *P. juliflora* was identified by the Food and Agriculture Organization of the United Nations as a valuable reforestation tree (Pasiecznik *et al.*, 2001) and today the global distribution of this species is truly remarkable (see Fig. 3.2). It has been introduced into 90 countries, has naturalized populations in 33 countries and is considered invasive in 48 countries (see Fig. 3.1; Shackleton *et al.*, 2014). The tree was introduced for multiple 'benefits', including as fuel source, fodder plants, shade trees and for 'environmental services' such as soil stabilization (Pasiecznik *et al.*, 2001; Wise *et al.*, 2012).

Some of the non-native ranges identified for *P. juliflora*, however, may be incorrect, stemming from the taxonomic issues associated with this species and the genus as discussed earlier (see Section 3.2). For instance, historical classification of *Prosopis* species may have led to the misidentification of some *P. juliflora* invasions (Palacios *et al.*, 2012). In North America, *P. glandulosa* and *P. velutina* were originally classified as *P. juliflora* var. *glandulosa* and *P. juliflora* var. *velutina*, respectively. Where these two species have been introduced outside their native ranges, variety names may have been dropped and introduced material wrongly classified as '*P. juliflora*' (Pasiecznik *et al.*, 2001). This error persists in many countries, particularly evident in regions with winter frosts, where *P. juliflora* is listed as invasive but is unlikely to survive frost conditions. Similarly, early taxonomists applied the binomial *P. chilensis* to several *Prosopis* species from the Americas, including *P. glandulosa*, *P. juliflora* and *P. pallida*, but this has since been corrected. Throughout Africa, however, the name *P. chilensis* was given to many early introductions of several *Prosopis* species, and these misidentifications continue (Pasiecznik *et al.*, 2001). For example, the source of most *P. juliflora* populations in Africa was planted by the Food and Agriculture Organization of the United Nations via their Danish International Development Agency-managed seed stocks, some of which were incorrectly labelled as *P. chilensis* (Pasiecznik *et al.*, 2001), or planted by NGOs who sourced seed from commercial suppliers from secondary sources outside the Americas, such as The Netherlands

These taxonomic misidentifications are aggravated by botanical and field guides using misapplied names (Pasiecznik *et al.*, 2001). Lastly, given the taxonomic issues and uncertainty within the *P. pallida–P. juliflora* complex, introductions of *P. juliflora* have often been misapplied to *P. pallida* and vice versa (Pasiecznik *et al.*, 2001).

3.2.3 Biogeography

It is not only taxonomic issues that have complicated the global introduction history of *P. juliflora*, but also biogeographic uncertainties (often in conjunction with taxonomic misidentifications). Given the scope of this book, we briefly discuss the introduction history of *P. juliflora* into eastern Africa here.

In Kenya, the first documented introduction of *Prosopis* (*P. juliflora* and *P. pallida*) dates to 1973 when these two species were introduced for the rehabilitation of quarries near the city of Mombasa (Mwangi and Swallow, 2005). Recent genetic analyses confirmed the identity of these two species in Kenya and Ethiopia (Castillo *et al.*, 2021a). These introductions involved seeds sourced from other non-native regions in Brazil and the Hawaiian Islands (Choge *et al.*, 2002). Both species were also introduced into the semi-arid districts of Baringo, Tana River and Turkana districts in the early 1980s, often planted together in trial plantations (Choge *et al.*, 2002). These were haphazard and seeds used for plantings were sourced from commercial suppliers without any reference to their origin(s) (Mwangi and Swallow, 2005).

The introduction history of *P. juliflora* into Ethiopia is equally muddled. The dates and origin(s) of introductions into the country is not known, with conflicting reports of when, and from where, introductions occurred (Mehari, 2015). Some authors suggest that the Ethiopian government first introduced *P. juliflora* in the 1970s, from India, into the country's Afar National Regional State (Tegegn, 2008; Ayanu *et al.*, 2015). Additional introductions into the region between the 1980s and 1990s have been reported but with no reference to their sources (Tessema, 2012).

In Tanzania *Prosopis*, sourced from South Africa, was planted in 1953 in the Mombo Arboretum from where the seeds were later harvested and distributed to other parts of the country. Whether the material introduced from South Africa is indeed *P. juliflora* remains unknown as there are numerous *Prosopis* species (and their hybrids) present in that country (Shackleton *et al.*, 2014). It is widely believed that invasive populations in Tanzania are *P. juliflora* that reached the country from Taveta in Kenya between 1988 and 1995 (Kilawe *et al.*, 2017). Further insights into the introduction history of *P. juliflora* and *P. pallida* in eastern Africa, inferred from molecular studies, are discussed below.

3.3 Molecular Insights Into the Taxonomy, Biogeography and Invasion Dynamics of the Genus *Prosopis*

A wealth of molecular studies on *Prosopis* has emerged since the 1950s (Fig. 3.3a). In this section we review this work and how it advanced our understanding of *Prosopis* invasions, particularly of *P. julifora*, the focal species of this book.

We conducted an online search in the ISI Web of Knowledge database (October 2022) using the search terms: (*Prosopis* OR Mesquite*) AND (genetic* OR molecular* OR genomic* OR gene* OR marker* OR phylogenetic* OR phylogeography OR heredit*). Our search retrieved 5852 studies, some of which were excluded based on their classification under Web of Science categories that are unrelated to our interest, such as 'peripheral vascular disease' or describing impacts and uses of *Prosopis* species. Many studies focused on plant growth performance trials or phenotypic comparisons among different *Prosopis* populations or species. These

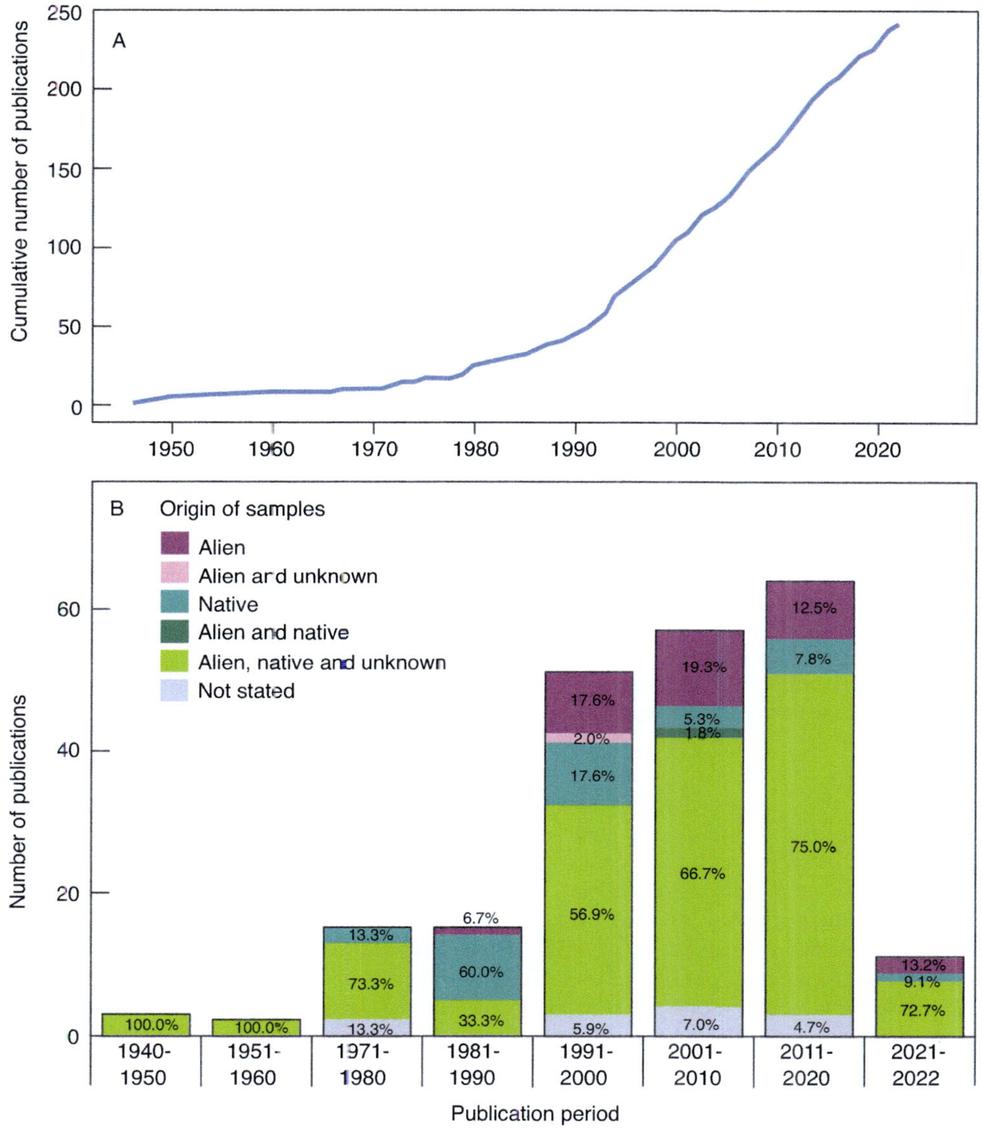

Fig. 3.3. Molecular and genetic studies on *Prosopis* shown as (A) the cumulative number of publications/year between 1946 and 2022 and (B) the geographic origin of *Prosopis* specimens used in studies.

studies were largely based on observational data from the field or common garden experiments. We only retained studies that used molecular approaches to investigate phenotypic traits for taxonomic identification of *Prosopis* species, trait heritability in plantation trials, genetic improvement studies or artificial selection and breeding. We also added 12 relevant studies not retrieved from the ISI search. These criteria resulted in a final collection of 242 papers.

The molecular data on *Prosopis* we retrieved included cytological (e.g. Cherubini, 1954;

Hunziker *et al.*, 1975; Naranjo *et al.*, 1984), chromatographical (e.g. Palacios and Bravo, 1981), DNA marker (e.g. Vázquez-Garcidueñas *et al.*, 2003; Mottura *et al.*, 2005; Landeras *et al.*, 2006; Palacios *et al.*, 2012) and protein-based (e.g. Solbrig and Bawa, 1975; Saidman and Vilardi, 1987; Freeman and Snyder, 2005; Bessega *et al.*, 2005, 2006) studies. Many of these were focused on *Prosopis* species in their native ranges, with few studies on the molecular aspects of invasive populations (Fig. 3.3b). Around 41% of the studies included *P. juliflora*, while most of the studies (ca. 83%) included species from the section Algarobia. About 12% of the studies investigated hybridization, while around 11% reported on *Prosopis* clones (most of these studies investigated the propagation of *Prosopis*). Approximately 54% of studies included comparisons of multiple *Prosopis* species.

3.3.1 Molecular insights into the systematics, occurrence of hybridization and taxonomy of *Prosopis*

As discussed in Section 3.2, taxonomic and biogeographic issues have plagued *Prosopis* for a long time. Molecular genetic studies have been instrumental in elucidating the systematics of, and in guiding a more stable taxonomy for, the genus. These studies have confirmed that the genus is polyphyletic (Hunziker *et al.*, 1975; Bessega *et al.*, 2006; Burghardt and Espert, 2007; Catalano *et al.*, 2008), for example by identifying *P. africana* (section Anonychium) as a monospecific genus (LPWG, 2017; Ringelberg *et al.*, 2022). Ringelberg *et al.*'s (2022) phylogenetic study separated the genus into three distinct lineages – one consisting of *P. africana*, a lineage including the remaining species from Africa and Asia, and another including all American taxa.

Based on the most comprehensive phylogenetic treatment of the genus to date, Hughes *et al.* (2022) proposed the disintegration of *Prosopis* and, in conjunction with morphological and geographical data, reclassified *Prosopis* into four genera (see Box 1.3, Chapter 1, this volume). Under this reclassification, the genus *Prosopis* only includes the three African species previously assigned to it, sections *Anonychium* and *Strombocarpa* are distinct genera (with *Anonychium*, i.e. *A. africana*, distributed from Senegal to Sudan and Ethiopia; *Strombocarpa* containing ten species restricted to the Americas) and the resurrected genus, *Neltuma*, containing species from the sections Algarobia and Molinicarpa (now comprising 43 species from the Americas).

Under the new classification, eight of the ten *Prosopis* species known to have invasive populations (Shackleton *et al.*, 2014) reside in the genus *Neltuma*. Although this reclassification provides some taxonomic stability, it does not solve the complicated delimitation of species that previously resided in section Algarobia of *Prosopis* (all now in the genus *Neltuma*), many of which are invasive. Nevertheless, the new phylogenetic data for these taxa (Hughes *et al.*, 2022; Ringelberg *et al.*, 2022) together with their reclassification (Hughes *et al.*, 2022) provide a framework for future assessments of *Prosopis* biogeography, trait evolution and diversification.

Hybrid and polyploid (the condition of having more than one set of chromosomes) alien plants are often successful invaders (Te Beest *et al.*, 2012). Hybridization can dramatically alter the genetic make-up of introduced populations and facilitate rapid evolution (Schierenbeck and Ellstrand, 2009). Hybrid genotypes often express trait values that transgress those of both parental species, referred to as heterosis, transgression or hybrid vigour (Te Beest *et al.*, 2012). Similarly, polyploidy may provide advantages to introduced populations associated with increased genetic diversity, altered gene expression, and enhanced physiological and environmental tolerance, amongst others (Te Beest *et al.*, 2012). These effects may not necessarily lead to higher fecundity (e.g. higher seed production) but can lead to enhanced growth or persistence under novel environmental conditions (Richardson *et al.*, 2015).

Hybridization and polyploidy in *Prosopis* have been largely studied in the native range of the genus. Early cytological studies suggested that most taxa, including hybrids, are diploid ($2n=2x = 48$; see Hunziker *et al.* (1975) and references therein). *Prosopis juliflora* was initially considered as having both diploid and tetraploid cytotypes (Cherubini, 1954; Hunziker *et al.*,

1975) but was subsequently shown to be entirely tetraploid (Trenchard *et al.*, 2008). It is thus the only polyploid member of the genus. Ploidal differences may facilitate reproductive isolation between different cytotypes as evidenced by the high genetic differentiation between *P. juliflora* and its congeners (Castillo *et al.*, 2021a). This also means that species that are difficult to distinguish morphologically from one another, such as *P. juliflora* and *P. pallida*, can be identified based on ploidal variation. For example, in eastern Africa, Castillo *et al.* (2021a) used flow cytometry analyses to differentiate between these two taxa, with important implications for understanding *Prosopis* invasions in the region (see Box 3.1 for more details).

Karyological studies in the native range of *Prosopis* have provided the first evidence for interspecific hybridization in the section Algarobia (now in the genus *Neltuma*; Hunziker *et al.*, 1975, 1977; Palacios and Bravo, 1981; Naranjo *et al.*, 1984; Table 3.1). These studies pointed to a general lack of genetic differentiation between species in this section and incomplete reproductive isolation between them. Prezygotic reproductive barriers (e.g. differences in phenology or visitation by pollinators) are thought to be weak in *Prosopis*, while postzygotic reproductive barriers (e.g. pollen unviability) may be more important (Palacios and Bravo, 1981; Naranjo *et al.*, 1984). For example, Naranjo *et al.* (1984) reported that *Prosopis* hybrids flowered profusely and prematurely, but also that they produced few fruits and low numbers of well-developed seeds. Incomplete reproductive isolation between closely related *Prosopis* species is supported by phylogenetic studies showing that species in the Algarobia section represent a recent radiation (Catalano *et al.*, 2008). Hybrids with intermediate morphological characteristics, compared to their parental species, have been found between *P. chilensis* and *P. flexuosa* (Verga, 1995; Joseau *et al.*, 2013; Teich *et al.*, 2015) and between *P. alba* and *P. nigra* (Vega *et al.*, 2020). Hybrids have also been identified based on morphological characters in Australia (between *P. glandulosa*, *P. juliflora*, *P. pallida* and *P. velutina*; Van Klinken and Campbell, 2009) and in the Hawaiian Islands (between *P. juliflora* and *P. pallida*; Gallaher and Merlin, 2010).

However, morphological descriptions of hybridization should also be treated with

Box 3.1. The different molecular ecologies of *Prosopis* invasions in East Africa and other parts of the world.

In eastern African countries such as Ethiopia, Kenya and Tanzania, tetraploid *P. juliflora* is the main invasive species, despite the presence of *P. chilensis* and *P. pallida* in these countries (Choge *et al.*, 2002; Muturi, 2012; Castillo *et al.*, 2021a). In Kenya, hybrids between *P. juliflora* and *P. pallida* were mostly found at sites where these two species were originally planted together, with few hybrids present far away from plantations (Castillo *et al.*, 2024). Interestingly, despite similar introduction histories of *P. juliflora* and *P. pallida* into Kenya, diploid *P. pallida* has not become invasive (Castillo *et al.*, 2024). As in Kenya, localized hybrids have been described between introduced *P. juliflora* and *P. pallida* from the Hawaiian Islands (Gallaher and Merlin, 2010). The localized distributions of *P. juliflora* × *P. pallida* hybrids probably reflect the fact that they are triploid and therefore sterile.

In contrast to eastern Africa, the invasion success of *Prosopis* in countries such as Australia and South Africa has been linked to extensive hybridization between congeners, where several diploid *Prosopis* species have been cointroduced (Van Klinken *et al.*, 2006; Zachariades *et al.*, 2011; see above).

Hybridization may create novel genotypes (Ellstrand and Schierenbeck, 2000) that may lead to an immediate increase in genetic diversity (Stebbins, 1959), the masking of deleterious alleles and fixed heterosis or hybrid vigour (Te Beest *et al.*, 2012) that usually increases plant performance and thus invasiveness. Polyploid alien plants are also often good invaders. For example, polyploids generally show elevated levels of stress tolerance and higher growth vigour through increased plant size, seed size, flower size, niche breadth and phenotypic plasticity (Te Beest *et al.*, 2012).

While hybridization enhances the invasiveness of *Prosopis* in countries such as Australia and South Africa, this does not appear to be the case in eastern Africa. Here, polyploidy and, in some contexts, rapid evolution appear to drive the invasion success of *Prosopis* (Castillo *et al.*, 2021a).

Table 3.1. Examples of putative *Prosopis* hybrids described from different native and non-native regions.

Putative hybrid combination	Study	Country/region	Range	Data type
P. affinis × *P. alba*	Hunziker *et al.* (1977)	Argentina	Native	Unknown
P. affinis × *P. nigra*	Naranjo *et al.* (1984)	Argentina	Native	Chromatography
P. affinis × *P. ruscifolia*	Hunziker *et al.* (1977)	Argentina	Native	Unknown
P. alba × *P. nigra*	Joseau *et al.* (2013)	Argentina	Native	Morphology
	Palacios and Bravo (1981); Vega and Hernández (2005); Ferreyra *et al.* (2007); Vega *et al.* (2020); Castillo *et al.* (2021a)			Morphology and genetic
P. chilensis × *P. juliflora*	Muturi (2012)	Kenya	Invaded	Genetic
P. alba × *P. chilensis*	Joseau *et al.* (2013);	Argentina	Native	Morphology
	Castillo *et al.* (2021a)			Morphology and genetic
P. alba × *P. flexuosa*	Ferreyra *et al.* (2007)	Argentina	Native	Morphology and genetic
P. alba × *P. rustifolia*	Palacios and Bravo (1981); Castillo *et al.* (2021a)	Argentina	Native	Morphology and genetic
P. alba × *P. hassleri*	Palacios and Bravo (1981)	Argentina	Native	Morphology and genetic
P. alba × *P. vinalillo*	Castillo *et al.* (2021a)	Argentina	Native	Morphology and genetic
P. chilensis × *P. flexuosa*	Joseau *et al.* (2013)	Argentina	Native	Morphology
	Verga, 1995; López Lauenstein *et al.* (2013); Teich *et al.* (2015); Castillo *et al.* (2021a)			Morphology and genetic
P. ruscifolia × *P. nigra*	Palacios and Bravo (1981)	Argentina	Native	Morphology and genetic
P. ruscifolia × *P. hassleri*	Palacios and Bravo (1981)	Argentina	Native	Morphology and genetic

Continued

Table 3.1. Continued

Putative hybrid combination	Study	Country/region	Range	Data type
Prosopis hybrids	Van Klinken and Campbell (2009)	Australia	Invaded	Morphology
P. juliflora × *P. pallida*	Gallaher and Merlin (2010)	Hawaii	Invaded	Morphology
	Castillo *et al.* (2021a)	Kenya	Invaded	Genetic
P. glandulosa × *P. velutina* × *P. laevigata*	Poynton (2009)	South Africa	Invaded	Morphology
	Mazibuko (2012)			Genetic

'Unknown' refers to cases where it was not possible to infer data type because the primary sources, e.g. old books, were not accessible.

caution as the traits of hybrids are not always intermediate to those of parental species. Genetic analyses identified hybrids between *P. juliflora* and *P. pallida* in Kenya (Castillo *et al.*, 2024). As stated before, hybrids involving tetraploid *P. juliflora* and diploid *Prosopis* species are unlikely to be widespread given that they will be sterile. For instance, in Kenya triploid *P. juliflora* × *P. pallida* hybrids were only present in low frequencies near the original plantations of the two species (Castillo *et al.*, 2021a).

In South Africa, hybrids between *P. glandulosa*, *P. laevigata* and *P. velutina* were identified based on morphological (Poynton, 2009) and genetic data (Mazibuko, 2012). Nevertheless, there have been instances where morphological data do not align with genetic data. For instance, morphological hybrids between *P. alba* × *P. nigra* (Vega *et al.*, 2020; Castillo *et al.*, 2021a), *P. alba* × *P. chilensis*, *P. alba* × *P. vinalillo*, and *P. chilensis* × *P. flexuosa* were identified as pure species using genetic analyses (Castillo *et al.*, 2021a).

It appears that certain environmental conditions may promote hybridization in *Prosopis* (Vega and Hernández, 2005), with hybrids often found in disturbed sites (Verga, 2005). Introgressive hybridization (i.e. hybridization followed by backcrossing), as has been noted for species in the section Algarobia (e.g. Ferreyra *et al.*, 2013), may also produce individuals that have morphological characteristics of one parental species while still carrying the genetic material of the other (e.g. Gaskin and Kazmer, 2009; Boswell *et al.*, 2016). This, combined with different selection pressures in different environments, may further contribute to high levels of intra- and interspecific morphological variation and the taxonomic confusion characteristic of *Prosopis*, even leading to the emergence of new ecotypes or species (Goulet *et al.*, 2017). For example, *P. vinalillo*, which recently diverged from *P. ruscifolia*, is thought to represent an unstable hybrid between *P. alba*, *P. ruscifolia* and *P. nigra* (Ferreyra *et al.*, 2004).

3.3.2 Studies on genetic improvement of *Prosopis*

Prosopis species have been widely introduced for forestry and agroforestry purposes and multiple studies have been conducted on their genetic improvement. Interestingly, most of these studies come from the native range (~68%). Despite this wealth of information, the problematic taxonomy of *Prosopis* complicates inferences that can be made from these studies, as the species identity of some taxa is often not reported, is uncertain or the species originally reported have been revised to a different species (Wojtusik *et al.*, 1993; Tewari *et al.*, 1997; Harsh *et al.*, 1998; Pasiecznik *et al.*, 2006). Similarly, the geographic origins of individuals used in these studies were often not reported or only provided as rough estimates (Tewari *et al.*, 1997; Harsh *et al.*, 1998).

Genetic improvement studies of *Prosopis* included those that characterized phenotypic traits or evaluated the performance of different provenances of the same species under common garden conditions, to examine the genetic basis and inheritance of traits. For example, in *P. cineraria*, high heritability in tree height and artificial selection yielded genotypes with improved height (Jindal, 1998). Similarly, *P. flexuosa* in Argentina showed high variability and heritability in growth and germination traits (Cony, 1996; Cony and Trione, 1996) and a good correspondence in growth performance between young seedlings and older trees suggests that trait selection can be achieved at the seedling stage (Mantovan, 2002).

Traits targeted for genetic improvement were also often linked to increased invasiveness of species such as *P. juliflora* (Pasiecznik *et al.*, 2001; Shiferaw *et al.*, 2004; Pasiecznik *et al.*, 2006; Castillo *et al.*, 2021b). In Kenya, the growth performance of (original) plantation and invasive seedlings of *P. pallida* and *P. juliflora* did not differ (Castillo *et al.*, 2021b), possibly because invasive trees originated from a few planted (and possibly) artificially selected trees. Native *Prosopis* provenances with the best performance have often been cloned and preserved in gardens to provide high-quality germplasm for planting and export (e.g. Ewens *et al.*, 2012). These long-term experiments aimed to create 'super trees' with high yield under a wide range of environmental conditions (Ewens *et al.*, 2012) and included progeny trails to select scions from spineless and erect trees to be grafted onto preselected trees (Wojtusik *et al.*, 1993; Shivkumar and Banerjee, 1998). In areas where *P. juliflora* is invasive and *P. pallida* not, it has been recommended to use the latter as a source of scion for grafting onto *P. juliflora* (Wojtusik *et al.*, 1993).

Another genetic improvement strategy is the identification of *Prosopis* families or individuals with high heritability for traits such as height, to negate the need for complex and time-consuming artificial selection and breeding programmes (Cony, 1996; Jindal, 1998). For example, for *P. alba*, progeny trials in contrasting environments were conducted in Argentina with approximately 200 common families to assess performance and plasticity of traits of forestry interest and for future 'genetic conservation' (López Lauenstein *et al.*, 2015; Verga, 2017). Similarly, the *Prosopis* National Germplasm Bank in Argentina prioritized nine species in 1995 for genetic conservation and future commercialization. Currently, there are 1650 accessions in its germplasm collection, corresponding to 1106 trees (Joseau *et al.*, 2021). These genetic improvement programmes aim to provide germplasm that is easy to grow and with desirable traits.

3.3.3 The evolutionary ecology of invasive *Prosopis juliflora*

It is now widely accepted that the invasiveness of alien species is not only a product of their ecologies (e.g. enemy release; Keane and Crawley, 2002; Heger and Jeschke, 2018), but also their ability to respond evolutionarily to conditions in the new environment (Le Roux, 2021). There are numerous studies showing how contemporary evolution influences the dynamics, spread and impacts of invasive species (Le Roux, 2021).

Genetic diversity is the currency of natural selection and therefore the amount of standing genetic variation present in introduced populations will influence their capacity to adapt to conditions in the new environment (Ward *et al.*, 2008). However, while high genetic diversity is advantageous for adaptation (Bijlsma and Loeschcke, 2012), low diversity may bring about non-adaptive evolutionary change, for example via strong founder events and genetic drift. The amount of genetic diversity in introduced populations depends on the species' evolutionary history in the native range and, following introduction (via single or multiple introduction events), on the demographic processes (e.g. founder events, genetic drift or spatial sorting) and selective pressures they experience (Keller and Taylor, 2008; Fig. 3.4).

Human-mediated processes in the native range may also impact the evolutionary capacity of species in their non-native ranges. For example, when using different provenances of a species for ecological restoration in the native range, humans may create genetically diverse and admixed populations that may subsequently be introduced to areas outside the

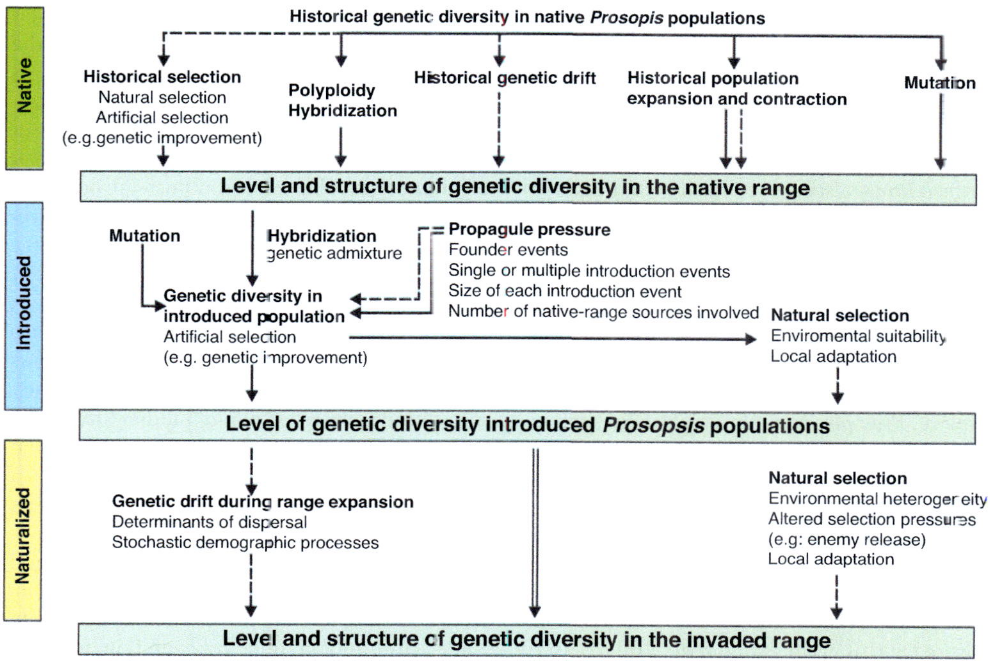

Fig. 3.4. Factors (bold font) associated with different stages of invasion that may have shaped the genetic characteristics of invasive *Prosopis* populations, including their evolutionary history in the native range, stochastics events and natural selection (modified from Keller and Taylor, 2008). Factors that are likely to decrease genetic diversity are indicated by solid arrows while those that are likely to increase genetic diversity are indicated by dashed arrows. The contribution of these factors to the dynamics of introduced *Prosopis* species is complicated by uncertain taxonomy and biogeography associated with the genus, with important evolutionary consequences (see main text and Box 3.1).

native range (e.g. Le Roux *et al.*, 2013). Some of the most invasive *Prosopis* species (i.e. *P. chilensis*, *P. flexuosa* and *P. juliflora*) have long histories of human usage in their native range (Roig, 1993; Cony, 1995, 1996) and have been the focus of many genetic improvement studies (see Section 3.3.2). *Prosopis* species were introduced for forestry and soil stabilization in many of their invasive ranges (Shackleton *et al.*, 2014). To maximize productivity, extensive selection of the best genotype(s) was often performed in breeding programmes (e.g. López Lauenstein *et al.*, 2015). Overall, this may have facilitated the introduction of genotypes that were well suited to conditions in non-native ranges or of so-called 'general-purpose genotypes' that can tolerate a wide range of environmental conditions (e.g. Parker *et al.*, 2003). While cultivation may in some instances increase genetic diversity and create novel genetic lineages (e.g. Thompson *et al.*, 2012), in most instances it is likely to lead to decreased genetic diversity.

In addition to the introduction history (e.g. source(s) and pathway(s)) of an alien species, propagule pressure (i.e. the number and size of introduction events) has important implications for genetic variation found in introduced populations and their evolutionary capacities. In eastern Africa, levels of genetic diversity in invasive *P. juliflora* and *P. pallida* populations were found to be similar, or in some instances higher, than levels of diversity found in native-range populations (Castillo *et al.*, 2021a). This supports the view that highly diverse introduction(s) probably founded these invasive populations (also see Pasiecznik *et al.*, 2001). These genetic characteristics also suggest that these populations may harbour high levels of trait variation for selection to act on (see Box 3.1 for more details).

Globally, *Prosopis* introductions often involved multiple species from diverse sources (Pasiecznik *et al.*, 2001). These cointroduced diploid species have often hybridized extensively, with consequences for levels of genetic diversity and, possibly, invasiveness, as discussed earlier. Hybridization between *Prosopis* species, coupled with the known taxonomic uncertainties in the genus, means that many studies simply refer to invasive populations as *Prosopis* spp., which complicates the study of the invasion biology of the genus. Molecular studies have provided some insights into these complexities (see Section 3.3.1).

Studies on the evolutionary ecology of invasive *Prosopis*, including features such as phenotypic plasticity, rapid post-introduction evolution and release from natural enemies, have been conducted to understand the ecological success of *Prosopis*. For example, in Kenya invasive *P. juliflora* trees were found to flower at only 17 months of age, while no offspring of the originally introduced plantation trees flowered at this age (Castillo *et al.*, 2021b). In its native range, *P. juliflora* generally only reaches reproductive maturity after 3 or 4 years of age (Vázquez-Yanes *et al.*, 1999). Earlier flowering in invasive *P. juliflora* populations in Kenya may thus be the result of post-introduction evolution and probably contributes to the species' invasion success in the region (Castillo *et al.*, 2021b). Invasive *P. juliflora* in Kenya also experiences significantly less seed damage compared to non-invasive *P. pallida*, suggesting that *P. juliflora* is potentially experiencing higher levels of enemy release than *P. pallida* (Castillo *et al.*, 2021b).

In general, most comparative studies between native and invasive ranges do not include the source(s) of invasive populations as these are often difficult to trace or may no longer exist. This makes the study of rapid post-introduction evolution of local adaptation in the context of invasion difficult. Castillo *et al.* (2021b) studied *Prosopis* invasions in the Afar region in Ethiopia and in Baringo County in Kenya where the founder trees that seeded invasions are still present in the environment. Their comparisons of plantation and invasive populations found that, despite the similar introduction history of *P. juliflora* and *P. pallida* in Kenya and the occurrence of hybridization between them, only *P. juliflora* have become invasive. This suggests that invasion success of *Prosopis* may not always be linked to hybridization. These authors also suggested that the higher ploidy of *P. juliflora* may enhance its invasiveness compared to diploid *P. pallida*. Common garden and reciprocal field transplant experiments in Kenya indicated that high levels of phenotypic plasticity and post-introduction evolution have contributed to the invasiveness of *P. juliflora*. High levels of phenotypic plasticity were absent from introduced, but non-invasive, *P. pallida*. Importantly, both *P. juliflora* and *P. pallida* were introduced into Kenya at the same time, to the same locations and with similar propagule pressure. This study also found invasive *P. juliflora* individuals to be genetically less diverse than founder plantation trees. Thus, the higher genetic diversity in founder individuals may have provided high initial adaptive capacity.

In another study, Castillo *et al.* (2024) found lower genetic diversity in founder *P. juliflora* trees in Ethiopia compared to founder trees in Kenya. In Ethiopia, ongoing gene flow between founding and invasive populations appears to have also led to the homogenization of genetic diversity between populations there (Castillo *et al.*, 2024). It is therefore unlikely that the same evolutionary dynamics are at play in Ethiopian and Kenyan *P. juliflora* populations.

Many studies on *Prosopis* performance under different abiotic and biotic conditions have been done, providing important insights into the ecological and evolutionary aspects underlying invasiveness in the genus. Many of these studies were done for genetic improvement via artificial selection. On the other hand, few studies that incorporate ecological and genetic data have been conducted to understand invasion dynamics. The *Prosopis* example from eastern Africa, discussed above, illustrates the utility of such studies (also see Box 3.1).

3.4 What Have We Learned From Genetic Studies on *Prosopis*?

For the implementation of management strategies such as biological control, sound taxonomic knowledge of invasive species is essential (Le Roux and Wieczorek, 2009).

The new genus *Neltuma* contains most of the species previously classified as *Prosopis* of economic importance and that have been widely introduced outside their native ranges and have become invasive (i.e. *P. chilensis*, *P. glandulosa*, *P. juliflora*, *P. nigra*, *P. pallida*, *P. velutina*; Shackleton *et al.*, 2014). For regulatory purposes, Castillo *et al.* (2021a) proposed that all diploid *Prosopis* taxa from the Americas be treated as a single taxonomic unit in national or regional lists in their non-native ranges, rather than as separate species.

The recent reclassification of *Prosopis* into four genera is unlikely to assist the taxonomic classification of invasive taxa (Hughes *et al.*, 2022). For example, most invasive *Prosopis* species now classified as *Neltuma* hybridize extensively in their non-native and native ranges. Correct identification of *Prosopis* species (and their hybrids), however, is important for studying the mechanisms underlying successful invasion and for management. An adequate understanding of the site-specific dynamics and/or generalities of invasive populations and, certainly, investigations of the efficacy of management approaches such as biological control require accurate taxonomic identification (Pysek *et al.*, 2013). For instance, the revised taxonomy of *Prosopis* (*sensu lato*) classifies *P. africana* (now *Anonychium africanum*) as a distant relative of invasive (*Neltuma*) species, which explains why biological control candidates such as *Evippe* sp. can develop on most invasive taxa but not on African *P. africana* (Kleinjan *et al.*, 2021). This also suggests that any new biological control agents against *Prosopis* in Africa are unlikely to have non-target impacts on this species.

We argue that future work on *Prosopis* should build on recent phylogenetic studies, to identify invasive populations using phylogeographic inferences based on extensive sampling of taxa and genomic data throughout their native and non-native distributions. This also opens the door for the development of high-density SNP arrays for reliable species identification (e.g. Mason *et al.*, 2015).

Differences in the drivers of *Prosopis* invasions in different parts of the world (see Box 3.1) serve as a powerful example of the context-dependency of invasion and illustrate the need for further research to evaluate the contributions of different processes to invasion success of *Prosopis* in different environments.

Acknowledgements

The authors acknowledge funding from the Swiss Programme for Research on Global Issues for Development (r4d), funded by the Swiss National Science Foundation (SNSF) and the Swiss Agency for Development and Cooperation (SDC), for the project 'Woody invasive alien species in East Africa: Assessing and mitigating their negative impact on ecosystem services and rural livelihood' (grant number: 400440_152085). We thank Klaas Vrieling and Rieks van Klinken for their valuable comments on an earlier draft of the manuscript.

References

Ayanu, Y., Jentsch, A. and Müller-Mahn, D. (2015) Ecosystem engineer unleashed: *Prosopis juliflora* threatening ecosystem services?. *Regional Environmental Change* 15, 155–167.

Bentham, G. (1875) VII. revision of the suborder mimoseae. *Transactions of the Linnean Society of London* 30, 335–664.

Bessega, C., Saidman, B.O. and Vilardi, J.C. (2005) Genetic relationships among American species of *Prosopis* (Leguminosae) based on enzyme markers. *Genetics and Molecular Biology* 28, 277–286.

Bessega, C., Vilardi, J.C. and Saidman, B.O. (2006) Genetic relationships among American species of the genus *Prosopis* (Mimosoideae, Leguminosae) inferred from ITS sequences: Evidence for long-distance dispersal. *Journal of Biogeography* 33, 1905–1915.

Bijlsma, R. and Loeschcke, V. (2012) Genetic erosion impedes adaptive responses to stressful environments. *Evolutionary Applications* 5, 117–129.

Boswell, A., Sing, S.E. and Ward, S.M. (2016) Plastid DNA analysis reveals cryptic hybridization in invasive dalmatian toadflax (*Linaria dalmatica*) populations. *Invasive Plant Science and Management* 9, 112–120.

Burghardt, A.D. and Espert, S.M. (2007) Phylogeny of *Prosopis* (Leguminosae) as shown by morphological and biochemical evidence. *Australian Systematic Botany* 20, 332–339.

Burkart, A. (1940) Materiales para una monografía del género *Prosopis* (Leguminosae). *Darwiniana* 4, 57–128.

Burkart, A.E. (1976) A monograph of the genus *Prosopis* (Leguminosae subfam Mimosoideae). *Journal of the Arnold Arboretum* 57(3), 219–249.

Castillo, M.L., Schaffner, U., van Wilgen, B.W. and Le Roux, J.J. (2021a) The contribution of phenotypic traits, their plasticity, and rapid evolution to invasion success: Insights from an extraordinary natural experiment. *Ecography* 44, 1035–1050.

Castillo, M.L., Schaffner, U., van Wilgen, B.W., Montaño, N.M., Bustamante, R.O. *et al.* (2021b) Genetic insights into the globally invasive and taxonomically problematic tree genus *Prosopis*. *AoB PLANTS* 13, 1–13.

Castillo, M.L., Mbaabu, P.R., Shiferaw, H. *et al.*, (2024) Following the footsteps of invasion: comparisons of founder and invasive genotypes of two independent invasions reveal site-specific demographic processes but similar determinants of dispersal. *NeoBiota* 93, 263–291.

Catalano, S.A., Vilardi, J.C., Tosto, D. and Saidman, B.O. (2008) Molecular phylogeny and diversification history of *Prosopis* (Fabaceae: Mimosoideae). *Biological Journal of the Linnean Society* 93, 621–640.

Cherubini, C. (1954) Números de cromosomas de algunas especies del género *Prosopis* (Leguminosae – Mimosoideae). *Darwiniana* 10, 637–643.

Choge, S.K., Ngunjiri, F.D., Kuria, M.W., Basaka, E.A. and Muthondeki, J.K. (2002) Status and impact of *Prosopis* in Kenya. Unpublished Technical Report. KEFRI, Nairobi, Kenya.

Conn, B.J. (2023) *Prosopis* L. Available at: https://profiles.ala.org.au/opus/weeds-australia/profile/Pros opis

Cony, M.A. (1995) Rational afforestation of arid and semiarid lands with multipurpose trees. *Interciencia* 2, 249–253.

Cony, M.A. (1996) Genetic variability in *Prosopis* Flexuosa D.C., a native tree of the monte phytogeographic province, Argentina. *Forest Ecology and Management* 87, 41–49.

Cony, M.A. and Trione, S.O. (1996) Germination with respect to temperature of two Argentinian *Prosopis* species. *Journal of Arid Environments* 33, 225–236.

Díaz Celis, A. (1995) *Los Algarrobos*. CONCYTEC, Lima, Perú.

Ellstrand, N.C. and Schierenbeck, K.A. (2000) Hybridization as a stimulus for the evolution of invasiveness in plants?. *Proceedings of the National Academy of Sciences of the United States of America* 97, 7043–7050.

Ewens, M., Gezan, S. and Felker, P. (2012) Five year field evaluation of *Prosopis* alba clones on ph 9–10 soils in Argentina selected for growth in the greenhouse at seawater salinities (45 ds m−1). *Forests* 3, 95–113.

Ferreyra, L.I., Bessega, C., Vilardi, J.C. and Saidman, B.O. (2004) First report on rapds patterns able to differentiate some Argentinean species of section algarobia (*Prosopis*, Leguminosae). *Genetica* 121, 33–42.

Ferreyra, L.I., Bessega, C., Vilardi, J.C. and Saidman, B.O. (2007) Consistency of population genetics parameters estimated from isozyme and RAPDs dataset in species of genus *Prosopis* (Leguminosae, Mimosoideae). *Genetica* 131, 217–230.

Ferreyra, L.I., Vilardi, J.C., Verga, A., López, V. and Saidman, B.O. (2013) Genetic and morphometric markers are able to differentiate three morphotypes belonging to section algarobia of genus *Prosopis* (Leguminosae, Mimosoideae). *Plant Systematics and Evolution* 299, 1157–1173.

Franklin, J. (2009) Mapping species distributions. In: *Spatial Inference and Prediction*. Cambridge University Press, Cambridge, UK.

Freeman, G.W. and Snyder, J.L. (2005) Identification of *Prosopis* l. seeds using high performance liquid chromatography. *Seed Technology* 27, 230–239.

Gallaher, T. and Merlin, M. (2010) Biology and impacts of pacific island invasive species. 6. *Prosopis pallida* and *Prosopis juliflora* (algarroba, mesquite, kiawe) (Fabaceae). *Pacific Science* 64, 489–526.

Gaskin, J.F. and Kazmer, D.J. (2009) Introgression between invasive saltcedars (*Tamarix chinensis* and *T. ramosissima*) in the USA. *Biological Invasions* 11, 1121–1130.

Goulet, B.E., Roda, F. and Hopkins, R. (2017) Hybridization in plants: Old ideas, new techniques. *Plant Physiology* 173, 65–78.

Harris, P.J.C., Pasiecznik, N.M., Smith, S.J., Billington, J.M. and Ramirez, L. (2003) Differentiation of *Prosopis juliflora* (Sw.) DC. and *P. pallida* (H.& B. ex. Willd.) H.B.K, using foliar characters and ploidy. *Forest Ecology and Management* 180, 153–164.

Harsh, L., Tewari, J.C. and Sharma, N. (1998) Performance of *Prosopis* species in arid regions of India. In: Felker, P. and Moss, J. (eds) *Prosopis: Semiarid Fuelwood and Forage Tree. Building Consensus for the Disenfranchised*. Center for Semi-Arid Forest Resources, Kingsville, Texas.

Heger, T. and Jeschke, J.M. (2018) Enemy release hypothesis. In: Jeschke, J.M. and Heger, T. (eds) *Invasion Biology: Hypotheses and Evidence*. CABI, Wallingford, UK, pp. 92–102.

Hirsch, H., Richardson, D.M., Pauchard, A. and Le Roux, J.J. (2021) Genetic analyses reveal complex introduction histories for the invasive tree acacia dealbata link around the world. *Diversity and Distributions* 27, 360–376.

Hughes, C.E., Ringelberg, J.J., Lewis, G.P. and Catalano, S.A. (2022) Disintegration of the genus *Prosopis* L. *PhytoKeys* 205, 147–189.

Hunziker, J.H., Poggio, L., Naranjo, C.A., Palacios, R.A. and Andrada, A.B. (1975) Cytogenetics of some species and natural hybrids in *Prosopis* (Leguminosae). *Canadian Journal of Genetics and Cytology* 17, 253–262.

Hunziker, J.H., Naranjo, C.A., Palacios, R.A. and Poggio, L. (1977) Chromosomal cytology and hybridization. In: Simpson, B.B. (ed.) *Mesquite, Its Biology in Two Desert Scrub Ecosystems*. Dowden, Hutchinson and Ross, Stroudsburg, Pennsylvania, pp. 56–60.

Hunziker, J.H., Saidman, B.O. and Naranjo, C.A. (1986) Hybridization and genetic variation of Argentine species of *Prosopis*. *Forest Ecology and Management* 16, 301–315.

Jindal, S. (1998) Combined selection for tree height improvement in *Prosopis cineraria*. In: Tewari, J.C., Pasiecznik, N.M., Harsh, L.N. and Harris, P.J.C. (eds) *Prosopis species in the arid and semi-arid zones of India*. Prosopis Society of India and the Henry Doubleday Research Association, Coventry, UK, pp. 69–72.

Johnston, M.C. (1962) The North American mesquites *Prosopis* sect. *Algarobia (Leguminosae). Brittonia* 14, 72–90.

Joseau, M.J., Verga, A.R., Pilar Díaz, M. del and Julio, N.B. (2013) Morphological diversity of populations of the genus *Prosopis* in the semiarid chaco of northern Cordoba and southern Santiago del Estero. *American Journal of Plant Sciences* 4, 2092–2111.

Joseau, M.J., Frassoni, J., Verzino, G., Rodriguez Reartes, S., Verga, A. *et al.* (2021) Avances en la conservación y obtención de material selecto del banco nacional de germoplasma de *Prosopis*, Córdoba, Argentina. *FAVE Sección Ciencias Agrarias* 20, 265–285.

Keane, R.M. and Crawley, M.J. (2002) Exotic plant invasions and the enemy release hypothesis. *Trends in Ecology and Evolution* 17, 164–170.

Keller, S.R. and Taylor, D.R. (2008) History, chance and adaptation during biological invasion: Separating stochastic phenotypic evolution from response to selection. *Ecology Letters* 11, 852–866.

Kilawe, C.J., Mbwambo, J.R., Kajembe, G.C., Mwakalukwa, E.E., Amri, A.M. *et al.* (2017) *Mrashia: Prosopis invading pastures and agricultural lands in Tanzania*. Insights report. The Woody Weeds Project. Morogoro, Tanzania.

Kleinjan, C.A., Hoffmann, J.H., Heystek, F., Ivey, P., Kistensamy, Y. *et al.* (2021) Developments and prospects for biological control of *Prosopis* (Leguminosae) in South Africa. *African Entomology* 29, 874.

Landeras, G., Alfonso, M., Pasiecznik, N.M., Harris, P.J.C. and Ramírez, L. (2006) Identification of *Prosopis juliflora* and *Prosopis pallida* accessions using molecular markers. *Biodiversity and Conservation* 15, 1829–1844.

Le Roux, J.J. (2021) *The Evolutionary Ecology of Invasive Species*. Academic Press, Cambridge, UK.

Le Roux, J. and Wieczorek, A.M. (2009) Molecular systematics and population genetics of biological invasions: Towards a better understanding of invasive species management. *Annals of Applied Biology* 154, 1–17.

Le Roux, J.J., Richardson, D.M., Wilson, J.R.U. and Ndlovu, J. (2013) Human usage in the native range may determine future genetic structure of an invasion: Insights from *Acacia pycnantha*. *BMC Ecology* 13, 1–8.

López Lauenstein, D., Vega, C., Luna, C. and Sagadin, M. (2015) Subprograma *Prosopis*. In: Marcó, M. and Llavallol, C. (eds) *Domesticación y Mejoramiento de Especies Forestales*. Ministerio de Agroindustria. UCAR (Unidad para el Cambio Rural), Ciudad Autónoma de Buenos Aires, Argentina, p. 113.

López Lauenstein, D.A., Fernández, M.E. and Verga, A.R. (2013) Drought stress tolerance of *Prosopis chilensis* and *Prosopis flexuosa* species and their hybrids. *Trees* 27, 285–296.

LPWG (2017) A new subfamily classification of the Leguminosae based on a taxonomically comprehensive phylogeny: The Legume Phylogeny Working Group (LPWG). *Taxon* 66, 44–77.

Mantovan, N.G. (2002) Early growth differentiation among *Prosopis flexuosa* D.C. provenances from the Monte phytogeographic province, Argentina. *New Forests* 23, 19–30.

Marchini, G.L., Arredondo, T.M. and Cruzan, M.B. (2018) Selective differentiation during the colonization and establishment of a newly invasive species. *Journal of Evolutionary Biology* 31, 1689–1703.

Mason, A.S., Zhang, J., Tollenaere, R., Vasquez Teuber, P., Dalton-Morgan, J. *et al.* (2015) High-throughput genotyping for species identification and diversity assessment in germplasm collections. *Molecular Ecology Resources* 15, 1091–1101.

Mazibuko, D.M. (2012) Phylogenetic relationships of *Prosopis* in South Africa: An assessment of the extent of hybridization, and the role of genome size and seed size in the invasion dynamics. MSc thesis, Stellenbosch University, Stellenbosch, South Africa.

Mehari, Z.H. (2015) The invasion of *Prosopis juliflora* and Afar pastoral livelihoods in the Middle Awash area of Ethiopia. *Ecological Processes* 4, 1–9.

Mottura, M.C., Finkeldey, R., Verga, A.R. and Gailing, O. (2005) Development and characterization of microsatellite markers for *Prosopis chilensis* and *Prosopis flexuosa* and cross-species amplification. *Molecular Ecology Notes* 5, 487–489.

Muturi, G.M. (2012) Ecological impacts of *Prosopis* invasion in riverine forest of Kenya. PhD thesis, Wageningen University, Wageningen, Netherlands.

Mwangi, E. and Swallow, B. (2005) Invasion of *Prosopis juliflora* and local livelihoods: Case study from the Lake Baringo area of Kenya. ICRAF Working Paper no. 3. Nairobi, Kenya, pp. 1–68.

Naranjo, C.A., Poggio, L. and Enus Zeiger, S. (1984) Phenol chromatography, morphology and cytogenetics in three species and natural hybrids of *Prosopis* (Leguminosae? Mimosoideae). *Plant Systematics and Evolution* 144, 257–276.

Palacios, R.A. (2006) Los mezquites mexicanos: Biodiversidad y distribución geográfica. *Boletín de La Sociedad Argentina de Botánica* 41, 99–121.

Palacios, R.A. and Bravo, L.D. (1981) Hibridación natural en *Prosopis* (Leguminosae) en la región chaqueña Argentina. *Darwiniana, Evidencias Morfológicas y Cromatográficas* 23, 3–35.

Palacios, R.A., Burghardt, A.D., Frías-Hernández, J.T., Olalde-Portugal, V., Grados, N. *et al.* (2012) Comparative study (AFLP and morphology) of three species of prosopis of the section Algarobia: *P. juliflora*, *P. pallida*, and *P. limensis*. Evidence for resolution of the '*P. pallida-P. juliflora complex*'. *Plant Systematics and Evolution* 298, 165–171.

Parker, I.M., Rodriguez, J. and Loik, M.E. (2003) An evolutionary approach to understanding the biology of invasions: Local adaptation and general-purpose genotypes in the weed *Verbascum thapsus*. *Conservation Biology* 17, 59–72.

Pasiecznik, N.M., Felker, P., Harris, P.J., Harsh, L., Cruz, G. *et al.* (2001) *The Prosopis juliflora–Prosopis pallida Complex: A Monograph*. HDRA, Coventry, UK.

Pasiecznik, N.M., Vall, A.O.M., Nourissier-Mountou, S., Danthu, P., Murch, J. *et al.* (2006) Discovery of a life history shift: precocious flowering in an introduced population of *Prosopis*. *Biological Invasions* 8, 1681–1687.

Poynton, R.J. (2009) *Tree Planting in Southern Africa. Other Genera.* Department of Agriculture, Forestry and Fisheries, Pretoria, South Afric.

Pysek, P., Hulme, P.E., Meyerson, L.A., Smith, G.F., Boatwright, J.S. *et al.* (2013) Hitting the right target: Taxonomic challenges for, and of, plant invasions. *AoB PLANTS* 5, lt042–plt042.

Richardson, D.M., Le Roux, J.J. and Wilson, J.R. (2015) Australian acacias as invasive species: Lessons to be learnt from regions with long planting histories. *Southern Forests* 77, 31–39.

Ringelberg, J.J., Koenen, E.J.M., Iganci, J.R., Queiroz, L.P., Murphy, D.J. *et al*. (2022) Phylogenomic analysis of 997 nuclear genes reveals the need for extensive generic re-delimitation in Caesalpinioideae (Leguminosae). *PhytoKeys* 205, 3–58.

Roig, F.A. (1993) Informe nacional para la selección de germoplasma en especies de *Prosopis* de la república Argentina. In: *Contribuciones Mendocinas a La Quinta Reunión Regional Para América Latina y El Caribe de La Red de Forestación Del CIID*. IADIZA-CRICYT, Mendoza, Argentina, pp. 1–37.

Ross, J.H. (1975) *Flora of Southern Africa*, Vol 16. Government Printer, Pretoria, p. 155.

Saidman, B.O. and Vilardi, J.C. (1987) Analysis of the genetic similarities among seven species of *Prosopis* (Leguminosae: Mimosoideae). *Theoretical and Applied Genetics* 75, 109–116.

Schierenbeck, K.A. and Ellstrand, N.C. (2009) Hybridization and the evolution of invasiveness in plants and other organisms. *Biological Invasions* 11, 1093–1105.

Shackleton, R.T., Le Maitre, D.C., Pasiecznik, N.M. and Richardson, D.M. (2014) *Prosopis*: A global assessment of the biogeography, benefits, impacts and management of one of the world's worst woody invasive plant taxa. *AoB PLANTS* 6, lu027.

Sherry, M., Smith, S., Patel, A., Harris, P., Hand, P. *et al*. (2011) RAPD and microsatellite transferability studies in selected species of *Prosopis* (section Algarobia) with emphasis on *Prosopis juliflora* and *P. pallida*. *Journal of Genetics* 90, 251–264.

Shiferaw, H., Teketay, D. and Nemomissa, S. (2004) Some biological characteristics that foster the invasion of *Prosopis juliflora* (sw.) DC. at Middle Awash Rift Valley area, north-eastern Ethiopia. *Journal of Arid Environments* 58, 135–154.

Shivkumar, A.M.R. and Banerjee, A.C. (1998) Genetic improvement of *Prosopis juliflora* – Ashwani kumar, Padmini. In: Tewari, J.C., Pasiecznik, N.M., Harsh, L.N. and Harris, P.J.C. (eds) *Prosopis Species in the Arid and Semi-Arid Zones of India*. Prosopis Society of India and Henry Doubleday Research Association, Coventry, UK, pp. 69–72.

Solbrig, O.T. and Bawa, K.S. (1975) Isozyme variation in species of *Prosopis* (Leguminosae). *Journal of the Arnold Arboretum* 56, 398–412.

Solbrig, O.T., Bawa, K. and Carman, N.J. Simpson, B.B. (ed.). (1977) *Mesquite, Its Biology in Two Desert Ecosystems*. Dowden, Hutchinson and Ross, Stroudsburg, Pennsylvania, pp. 44–60.

Stebbins, G.L. (1959) The role of hybridization in evolution. *Proceedings of the American Philosophical Society* 103, 231–251.

Te Beest, M., Le Roux, J.J., Richardson, D.M., Brysting, A.K., Suda, J. *et al*. (2012) The more the better? The role of polyploidy in facilitating plant invasions. *Annals of Botany* 109, 19–45.

Tegegn, G.G. (2008) *Experiences on Prosopis Management, Case of Afar Region*. Farm Africa, Addis Ababa, Ethiopia.

Teich, I., Mottura, M., Verga, A. and Balzarini, M. (2015) Asociación entre variabilidad genética y fenotípica con ajuste por autocorrelación espacial en *Prosopis*. *Journal of Basic and Applied Genetics* 26, 63–74.

Tessema, Y.A. (2012) Ecological and economic dimensions of the paradoxical invasive species – *Prosopis juliflora* and policy challenges in Ethiopia. *Journal of Economics and Sustainable Development* 3, 8.

Tewari, J.C., Sharma, N.K., Harsh, L N. and Felker, P. (1997) Early growth performance of latin American *Prosopis* species in Indian arid tract. *Indian Journal of Plant Genetic Resources* 10, 111–114.

Thompson, G.D., Bellstedt, D.U., Byrne, M., Millar, M.A., Richardson, D.M. *et al*. (2012) Cultivation shapes genetic novelty in a globally important invader. *Molecular Ecology* 21, 3187–3199.

Thompson, G.D., Bellstedt, D.U., Richardson, D.M., Wilson, J.R.U. and Le Roux, J.J. (2015) A tree well travelled: Global genetic structure of the invasive tree *Acacia saligna*. *Journal of Biogeography* 42, 305–314.

Trenchard, L.J., Harris, P.J.C., Smith, S.J. and Pasiecznik, N.M. (2008) A review of ploidy in the genus *Prosopis* (Leguminosae). *Botanical Journal of the Linnean Society* 156, 425–438.

van Klinken, R.D., Graham, J. and Flack, L.K. (2006) Population ecology of hybrid mesquite (*Prosopis* species) in Western Australia: How does it differ from native range invasions and what are the implications for impacts and management?. *Biological Invasions* 8, 727–741.

van Klinken, R.D. and Campbell, S.D. (2009) *Prosopis* L. species. In: Panetta, F.D. (ed.) *The Biology of Australian Weeds*, Vol. 3. R.G. and F.J. Richardson, Melbourne, Australia, pp. 238–273.

Vázquez-Garcidueñas, S., Palacios, R.A., Segovia-Quiroz, J., Frías-Hernández, J.T., Olalde-Portugal, V. *et al.* (2003) Morphological and molecular data to determine the origin and taxonomic status of *Prosopis chilensis* var. *riojana* (Fabaceae, Mimosoideae). *Canadian Journal of Botany* 81, 905–917.

Vázquez-Yanes, C., Batis-Muñoz, A.I., Alcocer-Silva, M.I., Gual-Díaz, M. and Sánchez-Dirzo, C. (1999) Árboles y arbustos potencialmente valiosos para la restauración ecológica y la reforestación. Reporte Técnico Del Proyecto J084. CONABIO - Instituto de Ecología, Mexico City, Mexico.

Vega, M.V. and Hernández, P. (2005) Molecular evidence for natural interspecific hybridization in *Prosopis*. *Agroforestry Systems* 64, 197–202.

Vega, C.D., Teich, I., Acosta, M.C., Lauenstein, D.L., Verga, A.R. *et al.* (2020) Morphological and molecular characterization of a hybrid zone between and in the Chaco region of northwestern Argentina. *Silvae Genetica* 69, 44–54.

Verga, A. (1995) Genetic study of *Prosopis chilensis* and *P. flexuosa* (mimosaceae) in the dry chaco of Argentina. PhD thesis, Universität Göttingen, Göttingen, Germany.

Verga, A. (2017) Estudio de casos: Avances en especies emblematicas: Algarrobos [Uruguay]. In: *Jornada Técnica – Domesticación y Diversificación de Especies Forestales de Alto Valor, Instituto Nacional de Investigación Agropecuaria*, pp. 15–16.

Verga, A. (2005) Recursos genéticos, mejoramiento y conservación de especies del género *Prosopis*. In: Norverto, C.A. (ed.) *Mejores Árboles Para Más Forestadores: El Programa de Producción de Material de Propagación Mejorado y El Mejoramiento Genético En El Proyecto Forestal de Desarrollo*. SAGPyA, Buenos Aires, Argentina, pp. 205–221.

Ward, S.M., Gaskin, J.F. and Wilson, L.M. (2008) Ecological genetics of plant invasion: What do we know? *Invasive Plant Science and Management* 1, 98–109.

Wise, R.M., van Wilgen, B.W. and Le Maitre, D.C. (2012) Costs, benefits and management options for an invasive alien tree species: The case of mesquite in the northern cape, South Africa. *Journal of Arid Environments* 84, 80–90.

Wojtusik, T., Felker, P., Russell, E.J. and Benge, M.D. (1993) Cloning of erect, thornless, non-browsed nitrogen fixing trees of Haiti's principal fuelwood species (*Prosopis juliflora*). *Agroforestry Systems* 21, 293–300.

Zachariades, C., Hoffmann, J.H. and Roberts, A.P. (2011) Biological control of mesquite (*Prosopis* species) (Fabaceae) in South Africa. *African Entomology* 19, 402–415.

Zimmermann, H.G. (ed.) (1991) Biological control of mesquite, *Prosopis* spp. (Fabaceae), in South Africa. *Agriculture, Ecosystems and Environment* 37, 175–186.

Appendix 3.1 Classification of Taxa within the Genus *Prosopis* by Burkart (1976)

Section	Series[a]	*Prosopis* species[a]	New classification[b]
Prosopis	c	*P. cineraria*	*P. cineraria*
		P. farcta	*P. farcta*
		P. koelziana	*P. koelziana*
Anonychium	Monotypic section	*P. africana*	*A. africanum*
Monilicarpa	Monotypic section	*P. argentina*	*N. argentina*
Strombocarpa	Strombocarpae	*P. strombulifera*	*S. strombulifera*
		P. reptans	*S. reptans*
		P. abbreviate	*S. abbreviate*
		P. torquata	*S. torquata*
		P. burkartii	*S. burkartii*
		P. palmeri	*S. palmeri*
		P. pubescens	*S. pubescens*
	Cavenicarpae	*P. ferox*	*S. ferox*
		P. tamarugo	*S. tamarugo*
Algarobia	Sericanthae	*P. sericantha*	*N. sericantha*
		P. kuntzei	*N. kuntzei*
	Ruscifoliae	*P. ruscifolia*	*N. ruscifolia*
		P. fiebrigii	*N. fiebrigii*
		P. hassleri	*N. hassleri*
		P. × *vinalillo*	*N.* × *vinalillo*
	Denudantes	*P. denudans*	*N. denudans*
		P. ruizleali	*N. ruizleali*
		P. castellanosii	*N. castellanosii*
		P. calingastana	*N. calingastana*
	Humilis	*P. humilis*	*N. humilis*
		P. rojasiana	*N. rojasiana*

Continued

Section	Series[a]	*Prosopis* species[a]	New classification[b]
	Pallidae	*P. rubriflora*	*N. rubriflora*
		P. pallida	*N. pallida*
		P. campestris	*N. campestris*
		P. affinis	*N. affinis*
		P. tamaulipana	*N. palmeri*
		P. elata	*N. elata*
		P. articulate	*N. articulata*
	Chilensis	*P. chilensis*	*N. chilensis*
		P. juliflora	*N. juliflora*
		P. flexuosa	*N. flexuosa*
		P. glandulosa	
		P. alba	
		P. nigra	*N. nigra*
		P. caldenia	*N. caldenia*
		P. pugionata	*N. pugionata*
		P. velutina	*N. velutina*
		P. alpataco	*N. alpataco*
		P. laevigata	*N. laevigata*
		P. laevigata var *andicola*	*N. andicola*

[a] Classification of *Prosopis* (*sensu lato*) following Burkart (1976).
[b] Reclassification of *Prosopis* (*sensu lato*) into the genera *Anonychium, Neltuma, Prosopis* and *Strombocarpa* following Hughes *et al*. (2022).

4 Distribution and Spread of *Prosopis juliflora* in Eastern Africa

Sandra Eckert[1,2*], Purity R. Mbaabu[3], Wai-Tim Ng[4] and Albrecht Ehrensperger[1]
[1]Centre for Development and Environment, University of Bern, Switzerland; [2]Institute of Geography, University of Bern, Switzerland; [3]Institute for Climate Change and Adaptation, University of Nairobi, Kenya; [4]Earth Observation Data Centre for Water Resources Monitoring, Vienna, Austria

Abstract
This chapter focuses on the distribution and spread of *Prosopis juliflora* trees in eastern Africa, particularly in Kenya and Ethiopia, and highlights the importance of accurate earth observation (EO) data for informing effective management strategies. There are hotspots of invasion in both Kenya and Ethiopia, and recent studies have quantified both the extent of invasion and the rate of spread. For example, case studies have revealed that over 1 million ha has become invaded in Ethiopia, notably in the Awash River basin. In Kenya – where our models estimate the total invaded area to be between 7000 and 12,000 km^2 – Turkana, Tana River, Isiolo and Baringo counties are particularly affected. Suitability models reveal that invasion by *P. juliflora* in eastern Africa is still at a very early stage and that it is likely to expand significantly if left uncontrolled. Around three-quarters of Kenya (453,000 km^2 out of 582,646 km^2) is at risk of being invaded if the spread is not effectively contained. In Ethiopia, 45% (492,000 km^2 out of 1,100,000 km^2) is at risk. This underscores the seriousness of the problem and the urgent need to take effective action to contain spread. The chapter also emphasizes the potential of EO data for research, monitoring and management, and provides a brief guide to the technology that is available to do so.

4.1 Introduction

Effective management of invasive *Prosopis* trees depends on the availability of up-to-date and accurate information on their current distribution, rate of spread and potential future distribution. Such spatial information allows for the design of meaningful management strategies, the identification and prioritization of management actions and the calculation of required budgets and other resources (Chapter 11, this volume). Ground-based methods are adequate for the generation of such data and information for relatively small areas, such as watersheds or village territories. However, they are unsuitable for large areas, for example entire districts, provinces or countries. At such scales, remotely sensed earth observation (EO) data are required to achieve meaningful results. Remotely sensed EO data include drone and aerial images (for intermediate scales) and satellite images, with which large areas can be covered. By using various algorithms and classification techniques, targeted information can be extracted from such images, for example a particular type of vegetation.

*Corresponding author: sandra.eckert@unibe.ch

© CAB International 2025. *The Ecology and Management of Invasive Prosopis Trees in Eastern Africa* (eds U. Schaffner *et al.*)
DOI: 10.1079/9781800623644.0004

Countries in eastern Africa have recently developed national strategies for the management of *Prosopis* trees and other invasive species. Ethiopia and Kenya adopted a National *Prosopis* Strategy in 2017 and 2024 respectively (Ministry of Livestock and Fisheries, 2017; Ministry of Environment and Forestry, 2022), and Tanzania adopted a National Invasive Species Strategy and Action Plan in 2019 (Anonymous, 2019). To translate these strategies into action plans, these countries require national maps showing the current distribution and suitable habitat for potential future spread, as well as the rate of spread of the targeted invasive species. Importantly, such mapping requires regular updates to monitor the success or failure of management. Thus, a standardized and replicable methodology needs to be put in place by national mapping agencies to enable such monitoring.

This chapter provides an overview of the available knowledge on the current distribution of the invasive tree *Prosopis juliflora*, rates of spread and suitable habitat at risk of becoming invaded in parts of eastern Africa. We focus mainly on Kenya and Ethiopia, where detailed studies have been carried out in the context of the Woody Weeds project (see Chapter 14, this volume, for a description of the Woody Weeds project). Unfortunately, no similar datasets are available in other countries of the region, where invasive *Prosopis* trees are a significant problem as well, such as Somalia or South Sudan. Besides national overviews, a few subnational case studies are provided here to illustrate the patterns of spread of *P. juliflora* in various contexts. Additionally, we give insights into the potential use of remote sensing and EO data, as well as recommendations to a non-specialist audience on how to integrate these tools into the management of invasive alien plant species.

4.2 Current Distribution and Spread

4.2.1 Eastern Africa

Although many different *Prosopis* species including *P. pallida*, *P. juliflora* and *P. chilensis* have been introduced and have become naturalized in eastern Africa (Chapter 3, this volume), only *P. juliflora* has become invasive. This species was also the most widely planted and is currently the most common in the region. In Kenya and Ethiopia, the original *Prosopis* tree plantations are still present which provides a unique opportunity to study their spread in space and time (Chapter 3, this volume).

Unfortunately, there is no regular and comprehensive monitoring of *P. juliflora* in the region based on field observations. This lack of monitoring at various scales implies that mapping and modelling techniques using EO data are required to achieve an approximation of the extent of *P. juliflora* invasions and their rate of spread in the region. In the next sections, we present results from studies in Kenya and Ethiopia that aimed to map or model the occurrence of this invasive tree species at various scales and points in time.

4.2.2 Kenya

4.2.2.1 Methodology

We conducted a national spatial assessment in 2018, which was further improved in 2020. The supervised classification used for this assessment was based on *Prosopis* presence points collected by various projects between 2008 and 2020 (S. Eckert, unpublished data). The point samples were characterized as (1) dense (pure) *Prosopis*, (2) sparse *Prosopis* with little to no other vegetation, and (3) *Prosopis* mixed with other dense woody vegetation. Samples of other land covers (grassland, shrubland, forest, bare land, rocks and water) were collected either in the field or based on the ESACCI African land cover product of 2016 (https://2016africalandcover20m.esrin.esa.int/). The samples were cross-checked using Google Earth Pro imagery, resulting in 3070 sample points.

The classification was performed at national scale and implemented in Google Earth Engine (GEE). The input data consisted of a dry season (2016) and a wet season (2018–2020) Sentinel-2 composite. Tasseled Cap components brightness, greenness and wetness indices (Shi and Xu, 2019) and two vegetation indices (normalized difference vegetation index and normalized burn ratio) were derived from the dry season composite. Additionally, four geospatial datasets were generated: distance to rivers, human footprint index 2009 (Venter *et al.*, 2018), and mean daytime and nighttime land surface temperatures (2000–2015) (Weiss *et al.*, 2014). This resulted in a final classification dataset of 25 bands. The *Prosopis* suitability map (see Fig. 4.5) was used to exclude areas from the classification that had a suitability below 0.7 (where *P. juliflora* is unlikely to spread).

All sample points were buffered with a 20m radius and mean values for the resulting circular polygons were calculated for all bands. The random forest classification algorithm was trained with 80% of the sampled values. With the remaining 20%, a fivefold accuracy assessment was conducted. Owing to the difficulty of distinguishing sparse *Prosopis* trees from *Prosopis* trees mixed with other vegetation, these two classes were merged. This resulted in a classification map with an overall accuracy of 84.5% and a kappa coefficient of 0.69. The detailed user's and producer's accuracies for each of the resulting three classes of the map are shown in Table 4.1.

4.2.2.2 Overall invasion patterns in Kenya

From the original plantations in Taita Taveta (1948), Mombasa (1973) and Rift Valley areas (early 1980s), *P. juliflora* spread across the arid and semi-arid lands of Kenya and invaded northern and eastern Tanzania. The national model estimates that *P. juliflora* has invaded an area of between 700,000 and 1,200,000 ha. Turkana (see case study one below) is the county with the largest area of dense *P. juliflora* invasion (approx. 56,000 ha), particularly in its north-western part and along the Turkwel and Kerio River corridors. Invasion progressed rapidly in Marigat Subcounty, between Lake Baringo and Lake Bogoria in Baringo County (see case study 2 below); from 882 ha in 1988 the invaded areas increased to 18,792 ha in 2016. The Lorian swamps and the corridor of the Ewaso Ng'iro River in Isiolo County form another hotspot of *P. juliflora* invasion (Fig. 4.1). The impact of this invasion, which totals around 20,000 ha in Isiolo County, is very serious owing to the crucial significance of these wetlands as habitats for wildlife and a source of water for livestock. Finally, around 30,000 ha are invaded by *P. juliflora* in Tana River County, particularly along the corridor and in the delta of the Tana River, as well as in important irrigation schemes and along seasonal streams (Fig. 4.2). According to different studies, *P. juliflora* continues to invade new areas in Kenya, but the spread rates vary depending on the characteristics of the area (Kyuma *et al.*, 2016; Mbaabu *et al.*, 2019).

4.2.2.3 Case study 1: Turkana county, Kenya

Turkana County is in north-western Kenya (Fig. 4.3). It has an arid to semi-arid climate, with the landscape dominated by xeric shrubland vegetation (Ng *et al.*, 2016a). *P. juliflora* was introduced in the early 1980s in Kakuma and Lokichogio from where it spread to other locations in the county. In the 1980s, stock theft and severe drought led to a loss of livestock, forcing some pastoralists to relocate to urban areas where they could get relief food aid, while others adopted irrigation farming along major rivers (Maundu *et al.*, 2009). Aid agencies encouraged people to plant drought-resistant trees, which fuelled extensive planting of *Prosopis* trees, including in refugee camps.

Using multitemporal Normalized Difference Vegetation Index (NDVI) change analysis on Landsat images combined with random field data collection within a 3 km buffer zone along the Turkwel River, Muturi *et al.* (2010) found that 39% of their sampled

Table 4.1. Accuracy assessment of the national-scale *Prosopis juliflora* occurrence model presented in Fig. 4.1. User's accuracy refers to the correctly classified samples as the share of all validation samples. Producer's accuracy refers to correctly classified samples as the share of all pixels of a class in a map.

Map classes	User's accuracy (%)	Producer's accuracy (%)
Dense *Prosopis*	81.73	77.36
Sparse and mixed *Prosopis*	54.80	77.07
Other land covers	95.18	87.88

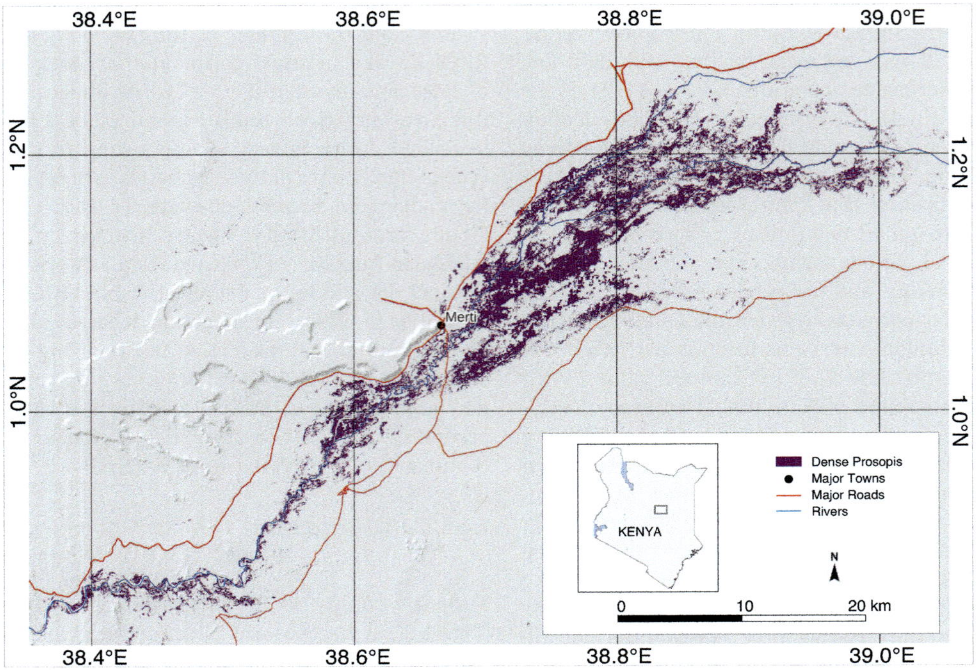

Fig. 4.1. The extent of invasion by *Prosopis juliflora* in the Lorian swamps and the corridor of the Ewaso Ng'iro River in Isiolo County, Kenya. Inset shows the location of this site in Kenya.

sites were invaded by *P. juliflora* in 2007, whereas there was no report of *Prosopis* trees in the same area in a previous study conducted in 1990. Ng *et al.* (2016a) performed a random forest land use and land cover classification on Sentinel-2 satellite data from 2016 and combined the classification with a large amount of field data covering a range of *P. juliflora* invasion density levels. Their detailed *P. juliflora* density map (see Fig. 4.3) for the Tarach watershed shows that approximately 88,200 ha (8% of the entire basin) was covered with *P. juliflora* at different densities. Presently, the trees are found at high densities in the Tarach basin, along the Turkwel River, at Kalokol on the shores of Lake Turkana, at Kakuma and Lokichogio and on the Lotikipi plains. The national *Prosopis* invasion assessment suggests that between 3.2% and 5.7% (approx. 206,000–360,000 ha) of the county is affected, making it one of the most highly invaded counties in Kenya (Ministry of Environment and Forestry, 2022).

4.2.2.4 Case study 2: Baringo county, Kenya

The Lake Baringo area is known as one of the *P. juliflora* invasion hotspots in Kenya. The species was introduced there in 1982 through the Fuelwood Afforestation Extension Project (Getahum, 1988; Kariuki, 1993). Most of the planting was done around Lake Baringo and a few other sites in upland areas, such as Yatya and Egerton University field station. At the time, the area around Lake Baringo was already heavily degraded and affected by soil erosion due to overgrazing. Thus, the planting project, conducted at around 40 sites covering more than 250 ha, aimed to mitigate firewood scarcity and desertification (Kariuki, 1993; Andersson, 2005). Some of the initial plantation sites are still present in Loruk, along the Kapthurin River, Salabani, Sandai, Chemeron Dam, Ng'ambo, Marigat and Eldume (Fig. 4.4).

Mbaabu *et al.* (2019) used a combination of dry and wet season Landsat satellite images for the

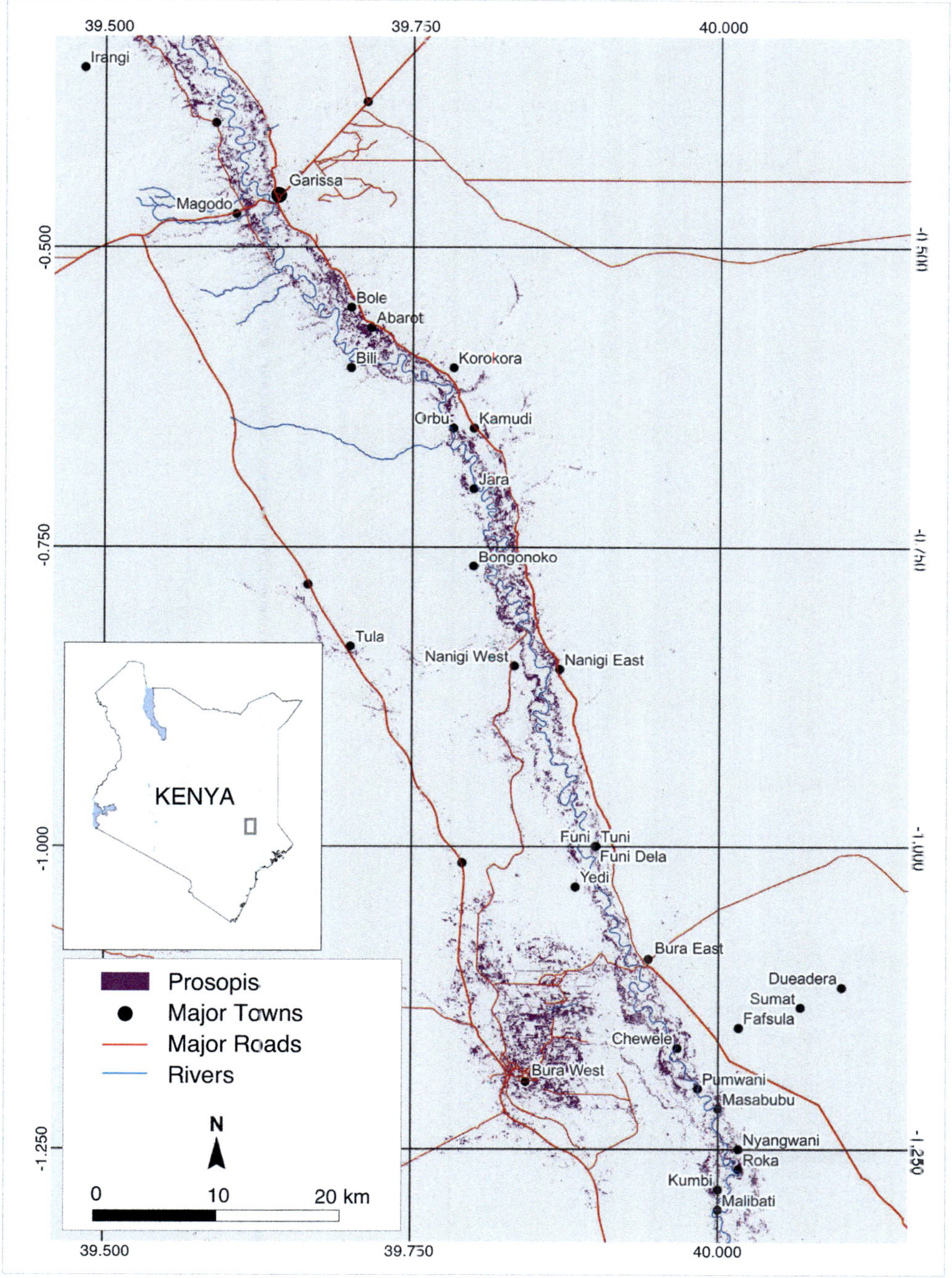

Fig. 4.2. The extent of invasion by *Prosopis juliflora* in Tana River County, particularly along the corridor and in the delta of the Tana River, Kenya. Inset shows the location of this site in Kenya.

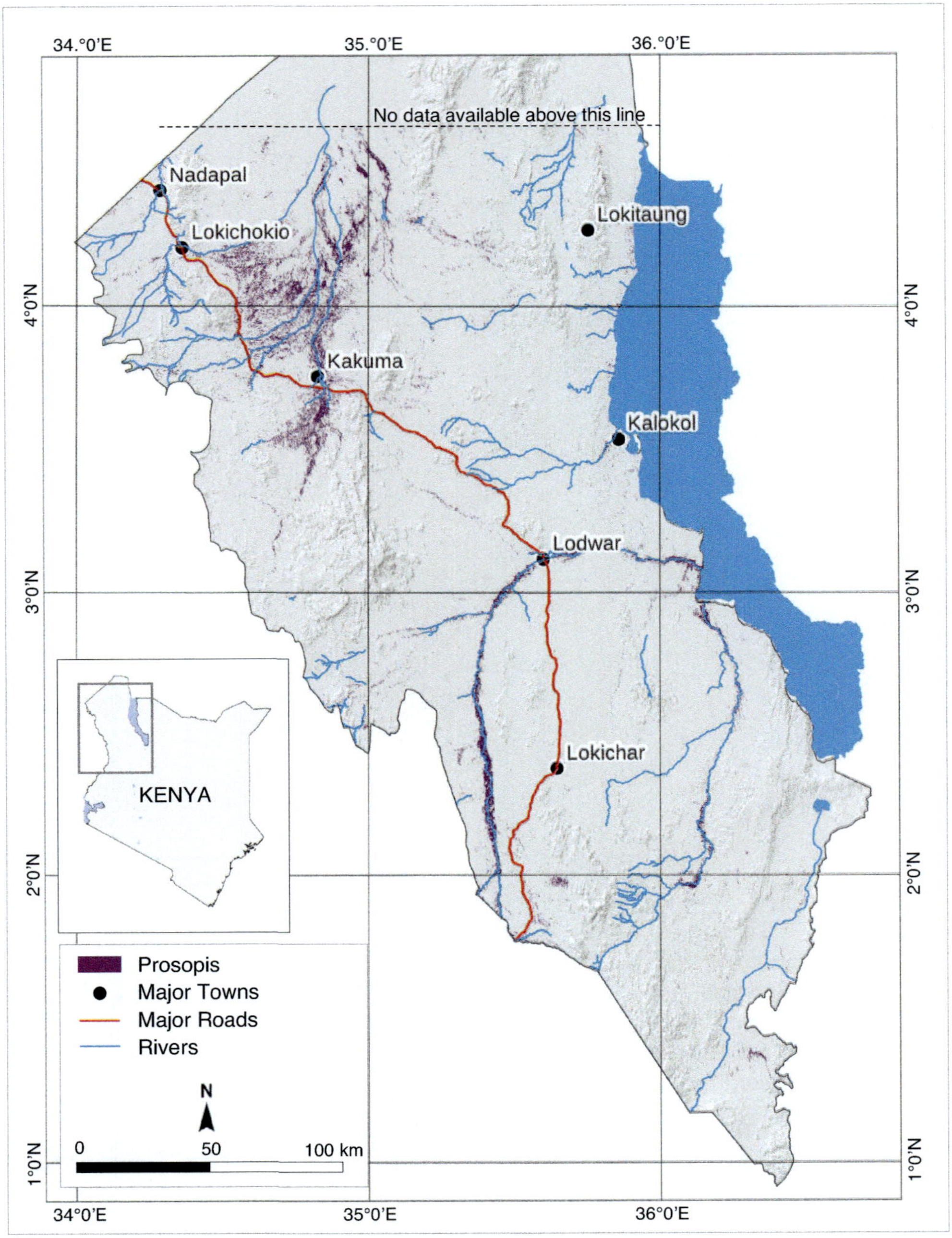

Fig. 4.3. *Prosopis juliflora* invasion in Turkana County, Kenya. Data are from S. Eckert 2018 and 2020, unpublished data. Inset shows the location of this site in Kenya.

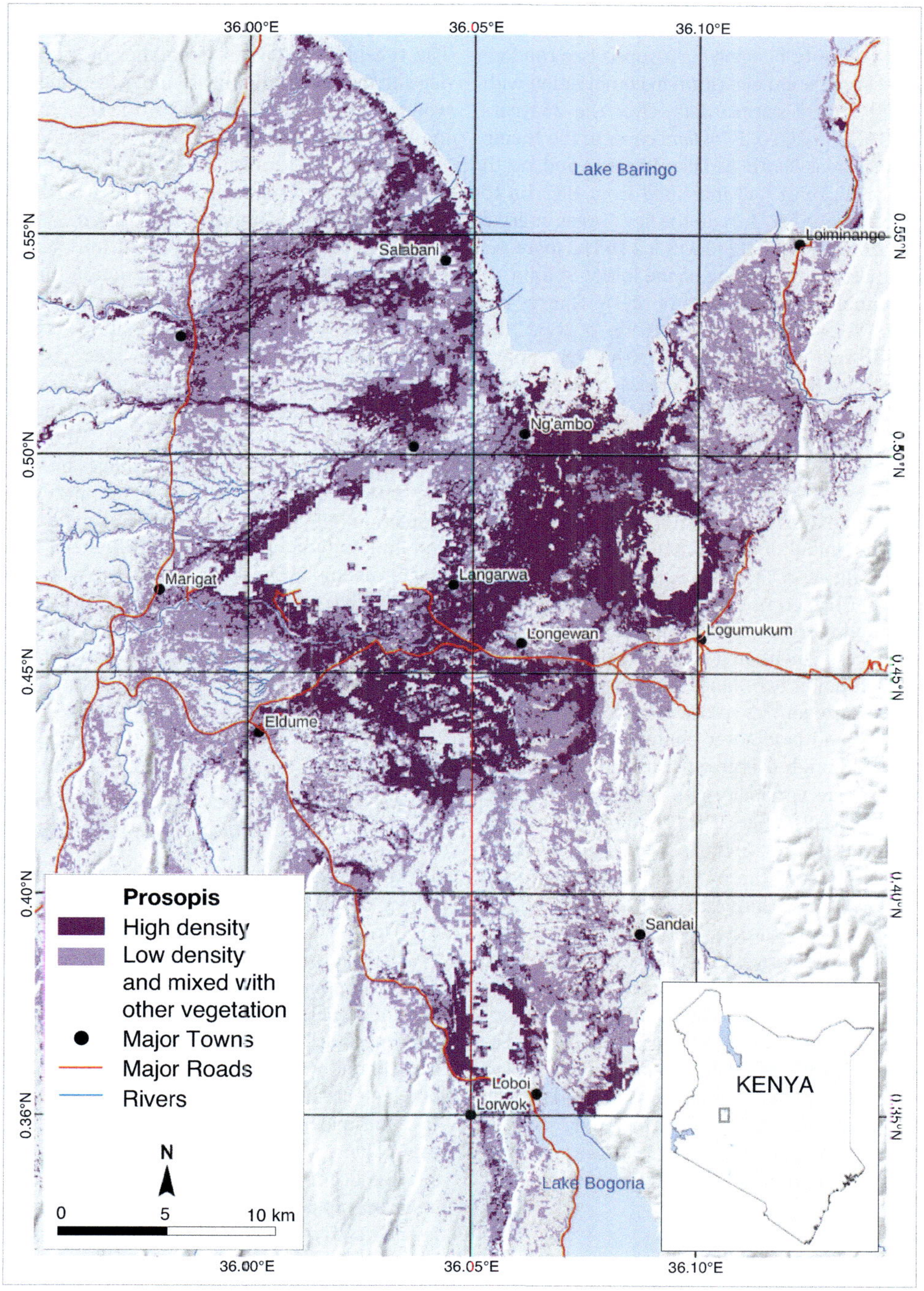

Fig. 4.4. The extent of *Prosopis juliflora* invasion in Marigat Subcounty, between Lake Baringo and Lake Bogoria. Data from Mbaabu *et al*. (2019). Inset shows the location of this site in Kenya.

years 1988, 1995, 2002, 2009 and 2016 to map the spread of *P. juliflora* in the Marigat Subcounty, Baringo County, by applying a supervised random forest classification algorithm in combination with ground control sample data. Over the 28 years from 1988 to 2016, *P. juliflora* cover in the Njemp flat (the areas south of Lake Baringo and north of Lake Bogoria) had increased from 882 ha to 18,792 ha (see Fig. 4.3). In the first 7-year interval from 1988 to 1995 there was a 2463 ha increase, i.e. almost a quadrupling of the initial 882 ha, at an annualized spread rate of 21%. During the second 7-year interval from 1995 to 2002, the invaded area increased by 5030 ha, reaching a total area of 8375 ha. The increase in the two subsequent 7-year intervals was 5211 and 5206 ha, with annualized spread rates of 7.2% and 4.7% respectively. The average spread rate over the full 28 years was 11.6%.

The analysis by Mbaabu *et al.* (2019) revealed that the initial invasion pathways in Baringo were along rivers (Molo, Loboi, Perkerra, Lembus, Endau, Chemeron, Sandai), irrigation canals, roads and marshes connecting Lake Baringo and Lake Bogoria. Invasion progressed from these areas to surrounding bare lands south and south-west of Lake Baringo. This spread resulted in significant land use and land cover changes in the affected areas. Although overgrazing, agricultural expansion, climate variability and fire are also drivers of land cover change, it was estimated that from 1988 to 2016, *Prosopis* trees were responsible for 41% (3452 ha) of the loss of native *Vachellia tortilis* trees, 37% (2675 ha) of the loss of grassland and 28% (502 ha) of the loss of cropland. Their impact became particularly noticeable between 2009 and 2016, when 74% of the irrigated cropland in the Njemps flat was invaded by *P. juliflora*, particularly in important irrigation schemes (Perkerra, Loboi, Eldume, Lororo, Sandai) (Mbaabu *et al.*, 2019). Presently, more than ten sublocations around Lake Baringo are densely invaded, including parts of protected areas such as the Lake Bogoria Reserve and Ruko Community Wildlife Conservancy.

4.2.3 Ethiopia

In Ethiopia, *P. juliflora* was introduced and planted mainly in the Afar National Regional State in the north-east of the country, bordering Eritrea and Djibouti. Afar has an arid to semi-arid climate and is the hottest part of Ethiopia. The vegetation consists of patches of scattered dry shrubs, woodland, bushland, grassland and wooded grassland (Engda, 2009). One of the main invasion hotspots in Afar is the Awash River basin. Pastoralism is the main economic activity but agropastoralism is on the rise due to sedentarization (Schmidt and Pearson, 2016), as well as private and government investments in large-scale agricultural production in the Middle Awash Valley.

There is very little information regarding the location and number of sites where *P. juliflora* was originally introduced 50 years ago, primarily for soil conservation (Ayanu *et al.*, 2015; Kebede and Coppock, 2015). Additional plantations were established in villages between the 1980s and 1990s for shade and wind protection, and as a source of firewood, fencing and building material (Ayanu *et al.*, 2015). The first problems arose soon thereafter, when the tree started to invade grasslands, riverbanks and roadsides. *P. juliflora* has become a source of conflict among pastoralist groups due to the increasingly difficult access to grazing areas (Kebede and Coppock, 2015; Shiferaw *et al.*, 2019a; Chapter 7, this volume).

4.2.3.1 Case study 3: Awash basin, Ethiopia

Shiferaw *et al.* (2019b) used Landsat satellite images of the dry season in 1986, 2000 and 2017 to assess the spread of *P. juliflora* in Middle Awash (a subsection of the Afar Region) using the random forest classification algorithm and a large amount of field data (Fig. 4.5). They found that cover increased from 241 ha in 1986 to 148,218 in 2000 and 965,166 ha in 2017, or 14% of the total land area in Afar (Shiferaw *et al.*, 2019a, c). As in Baringo, invasion by *P. juliflora* was concentrated along major rivers, floodplains and roads and significantly affected the ecosystems of these areas by replacing native vegetation. Over the 31-year study period, 340,000 ha of grassland became invaded, which severely affected the livelihoods of many pastoralists especially in the Awash Basin (Chapter 8, this volume). *P. juliflora* also invaded the Allideghi Wildlife Reserve, thus threatening biodiversity, livelihoods and essential ecosystem services.

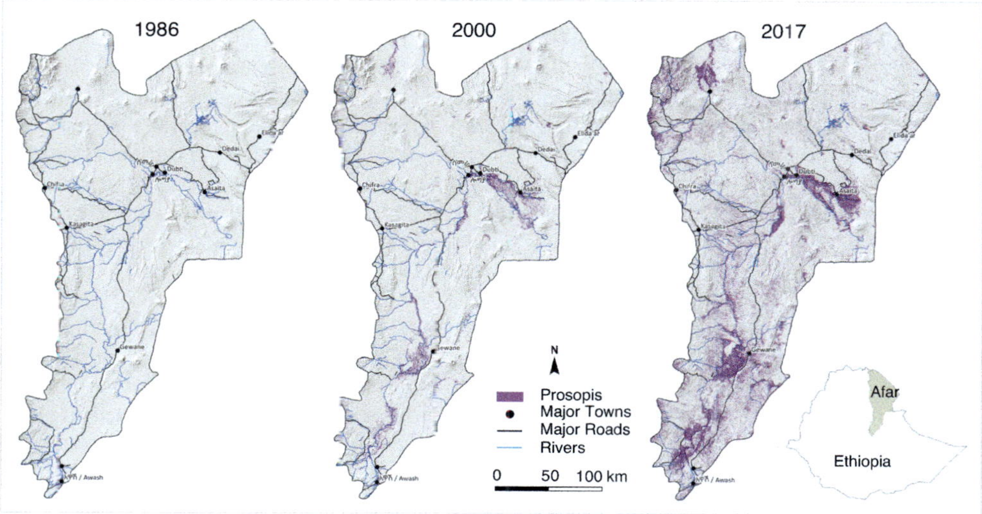

Fig. 4.5. Spread of *Prosopis juliflora* (shaded purple) between 1986 and 2017 in the central and southern parts of Afar National Regional State, Ethiopia. Map redrawn from Shiferaw *et al*. (2019a). Inset shows the location of this site in Ethiopia.

The rapid spread of *P. juliflora* in Afar is correlated with the availability of water and the presence of infrastructure networks such as roads and water canals, settlements and recent agricultural expansion (Shiferaw *et al*., 2019b).

4.2.4 Conclusions on current spread

The case studies above show that the local context strongly influences the pattern, pace and direction of *P. juliflora* invasions. Pathways of spread such as rivers, roads or livestock routes act as accelerators along which invasions spread rapidly in a linear fashion. On the other hand, natural barriers such as lakes or escarpments effectively act as obstacles to spread (see Section 4.3 below), leading to sharp invasion edges. Finally, in open terrain that lacks roads, rivers or pathways of livestock movement, such as in the north-western part of Turkana County, *P. juliflora* tends to spread along the leading edge of the invasion at a much lower rate than along linear pathways. However, the case of Afar shows that even such areas can become invaded at high densities over time.

4.3 Areas at Risk of Future Invasion

The rate at which *P. juliflora* is spreading in Kenya and Ethiopia, as detailed in the preceding section, is concerning. It is therefore important for future preparedness and planning to quantify the extent and location of potentially suitable habitat for this invasive species in eastern Africa – that is, the territory it might inhabit if no effective control measures are undertaken.

4.3.1 Methodology

Eckert *et al*. (2020) analysed the suitable habitat that could potentially be invaded by *P. juliflora* in eastern Africa. They also assessed how these habitats might change in the future considering various climate change scenarios. The authors used global occurrence points and seven spatial climate variables (four temperature and three precipitation variables) to model the global niche of the species. Further, they used a set of occurrence points in eastern Africa and 13 spatial variables (four temperature, four precipitation, two topographic, two hydrological and one anthropogenic), as well as two vegetation indices, to model the regional niche of *P. juliflora*.

Here, we provide a summary of their findings concerning the suitable range for *P. juliflora*, as determined by the regional model under current climatic conditions. We offer a binary rendition of the suitability layer, where '1' denotes suitability and '0' denotes unsuitability across Ethiopia, Kenya and Tanzania. This binary layer was derived from the original layer, which featured continuous values ranging between 0 and 1. A threshold was set at 0.6425 (equivalent to 64.25% suitability); values below this threshold were assigned '0' while those equal to or exceeding the threshold were assigned '1'. The model exclusively indicates the likelihood of *P. juliflora* establishing itself in each location, without providing insights into its behaviour in suitable locations, such as invasiveness, dominance over other species or density.

4.3.2 Spatial characteristics of suitable area

Among the regions where *P. juliflora* could potentially establish within the current climatic conditions in Ethiopia, Kenya and Tanzania (Fig. 4.6), Kenya stands out as having the largest proportion of suitable areas: around three-quarters of its territory (453,000 km^2 out of 582,646 km^2), i.e. almost all its arid and semi-arid lands, are at risk of being invaded if spread is not contained. About half the territory of Tanzania is suitable for the growth of *P. juliflora* (484,000 km^2 out of 945,000 km^2). These areas include vast stretches along the coast and the Kenyan border, parts of the Ruvuma and Mtwara regions in the south, as well as Mwanza, Geita, Simiyu, Shinyanga and Tabora regions south of Lake Victoria. Finally, about 45% of Ethiopia (492,000 km^2 out of 1,100,000 km^2), i.e. a major part of the Ogaden and Afar regions, is suitable for the growth of *P. juliflora*. A comparison between Fig. 4.5 and the currently invaded areas outlined in Section 4.2 above shows that invasion by *P. juliflora* in several parts of the region is still at a very early stage and that it is likely to expand significantly if left uncontrolled (Eckert *et al.*, 2020).

4.3.3 Implications for management

The *P. juliflora* occurrence and suitability models presented above can provide vital information to support the implementation of national strategies such as the Kenyan National Prosopis Strategy or the Tanzanian National Invasive Species Strategy and Action Plan. They have direct practical relevance for the prioritization of management efforts, the estimation of management costs and related allocation of funds. As *P. juliflora* is likely to expand substantially if left uncontrolled, the preparation of such maps at regular intervals is necessary for the monitoring of management effectiveness, the detection of new invasions and consequently for the adaptation of management plans. Details on how the models and resulting maps and data can be used to design and implement spatially explicit management strategies at county level are provided in Chapter 11 (this volume).

4.4 Considerations on the Use of Remote Sensing

4.4.1 The importance of remote sensing

Remote sensing with EO data is crucial for the surveillance and management of invasive species like *P. juliflora*. EO data enable the creation of accurate maps illustrating the species' distribution across vast territories at different times, which allows for rapid, frequent and cost-effective assessments. Furthermore, EO data and resulting maps provide detailed, quantitative insights, making them invaluable for disseminating information on distribution and spread rates. They can serve as baseline data for mapping spatial patterns and modelling areas at risk of invasion, as well as for assessing and monitoring control and rehabilitation efforts.

This section provides methodological considerations for the use of remotely sensed EO data for *P. juliflora* research, monitoring and management. These considerations are intended for a non-technical audience and focus on guiding important decisions regarding data selection, while also providing references for further reading.

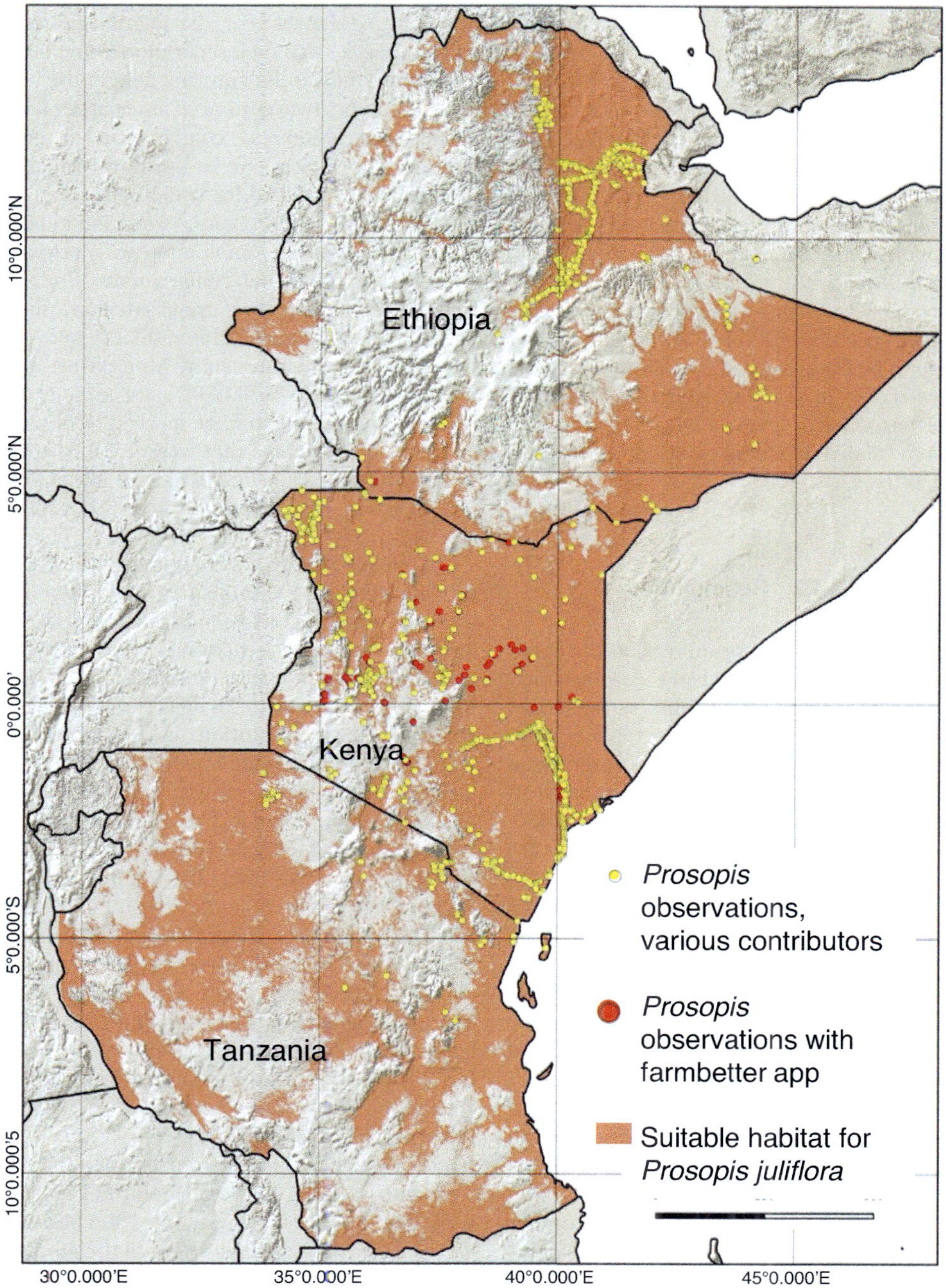

Fig. 4.6. Areas that are suitable for the growth of *Prosopis juliflora* in Ethiopia, Kenya and Tanzania, under current climatic conditions. Results from a regional model based on 15 spatial variables (Eckert *et al.*, 2020). The continuous values of the model (between 0 and 1) were transformed to binary values (either 0 or 1 to represent suitable or unsuitable areas) in the present map with a threshold at 0.6425 to differentiate the binary values.

Prosopis juliflora trees can be young or old, small or large, standing alone, in small thickets or expansive patches, sometimes as the only tree species and sometimes mixed with other tree species. These many possible forms of *P. juliflora* occurrence are a challenge for research, monitoring or planning efforts relying on the use of remotely sensed EO data. For example, the identification of isolated small trees requires data with a very high spatial resolution, while spotting *P. juliflora* trees from space when they are mixed with other species requires data that allow the spectral reflectance of the various tree species to be distinguished. Below, we list a few characteristics of remotely sensed EO data that are of particular importance in the context of research aimed at monitoring or managing *Prosopis* trees.

4.4.2 Resolution

Resolution plays an important role in the application domain of remote sensing sensors, which can vary depending on the satellite's orbit height and sensor design. There are four types of resolution for any dataset: spatial, spectral, temporal and radiometric (Table 4.2). Trade-offs between these resolutions greatly influence the ability to accurately detect, map or monitor invasive species.

Spatial resolution is defined by the size of each pixel within a digital image and the area on Earth's surface represented by that pixel. In other words, the smallest object visible on a remotely sensed image is determined by the spatial resolution at which the data were captured. The most widely used sensors have a spatial resolution ranging from 1 to 1000m (see Table 4.2). About half, including the widely used Landsat, SPOT and Sentinel sensors, have a resolution of between 10 and 30 m. However, there is a trend towards higher spatial resolution, to capture more details. This will improve the potential to identify young *P. juliflora* trees or sparse covers that are often found at the invasion front.

Spectral resolution is a measure of the ability of a sensor to resolve features into separate spectral bands on the electromagnetic spectrum. Many sensors have 3–10 bands (see Table 4.2) capturing information in the visible and infrared ranges. For the identification of invasive species and their separation from native vegetation, the bands capturing data in the red, near- and shortwave-infrared are particularly useful. Spectral indices – combinations of spectral reflectance from two or more wavelengths – have been developed for better separation of very similar vegetation types. Vegetation indices are often applied in combination with original wavelengths to better differentiate *Prosopis* from other land covers. New satellites, such as the FLuorescence EXplorer (FLEX) and the hyperspectral Environmental Monitoring and Analysis Program (ENMAP) satellite sensors, measure reflectance in finer spectral resolution or in spectral domains that were not measured by satellite sensors before, which might enable an improved identification of invasive plant species.

Temporal resolution is the time it takes for a satellite to revisit the same area. Geostationary satellites match the rotation of Earth and therefore have very fine temporal resolution. Polar orbiting satellites have a temporal resolution that varies from one to 16 days (see Table 4.2). High temporal resolution is favourable in tropical areas with frequent cloud cover. Nanosatellite constellations (e.g. PlanetLabs) have recently become popular. These consist of many very small and relatively inexpensive satellites, thus having an extremely high temporal resolution and very high spatial resolution, while lacking in spectral resolution and having a short lifespan in lower orbits. They can capture data at almost any given time due to their abundance, but they are unsuitable for invasive species monitoring unless the target species is well identifiable and separable from native vegetation. Very high temporal resolution is not required to identify *P. juliflora*. Rather, it is important to have images from both the dry and wet seasons owing to the phenological characteristics of *P. juliflora* (see Section 4.4.5 below).

Radiometric resolution refers to the amount of information in each pixel, specifically the number of bits representing the recorded energy. The level of detail each pixel can store is influenced by the sensor's ability to capture fine details of the real world. Each bit increases the potential radiometric resolution exponentially by a factor of 2.

Table 4.2. Overview of most widely known optical satellite sensors, excluding thermal bands. For time in operation, the start and end year is shown; if no end year is given, the sensors are still operational.

	Spatial resolution		Spectral resolution	Number of bands	Temporal resolution	Radiometric resolution	Time in operation
	Spectral bands	Panchromatic band					
Sensor	(m)	(m)	(nm)		(days)	(bit)	(years)
MODIS Terra	250/500/1000	–	400–1440	36	1–2	12	1999–2022
ASTER	15/30	–	520–2430	10	16	8	1999
Landsat-5 TM	30	–	450–2350	6	16	8	1984–2011
Landsat-7 ETM+	30	15	450–2350	7	16	8	1999–2022
Landsat-8 OLI	30	15	430–2290	7	16	12	2013
Landsat-9 OLI-2	30	15	230–2290	8	15	14	2021
Sentinel-2A/B MSI	10/20/60	–	443–2190	13	10	12	2015
SPOT-1–5	5/20	2.5–10	510–1750	4/5	3	8	1986–2015
GeoEye-1	1.65	0.41	450–920	4	2–3	11	2008–2015
IKONOS	4	1	445–853	4	3	11	1999–2014
Quickbird	2.4	0.6	450–900	5	3–4	11	2001–2015
Pléiades	2	0.5	450–915	4	1	12	2011
RapidEye	5		440–850	5	1	12	2008
WorldView-2/3	2/1.24/3.7	0.5/0.3	400–2245	9/17	1	11/14	2009
PlanetScope	3	–	431–888	8	1	12	2014
Skysat	1/0.72	0.86/0.58	450–900	5	4–5	16	2013

4.4.3 Available period

Maintenance and expansion of EO satellite programmes is an ongoing task, which ensures a stable stream of comparable satellite data over time. Prominent examples are the Landsat satellite series launched in the 1970s (see Table 4.2) and the Sentinel-2 satellites that were launched in 2014 and will be continued for several years to come. Such long-term data series are a requirement for the spatial-temporal assessment of a plant invasion, including the identification of pathways of spread and the calculation of the rate of spread. Furthermore, such information is necessary for the calculation of the impacts of an invasion, and it forms the baseline for the planning and monitoring of effective land and invasion management.

4.4.4 Processing capacity

Given the increased amount and range of remotely sensed data captured daily and the increased processing and analysis complexity, remote sensing data have become a part of the big data paradigm. Desktop computers have become unsuitable to store and process available time series of EO data and cloud-based, paid data storage, as well as processing services, are almost mandatory when working with EO data. In parallel, novel algorithms have emerged to analyse the growing amount of data, such as artificial intelligence, machine learning and time series analysis.

4.4.5 Relevant characteristics of *Prosopis juliflora*

Pathways of spread and phenology are of particular importance in relation to the monitoring of *P. juliflora* using remotely sensed EO data. *P. juliflora* can establish near settlements, along water bodies and river courses, on moister than average land, as well as uncultivated, bare and disturbed grasslands. It spreads along roads, as well as wildlife and livestock migration routes. Spatial elements that influence these patterns of spread can be identified in the field or in existing spatial data and used to guide the collection of ground control points or to inform the classification of land cover – including *P. juliflora* – with EO data.

In eastern Africa, *P. juliflora* is evergreen, while the native African trees are mainly drought-deciduous. Consequently, sensor data captured during or towards the end of the dry season are optimal for differentiating between *P. juliflora* and native tree species. This important difference in phenology was used to successfully differentiate *P. juliflora* from native *Vachellia* tree species in eastern Africa. Ng *et al.* (2016a) and Mbaabu *et al.* (2019) have made use of this phenological variation and consequently worked with a combination of dry and wet season satellite datasets in their studies in Baringo, Kenya. However, this phenological trait of *Prosopis* trees is specific to eastern Africa and cannot simply be transferred to other regions. For example, invasive *Prosopis* trees as well as native trees are drought-deciduous in South Africa.

Several scholars have contributed to advancing EO methodology for the accurate and reliable identification and mapping of *Prosopis* trees in Africa. In Table 4.3 we provide a list of key peer-reviewed publications pursuing this objective and summarize the classification type and EO data used by the authors, as well as the areas of interest and the accuracies that they achieved with their approach.

4.5 Conclusions

Huge areas are already affected by *P. juliflora* in eastern Africa. In the Awash basin in Ethiopia, the species has invaded over a million hectares. In Isiolo, Turkana and Baringo in Kenya, the invaded areas cover more than 3% of these counties' total territory. The maps in Figs 4.1–4.4 show that invasions tend to follow rivers, roads and livestock routes, and that *P. juliflora* invades particularly suitable areas at a very high rate. This is the case, for example, along the Kerio, Turkwel and Tana rivers, in the fertile flatlands between the lakes of Bogoria and Baringo and in the Tana River delta.

Differences in the rate and patterns of spread can be caused by a number of factors, including (1) the number, size and location of

Table 4.3. Key peer-reviewed publications on remote sensing-based mapping of *Prosopis* trees in Africa. The table lists classification type, Earth Observation (EO) data, mono-, bi- or multitemporality, number of land cover/land use classes produced and accuracy achieved.

Reference	Region of interest	Classification type	EO data	LU/LC classes	Accuracy
Dubow (2011)	Garissa, Kenya	Subpixel classifier	Bitemporal Landsat	2	NA
van den Berg *et al.* (2013)	Northern Cape, South Africa	Thresholding	Multitemporal Landsat	5	1849 points (OA: 0.73)
Amboka and Ngigi (2015)	Baringo, Kenya	Maximum likelihood	Multitemporal Landsat NDVI	3	250 random points (OA: 0.88–0.90)
Rembold *et al.* (2015)	Somaliland	Maximum likelihood	Landsat and bitemporal NDVI	3	400 random points (OA: 0.77)
Wakie *et al.* (2014)	Afar, Ethiopia	Maxent	MODIS NDVI and multitemporal EVI	2	K: 0.85 AUC: 0.94
Meroni *et al.* (2017)	Hargeisa, Somaliland	Object- and pixel-based RFC	Bitemporal Landsat	7	10-fold cross-validation (OA: 0.82–0.84)
Ng *et al.* (2016b)	Hargeisa, Somaliland	Object- and pixel-based RFC	Bitemporal Landsat	16	10-fold cross-validation (OA: 0.67–0.78)
Ng *et al.* (2016a)	Tarach basin, Kenya	Pixel-based RFC	Monotemporal Sentinel-2	2	20 selected points (OA: 0.89)
Ng *et al.* (2017)	Marigat, Baringo, Kenya	Object-based RFC	Monotemporal Sentinel-2	4	OOB and 200 points (OA: 0.79–0.96)
Ng *et al.* (2017)	Marigat, Baringo, Kenya	Object-based RFC	Monotemporal Pléiades	4	OOB and 200 points (OA: 0.88–0.96)
Adam *et al.* (2017)	Northern Cape, South Africa	SVM and RFC	Monotemporal WorldView-2	6	164 points (OA: 0.86, K: 0.83–0.84)
Shiferaw *et al.* (2019c)	Afar National Regional State, Ethiopia	Pixel-based RFR (regression)	MODIS Temp, Landsat, other environmental variables	1	OOB and 10-fold cross-validation (OA: 0.92; K: 0.8)
Shiferaw *et al.* (2019b)	Afar National Regional State, Ethiopia	Pixel-based GLM, GBM, GBM-BRT, RFR, SVM, DNN and ensemble classifiers	MODIS Temp, Landsat, other environmental variables	1	20/80 split (AUC: 0.60–0.97, K: 0.32–0.80)

Continued

Table 4.3. Continued

Reference	Region of interest	Classification type	EO data	LU/LC classes	Accuracy
Shiferaw *et al.* (2019a)	Middle Awash, Ethiopia	Pixel-based RFC	Multitemporal Landsat	9	30/70 split and 10-fold cross-valida-tion (K: 0.88–0.92)
Mbaabu *et al.* (2019)	Baringo County, Kenya	Pixel-based RFC	Bitemporal Landsat	8	OOB and 10-fold cross-validation (OA: 0.98, K: 0.93–0.96)
Ahmed *et al.* (2021)	Lower Awash River basin, Ethiopia	Ensemble (BioClim, SVM, BRT, GAM, GLM, RFC)	Sentinel-2 and indices	2	K: 0.34–0.91, TSS: 0.37–0.93, AUC: 0.69–0.99
Ahmed *et al.* (2022)	Lower Awash basin, Ethiopia	RFC	Sentinel-2 and EV	2	TSS: 0.74–0.93, AUC: 0.91–0.98

LU/LC, land use/land cover; Classification types: RFC, random forest classifier; SVM, support vector machine; RFR, random forest regression; GBM, gradient boost machine; BRT, boost regression tree; DNN, deep neural network; Accuracy: K, kappa; OA, overall accuracy; OOB, out-of-bag accuracy; AUC, area under the curve; NDVI, Normalized Difference Vegetation Index; EVI, Enhanced Vegetation Index; GLM, General Linear Model; GAM, General Additive Model.

initial plantations, (2) the presence of pathways that can accelerate spread in a linear fashion (for example, roads and rivers), (3) the presence of obstacles to invasion (for example, Lake Baringo or the Rift Valley escarpments bordering Marigat Subcounty in Baringo), (4) the availability of suitable habitat, or (5) the level of management that has been undertaken in the area.

Prosopis juliflora invasion in eastern Africa is still at an early stage when compared to the potentially suitable habitats in this region, which cover 45%, 51% and 77% of Ethiopia, Tanzania and Kenya respectively. The areas most at risk are the arid and semi-arid lowlands such as the Afar and Ogaden lowlands in Ethiopia, the entire northern and eastern regions of Kenya and a mosaic of areas throughout Tanzania. This finding emphasizes the enormous risk that the countries in eastern Africa are facing and that their national governments have recently recognized. For example, in March 2024 the Kenyan Cabinet declared *P. juliflora* a national disaster and approved a National *Prosopis* Strategy for a coordinated management of the invasive plant at various planning and decision-making scales (Chapter 15, this volume).

The potential scale of *P. juliflora* invasion in eastern Africa underscores the importance of finding more effective ways to contain their spread and avoid the catastrophic consequences that would accompany such an invasion. This includes a focus of management on preventing further spread (see Part 3, this volume), i.e. by focusing on the leading edges of the advancing invasion. Indeed, unless biological control (Chapter 10, this volume) is integrated into a management strategy, effectively preventing the ongoing spread of *P. juliflora* may not be achievable. Further, there is a need for careful prioritization of preventive measures along pathways of spread, which implies a thorough understanding of the socio-economic and environmental context. At larger scales, there is a need for a coherent transboundary cooperation among neighbouring counties or nations.

Finally, this chapter illustrates the potential of remote sensing and EO data for studying, monitoring and managing *P. juliflora* invasions. We have summarized important technical considerations, particularly in terms of data selection (based on various types of resolution), for these specific applications. Considering the complexity of these tools and methods, as well as their need for massive storage capacity and computing power to secure regular monitoring at a large scale, it is advisable to set up adequate capacities for this task at national or even regional scale in eastern Africa. This includes infrastructure, a data access strategy and in-country capacity development. For the finer-grained monitoring of *P. juliflora* invasions at smaller scales – for example counties, wards or landscapes – EO data from satellite sensors need to be combined with the collection of field data and/or the use of other remote sensing devices such as unmanned aerial vehicles (Chapter 11, this volume).

Acknowledgements

This work was supported by the Swiss Programme for Research on Global Issues for Development (r4d), funded by the Swiss National Science Foundation and the Swiss Agency for Development and Cooperation, through the project 'Woody invasive alien species in East Africa: Assessing and mitigating their negative impact on ecosystem services and rural livelihood' (Grant Number: 400440_152085).

References

Adam, E., Mureriwa, N. and Newete, S. (2017) Mapping *Prosopis glandulosa* (Mesquite) in the semi-arid environment of South Africa using high-resolution Worldview-2 imagery and machine learning classifiers. *Journal of Arid Environments* 145, 43–51.

Ahmed, N., Atzberger, C. and Zewdie, W. (2021) Species distribution modelling performance and its implication for Sentinel-2-based prediction of invasive *Prosopis juliflora* in lower Awash River Basin, Ethiopia. *Ecological Processes* 10, 18.

Ahmed, N., Atzberger, C. and Zewdie, W. (2022) The potential of modeling *Prosopis juliflora* invasion using Sentinel-2 satellite data and environmental variables in the dryland ecosystem of Ethiopia. *Ecological Informatics* 68, 101545.

Amboka, A.A. and Ngigi, T.G. (2015) Mapping and monitoring spatial-temporal cover change of *Prosopis* species colonization in Baringo Central, Kenya. *International Journal of Engineering Science Invention* 4, 50–55.

Andersson, S. (2005) Spread of the introduced tree species *Prosopis juliflora* (Sw.) DC in the Lake Baringo Area, Kenya. Thesis, Swedish University of Agricultural Sciences, Umeå, Sweden.

Anonymous (2019) *National invasive species strategy and action plan (NISSAP) (2019-2029)*. Office of the Vice President, Dodoma, Tanzania.

Ayanu, Y., Jentsch, A., Müller-Mahn, D., Rettberg, S., Romankiewicz, C. *et al.* (2015) Ecosystem engineer unleashed: *Prosopis juliflora* threatening ecosystem services?. *Regional Environmental Change* 15, 155–167.

Dubow, A.Z. (2011) Mapping and managing the spread of *Prosopis juliflora* in Garissa county, Kenya. Thesis, Kenyatta University, Nairobi, Kenya.

Eckert, S., Hamad, A., Kilawe, C.J., Linders, T.E., Ng, W.T. *et al.* (2020) Niche change analysis as a tool to inform management of two invasive species in Eastern Africa. *Ecosphere* 11(2), e02987.

Engda, G. (2009) Spatial and temporal analysis of *Prosopis juliflora* (Swarz) DC invasion in Amibara Woreda of the Afar NRS. Thesis, Addis Ababa University, Ethiopia.

Getahum, A. (1988) Agroforestry for development in kenya. In: Kilewe, A.M., Kealey, K.M. and Kebaara, K.K. (eds) *Proceedings of the Second Kenya National Seminar on Agroforestry, 7–16 November 1988*, International Council for Research in Agroforestry, Nairobi, Kenya, pp. 85–109.

Kariuki, P. (1993) A social forestry project in Baringo, Kenya: A critical analysis. Thesis, University of Queensland, Brisbane, Australia.

Kebede, A.T. and Coppock, D.L. (2015) Livestock-mediated dispersal of *Prosopis juliflora* imperils grasslands and the endangered Grevy's zebra in Northeastern Ethiopia. *Rangeland Ecology and Management* 68, 402–407.

Kyuma, R.K., Wahome, R.G., Kinama, J.M. and Wasonga, V.O. (2016) Temporal relationship between climate variability, *Prosopis juliflora* invasion and livestock numbers in the drylands of Magadi, Kenya. *African Journal of Environmental Science and Technology* 10, 129–140.

Maundu, P., Kibet, S., Morimoto, Y., Imbumi, M. and Adeka, R. (2009) Impact of *Prosopis juliflora* on Kenya's semi-arid and arid ecosystems and local livelihoods. *Biodiversity* 10, 33–50.

Mbaabu, P.R., Ng, W.T., Schaffner, U., Gichaba, M. and Olago, D. (2019) Spatial evolution of *Prosopis* invasion and its effects on LULC and livelihoods in Baringo, Kenya. *Remote Sensing* 11, 1217.

Meroni, M., Ng, W.T., Rembold, F., Leonardi, U., Atzberger, C. *et al.* (2017) Mapping *Prosopis juliflora* in West Somaliland with Landsat 8 satellite imagery and ground information. *Land Degradation and Development* 28, 494–506.

Ministry of Environment and Forestry (2022) *National Strategy and Action Plan for the Management and Control of the Invasive Prosopis juliflora Tree Species. 2022 – 2030*. Ministry of Environment and Forestry, Nairobi, Kenya.

Ministry of Livestock and Fisheries (2017) *National Strategy on Prosopis juliflora Management*. Ministry of Livestock and Fisheries, Addis Ababa, Ethiopia.

Muturi, G.M., Mohren, G.M.J. and Kimani, J.N. (2010) Prediction of *Prosopis* species invasion in Kenya using geographical information system techniques. *African Journal of Ecology* 48, 628–636.

Ng, W.T., Immitzer, M., Floriansitz, M., Vuolo, F., Luminari, L. *et al.* (2016a) Mapping *Prosopis* spp. within the Tarach Water Basin, Turkana, Kenya using Sentinel-2 Imagery. In: Neale, M.U. and Maltese, A. (eds) *Proceedings of the 2016 SPIE Remote Sensing, Edinburgh, UK, 25 October 2016*, p. 99980L.

Ng, W.T., Meroni, M., Immitzer, M., Böck, S., Leonardi, U. *et al.* (2016b) Mapping *Prosopis* spp. with Landsat 8 data in arid environments: Evaluating effectiveness of different methods and temporal imagery selection for Hargeisa, Somaliland. *International Journal of Applied Earth Observation and Geoinformation* 53, 76–89.

Ng, W.T., Rima, P., Einzmann, K., Immitzer, M., Atzberger, C. *et al.* (2017) Assessing the potential of Sentinel-2 and Pléiades data for the detection of *Prosopis* and *Vachellia* spp. in Kenya. *Remote Sensing* 9, 74.

Rembold, F., Leonardi, U., Ng, W.T., Gadain, H., Meroni, M. *et al.* (2015) Mapping areas invaded by *Prosopis juliflora* in Somaliland on Landsat 8 imagery. *Proceedings of SPIE* 9637, 295–306.

Schmidt, M. and Pearson, O. (2016) Pastoral livelihoods under pressure: Ecological, political and socio-economic transitions in Afar (Ethiopia). *Journal of Arid Environments* 124, 22–30.

Shi, T. and Xu, H. (2019) Derivation of tasseled cap transformation coefficients for Sentinel-2 MSI at-sensor reflectance data. *IEEE Journal of Selected Topics in Applied Earth Observations and Remote Sensing* 12, 4038–4048

Shiferaw, H., Bewket, W., Alamirew, T., Zeleke, G., Teketay, D. *et al.* (2019a) Implications of land use/land cover dynamics and *Prosopis* invasion on ecosystem service values in Afar region, Ethiopia. *Science of the Total Environment* 675, 354–366.

Shiferaw, H., Bewket, W. and Eckert, S. (2019b) Performances of machine learning algorithms for mapping fractional cover of an invasive plant species in a dryland ecosystem. *Ecology and Evolution* 9, 2562–2574.

Shiferaw, H., Schaffner, U., Bewket, W., Alamirew, T., Zeleke, G. *et al.* (2019c) Modelling the current fractional cover of an invasive alien plant and drivers of its invasion in a dryland ecosystem. *Scientific Reports* 9, 1576.

van den Berg, E.C., Kotze, I. and Beukes, H. (2013) Detection, quantification and monitoring of *Prosopis* in the Northern Cape province of South Africa using remote sensing and GIS. *South African Journal of Geomatics* 2, 68–81.

Venter, O., Sanderson, E.W., Magrach, A., Allan, J.R., Beher, J. *et al.* (2018) *Last of the Wild Project, Version 3 (LWP-3): 2009 Human Footprint, 2018 release.* NASA Socioeconomic Data and Applications Center (SEDAC), New York.

Wakie, T.T., Evangelista, P.H., Jarnevich, C.S. and Laituri, M. (2014) Mapping current and potential distribution of non-native *Prosopis juliflora* in the Afar region of Ethiopia. *PLoS One* 9(11), e112854.

Weiss, D.J., Atkinson, P.M., Bhatt, S., Mappin, B., Hay, S.I. *et al.* (2014) An effective approach for gap-filling continental scale remotely sensed time-series. *ISPRS Journal of Photogrammetry and Remote Sensing* 98, 106–118.

Part 2: Impacts

This Part includes five chapters that provide an overview of the impacts of invasive *Prosopis* trees on aspects of the environment and people's livelihoods. The environmental issues that have received the most research attention include biodiversity, water resources and rangeland productivity, all of which affect people's wellbeing. These impacts are often quantified and reported separately but of course, they interact in many ways that have seldom been considered in integrated research projects. The chapters in this section review impacts across individual biophysical and social domains, and then attempt to draw them together in a more holistic manner using the One Health approach.

In Chapter 5, van Wilgen *et al.* review the impacts of *Prosopis* tree invasions on biodiversity, drawing on an increasing body of evidence that has recorded these aspects in southern and eastern Africa. These invasions can dramatically change the composition and structure of native vegetation, eliminating grasses and forbs and converting open savannas to closed, impenetrable thickets. These impacts translate into net financial losses to the rural communities which depend on the ecosystem services that are underpinned by biodiversity.

In Chapter 6, Shiferaw *et al.* explore the impacts of invasions by *Prosopis* trees on groundwater resources in the semi-arid ecosystems of eastern Africa. The trees are well adapted to survive in water-scarce areas. They have structural and physiological features (such as deep root systems and an ability to redistribute water in the soil profile) that allow them to outcompete native trees, resulting in significant losses of groundwater resources. These losses contribute further to the increasing impacts of climate change on nature and people, and the financial burdens borne by rural communities.

In Chapter 7, Kibet and van Wilgen evaluate impacts on rangeland condition and carrying capacity. A substantial proportion of the region's economy is derived from pastoralism and tourism, which in turn depend heavily on healthy, productive rangelands that support both domestic livestock and wild herbivores. Invasions reduce carrying capacity by eliminating the grasses and forbs that grazing herbivores depend on. In addition, the excessive consumption of *Prosopis* pods can lead to direct injury and toxic effects, while thorns cause serious injuries to animals. Pastoral communities face many threats to their way of life, and invasion of their shrinking rangelands by *Prosopis* trees is one of them that is covered here.

In Chapter 8, Bekele *et al.* address how these impacts affect people. They highlight how people originally benefited from the introduction of these trees, but how they were increasingly negatively affected as the trees became invasive and spread across the landscape. Perceptions of the value of the trees began to change, but also to diverge as those who benefit from the trees (mostly charcoal producers) retain positive attitudes, while those who suffer the consequences become increasingly concerned about the impacts. Overall, the invasions come at a considerable net cost.

In Chapter 9, Schaffner *et al.* address the interdependence of human, animal, plant and ecosystem health. They apply the One Health approach, informed by a wide range of impacts documented in the literature, to examine the multidimensional impacts of *Prosopis juliflora* by (i) assessing its impacts on ecosystem service multifunctionality, considering indicators of ecosystem services relating to all four dimensions of One Health, and (ii) contrasting the positive and negative aspects of *Prosopis juliflora* across the four dimensions to calculate a One Health score for different landscapes with varying degrees of invasion. Both approaches indicate that pristine or restored environments deliver positive benefits, but that heavily invaded sites are characterized by substantial changes in human, animal, plant and environmental health, a situation that is predominantly negative and unsustainable. Such holistic assessments should be very useful in demonstrating the overall effects of invasions, and could be used to advocate strongly for the funding needed to address the problem.

5 The Impact of Invading *Prosopis* Species on Biodiversity and Ecosystem Services

Brian W. van Wilgen[1]*, Theo E.W. Linders[2] and Ketema Bekele[3]
[1]Centre for Invasion Biology, Department of Botany and Zoology, Stellenbosch University, South Africa; [2]Management, Nature and Landscape Department, Central Government Real Estate Agency of the Dutch Ministry of the Interior and Kingdom Relations, The Hague, The Netherlands; [3]School of Agricultural Economics, Haramaya University, Dire Dawa, Ethiopia

Abstract

This chapter reviews the impacts of *Prosopis* invasions on biodiversity and ecosystem services in eastern Africa and addresses the potential for their restoration following control. Invasion by *Prosopis* trees affects the composition, structure and functioning of invaded ecosystems because the trees are superior competitors for water, light and nutrients. *Prosopis* trees form impenetrable stands that alter habitats and exclude wildlife, and they cause harm by being toxic and causing physical injury to animals and people. Studies in eastern and southern Africa have demonstrated that invasions reduce the species richness of native tree, shrub and grass communities, with knock-on impacts on native bird, mammal and invertebrate communities. Invasion brings about dramatic changes in structure, from ecosystems with scattered trees and abundant grasses to closed-canopy, impenetrable stands of trees with little or no grass below them. Ecosystem functioning and the ability of ecosystems to deliver services such as livestock production, water supply, traditional medicinal remedies and carbon storage are also affected, both directly and indirectly. Control of invasions can restore some of these services, and such action is justified because of the positive returns on investment in control. Such control would achieve more if the diverse and productive herbaceous communities that are important for ecosystem functioning could also be restored. Indigenous peoples and local communities are generally aware of the impacts of invasions on biodiversity, although a smaller minority are either unaware or their perceptions tend to be positive due to perceived benefits. The implications of these findings are that governments, aid agencies and communities would benefit from investing in control to protect biodiversity and ecosystem services by preventing the further spread of invasive *Prosopis* trees, implementing targeted control efforts where the species has become established, and actively restoring biodiversity to areas where the invasive trees have been brought under control.

5.1 Introduction

Biological diversity (typically shortened to 'biodiversity') is a term that describes the variety and variability of life on Earth. The shortened term biodiversity was first coined by W. G. Rosen in 1985 (Wilson, 1988) and has since become widely used in the scientific literature, as well

*Corresponding author: bvanwilgen@sun.ac za

© CAB International 2025. *The Ecology and Management of Invasive Prosopis Trees in Eastern Africa* (eds U. Schaffner *et al.*)
DOI: 10.1079/9781800623644.0005

as in policy documents (e.g. IPBES, 2018). The Earth's biodiversity is not evenly distributed across the globe, with some regions being markedly more diverse than others (Mittermeier *et al.*, 2004). Biological invasions constitute a significant threat to global biodiversity (Pyšek *et al.*, 2020), often operating in conjunction with other elements of global change to create large impacts on the composition, structure and functioning of ecosystems (Sala *et al.*, 2000).

The developing science of biological invasions did not initially have a focus on impacts and was developed around questions of which species would become invasive, which ecosystems would be prone to invasions and how science could inform effective management (Drake *et al.*, 1989). As a result, impact-related studies lagged behind other aspects of the problem but more recently, Pyšek and Richardson (2010) noted that the proportional contribution of impact-related studies to the overall invasion literature increased steadily between 1990 and 2009 (from c. 14% of studies to c. 27% of studies).

Although there have been several early anecdotal accounts of the impacts of invasive *Prosopis* trees in Africa, it was only after the start of the 21st century that careful studies documenting observed impacts in the field began to emerge. Initially, these studies were carried out in South Africa, where *Prosopis* trees have been recognized as a problematic invasive species since the 1970s and where attempts to control them were initiated in the 1980s (Zimmermann, 1991). Several studies have now also been conducted in, *inter alia*, Ethiopia and Kenya, where *Prosopis* trees were introduced more recently, and where they only started to be recognized as a problem in the late 1990s (Chapter 2, this volume).

In this chapter, we review the current state of understanding regarding the impacts that *Prosopis* invasions have had on biodiversity and ecosystem services in eastern Africa. We examine the determinants of impact (i.e. the range, abundance and per capita effect of individuals of a species; Parker *et al.*, 1999), and then review available studies documenting the effects of these impacts on aspects of biodiversity (i.e. the structure, composition and functioning of ecosystems) using the framework of Noss (1990). Our review also covers impacts recorded elsewhere in Africa (predominantly in South Africa), as these often provide additional information in support of general trends and allow for evaluation of the magnitude of impacts with a greater degree of confidence. We also examine the potential for the restoration of species composition and ecosystem functioning, and the benefits that would arise from such interventions.

5.2 Biodiversity and Ecosystem Services in Eastern Africa

Mittermeier *et al.* (2004) recognized 40 'hotspots' of biodiversity globally, seven of which are in Africa, and three of which are in eastern Africa. Hotspots are important because they cover only 2.3% of the world's land area, yet they contain more than half of the world's plant species and a third of its land vertebrates. Parts of Ethiopia, Somalia, Djibouti and Kenya fall within the Horn of Africa hotspot, which contains 2800 endemic plant species. The Eastern Afromontane hotspot is found in Burundi, Ethiopia, Kenya, Rwanda, Tanzania and Uganda and contains 2400 endemic plant species. The Coastal Forests of Eastern Africa hotspot occurs along the coast of Tanzania, Kenya and Somalia, reaching about 200 km inland, and contains 1750 endemic plant species. Together, these three hotspots contain almost 7000 endemic plant species, although many more remain to be discovered (Spawls and Mathews, 2012), and much still needs to be done to obtain a more comprehensive account of their distribution and value. Examples quoted by Tsinda *et al.* (2016) include Ethiopia, which has 31 endemic species of mammals and is also a global centre of avian diversity, with 856 bird species including 20 endemics. Kenya is perhaps the country in eastern Africa that is most diverse, because of the variety of landscapes, ecosystems and habitats, and the convergence of at least seven biogeographic units (Spawls and Mathews, 2012). Tanzania is also one of the globe's mega-biodiversity countries with many endemic species. The country ranks second highest in terms of the number of vertebrates and plant species in sub-Saharan Africa and possesses about 74% of all plant species found in eastern Africa (Rija *et al.*, 2013).

According to the regional assessment report for Africa on biodiversity and ecosystem services (IPBES, 2018), Africa has an extraordinary richness in biodiversity and a wealth of indigenous and local knowledge that together comprise a strategic asset for sustainable development in the region. The report notes that:

> Africa's rich and diverse ecosystems generate flows of goods and services that are essential in providing for the continent's food, water, energy, health and secure livelihood needs. More than 62 per cent of the population depend directly on these services in rural areas, while the urban and peri-urban population supplement their incomes, as well as their energy, medicine and other essentials, from ecosystem-based resources. Tangible and intangible assets such as food, water, medicinal plants, sacred rituals, as well as religious and cultural spaces, underpin nature's contributions to the economy and are central to a multitude of other livelihood strategies.

The IPBES report stresses a few key messages, including that Africa's biodiversity is essential for the continent's food, water, energy, health and secure livelihood needs; that its true value is underappreciated; that there are opportunities to capitalize on these resources and to sustainably manage them; that certain African ecosystems are of great ecological, biological and cultural importance; and that Africa has important and unique genetic diversity.

Tourism, especially ecotourism which is underpinned by biodiversity, is one of the most significant sectors of the economy of the East African Community region. It contributes about 17% to export earnings and about 10% to gross domestic product, generating about 7% of employment in the region. Moreover, tourism has important linkages with other sectors of the economy including, among others, agriculture, manufacturing, insurance and finance (https://www.eac.int/tourism). While the full impact of invasive species on the perceptions of tourists is poorly understood (Shackleton *et al.*, 2022), evidence is emerging that invasion by alien plants (including *Prosopis* trees) has measurable negative effects on tourism. In Kenya, increasing cover of invading *Prosopis* trees was found to have a significantly negative impact on tourism potential (Linders *et al.*, 2021). Research in South Africa has also shown that tourists in some protected areas are aware of the problem, and most (78%) were willing to pay increased entrance fees to control invasive species (Nikodinoska *et al.*, 2014). The prevention of invasion of ecosystems by *Prosopis* trees, and their restoration following effective control, can therefore be justified in economic terms, as well as in terms of protecting the livelihoods of rural communities that are heavily dependent on biodiversity and ecosystem services.

5.3 Determinants of Spread and Impact

5.3.1 A framework for assessing impact

Parker *et al.* (1999) proposed an equation to describe the determinants of the impacts of biological invasions, expressed as $I = R \times A \times E$, where impact (I) is the product of the geographical range of the invader (R), its abundance or density (A), and the per-capita effect (E) of an individual invader. This provides a useful framework for understanding how impacts come about, noting that the individual determinants have to be considered in combination to estimate impacts over larger areas (Box 5.1). In this section, we discuss the elements of this framework, while Section 5.4 reviews the growing body of evidence that exists to assign values to individual determinants.

5.3.2 Increases in geographical range

The range occupied by an invasive species is the first of three determinants of impact (Parker *et al.*, 1999), and invasive *Prosopis* species increase their range through aggressive seed dispersal. The trees produce one crop of pods annually, and each seed pod holds 5–20 seeds, with potentially hundreds of thousands of seeds produced per mature plant (Pasiecznik *et al.*, 2001) (Fig. 5.1). For example, flowering in most *Prosopis* species introduced to South Africa began within 2–6 years from germination, and mature trees yielded 90–140 kg of pods in a single season (Poynton, 1990). Documented evidence of increases in geographical range

Box 5.1. Estimating the impact of invasion by *Prosopis* trees on biodiversity and ecosystem services.

In order to quantify the impact of invasive alien species on ecosystems, it is necessary to be able to quantify the area occupied by the species (i.e. its range), the abundance of the species within its range (i.e. cover or biomass per unit area for plants, or the number of individuals for animals), and the per-capita effect that each individual of the invasive species, or each unit of area occupied by invasive plants, has on the ecosystem. Impact has been expressed by Parker *et al.* (1999) as the product of these three determinants as $I = R \times A \times E$, where impact (I) is the product of the geographical range of the invader (R), its abundance or density (A), and the per-capita effect (E) of an individual invader. When estimating the impacts of invasion on biodiversity and ecosystem services, the effects need to be further considered in terms of the attributes of biodiversity, namely composition, structure and function (Noss, 1990). The methods used to quantify the determinants of impact and their outcomes as summarized in this chapter are listed in the table below. Chapters indicate where some of these issues are dealt with in more detail in this volume.

Methods used to estimate range	Methods used to estimate abundance	Methods used to estimate per-capita effects	Outcomes in terms of impacts
Remote sensing (Chapter 4, this volume) Ground surveys (Witt et al., 2018)	Studies designed to quantify a range of cover and biomass categories	*Effects on ecosystem composition*: Enumeration of native species diversity and abundance across taxonomic groups *Effects on ecosystem structure*: Estimation of biomass and density *Effects on ecosystem functioning*: Carbon storage potential; changes to hydrological cycle (Chapter 6, this volume); models of indirect effects; toxic and allelopathic effects	Reductions in native species richness and abundance (this chapter) Reductions in rangeland carrying capacity (Chapter 7, this volume) Decreases in ecotourism potential Decreases in water resources (Chapter 6, this volume) Loss of income for indigenous peoples and local communities (Chapter 8, this volume) Toxicity and direct injury to animals and people (Chapter 9, this volume) Loss of access to grazing and medicinal plants

Fig. 5.1. The seeds of *Prosopis* trees are produced in large numbers (A) and consumed by domestic livestock and wild ungulates, which disperse them over long distances in their dung, leading to the establishment of new populations (B). Photos courtesy of Ketema Bekele (A) and Purity Mbaabu (B).

are provided in more detail in Chapter 4 (this volume).

The major dispersal agents are livestock and wildlife that consume the pods and spread them in their dung (Sawal *et al.*, 2004). A general trend emerging from studies in South Africa showed that rapid but episodic seed dispersal also takes place along ephemeral rivers in years of above-average rainfall, where they are transported long distances by infrequent floodwaters (van den Berg *et al.*, 2013). In addition, humans actively spread the species by establishing new populations, thus creating new foci for invasion. Rapid spread in new environments may well be facilitated by the absence of natural seed predators that limit seed production in the native environment. This general phenomenon in invasion biology has been termed the 'enemy release hypothesis' by Keane and Crawley (2002), and biological control in the form of seed-attacking insects has been attempted against *Prosopis* species in South Africa to counter this phenomenon (Kleinjan *et al.*, 2021: see also Chapter 12, this volume).

5.3.3 Increases in abundance

The abundance of an invasive species within its invaded range is the second determinant of impact (Parker *et al.*, 1999). Several *Prosopis* species are able to rapidly dominate in newly colonized areas due to their superior competitive abilities (Fig. 5.2) (see sections 5.4.2.1 and 5.4.2.2 for examples of where this has been documented). *Prosopis* trees also become dominant over several years (e.g. Mbaabu *et al.*, 2019; Shiferaw *et al.*, 2019a; see also Chapter 4, this volume) by competing successfully for light and nutrients. Other factors that assist *Prosopis* trees to increase in abundance include their ability to fix nitrogen (reducing any competitive advantage that C_4 grasses may have had; Ansley *et al.*, 2019), while allelopathy has been suggested as one of the mechanisms that allows *Prosopis* species to become dominant in sites where it has established (Kaur *et al.*, 2012; Noha *et al.*, 2021; Bibi *et al.*, 2023).

5.3.4 Per-capita effects of invading *Prosopis* trees

The third determinant of impact is the per-capita effect that individual trees have on ecosystem composition, structure and functioning (Parker *et al.*, 1999). As superior competitors, they displace native plants, changing the structure and composition of the vegetation and its suitability as habitat for native animals. The evidence for per-capita impacts on composition and structure is reviewed in sections 5.4.2 and 5.5.4.

Fig. 5.2. *Prosopis* trees were originally brought to Africa to provide fodder and shade for livestock. Panel A shows a non-invasive *Prosopis* tree planted along a road in South Africa to provide shade. However, some species rapidly became invasive and dominated the sites that they invaded, resulting in substantial loss of native biodiversity (B). These dense stands are also impenetrable to wildlife and livestock, so that the original intended benefits are lost. Photos courtesy of Brian van Wilgen (A) and René Eschen (B).

5.4 Evidence for the Impact of *Prosopis* Invasions on Biodiversity

5.4.1 A growing body of evidence

Noss (1990) proposed that biodiversity could be described in terms of three attributes (composition, structure and function), and that each of these attributes could be examined at different scales (regional-landscape, community-ecosystem, population-species or genetic make-up). Studies on the impacts of invasion can provide values to assign to effects (E in Parker *et al.*'s equation $I = R \times A \times E$; see Section 5.3.1). There are many studies in the published literature that mention the impacts of invasive *Prosopis* trees on the composition, structure and function attributes of ecosystems. Many of these are unquantified observations (e.g. Ayanu *et al.*, 2015) or are based on the opinions of local people (e.g. Berhanu and Tesfaye, 2006; Wakie *et al.*, 2016).

There is, however, a growing body of literature that reports on the measurement of impacts in the field in Africa. We located 23 papers from five African countries that documented the impacts of invasion on various attributes of biodiversity and ecosystem services, based on quantitative field research. Papers were located in Google Scholar using the search terms '*Prosopis*', 'Africa' and 'impact', and the papers identified were narrowed down to those that reported impacts that were measured in the field. We identified further relevant studies by examining the sources cited in the selected papers, as well as in papers published by the Woody Weeds research team (https://woodyweeds.org/?page_id=2), and those cited in a recent review of impacts in South Africa (van Wilgen *et al.*, 2022).

The first study (Steenkamp and Chown, 1996) was published in 1996, but it was only after 2010 that research efforts on this topic increased noticeably (Fig. 5.3). In this section, we provide a review of these studies in terms of three attributes of biodiversity (the composition, structure and functioning of natural ecosystems).

5.4.2 Impacts on the composition of ecosystems

Prosopis species have been shown to be aggressive invaders that displace native plant species through competition or change ecosystems in ways that make them unsuitable as habitats for native animal species. The invasions have affected a wide range of native species communities, including plants, mammals, birds, invertebrates and microbial organisms, and these are summarized below.

5.4.2.1 Native trees and shrubs

Invasion by *Prosopis* trees has marked effects on the cover and richness of woody plants in Africa. In the Afar region of Ethiopia, invasion

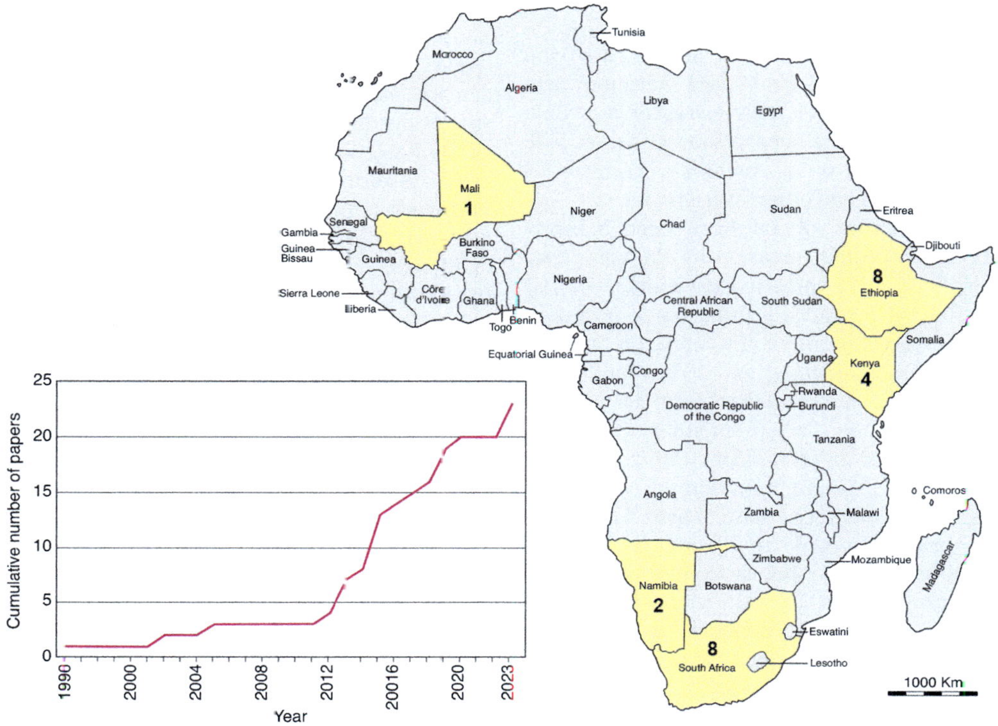

Fig. 5.3. Cumulative number of papers published between 1996 and 2023 which describe the impacts of invasion by *Prosopis* trees on elements of biodiversity in Africa. Map shows the number of studies conducted by country.

by *P. juliflora* reduced plant species richness, including several species of medicinal plants, by 29% on average (Linders *et al.*, 2019), and evidence suggested that allelopathy affected the germination and growth of selected native tree species (Getachew *et al.*, 2012). Invasion of open grazing lands by *P. juliflora* in the Afar region reduced the diversity of seed banks (both woody and herbaceous plants) from 19.2 to 11.0% (as measured by the Shannon index; Shiferaw *et al.*, 2019a).

In the Turkana district in Kenya, invasion of a riverine forest ecosystem by *P. juliflora* resulted in a gradual conversion of herbaceous-rich *Vachellia tortilis* woodland to herbaceous-poor *P. juliflora* woodland or thickets (Muturi *et al.*, 2013). In Namibia, *Prosopis* trees were found to have largely replaced the natural vegetation (including native trees, shrubs and herbaceous plants) along the Fish River, with only remnants of the original vegetation remaining (Strohbach *et al.*, 2015). In the Northern Cape Province, South Africa, significant decreases in native woody species density, basal area, richness and diversity were associated with increases in the basal area of *Prosopis* stands (Fig. 5.4 A,C) (Shackleton *et al.*, 2015a, b). A decrease of 79.7% in the density of native woody species in sites invaded by *P. velutina* in South Africa's Northwest Province has also been recorded (Tiawoun *et al.*, 2023).

Prosopis root systems consist of a very deep taproot (Villagra *et al.*, 2010), combined with extensive lateral roots. In times of drought, *Prosopis* trees are able to access groundwater that is beyond the reach of the native tree *Vachellia erioloba* (Schachtschneider and February, 2013). In the study in the Northern Cape Province, South Africa, the canopy dieback in *V. erioloba* trees was 50% higher on plots invaded by *Prosopis*, due to direct competition for groundwater.

5.4.2.2 Native herbaceous plants

Herbaceous plants, notably grasses, are vital components of Africa's arid savanna ecosystems, because of their value for providing grazing and other food resources for a wide range of herbivorous mammals, granivorous birds, termites and other invertebrates.

Herbaceous plants have been severely impacted through invasion by *Prosopis* trees. Studies on survey plots in Kenya indicated that invasion of degraded grazing lands decreased plant species richness by 29% on average (Fig. 5.4B) (Linders *et al.*, 2019), and plant species richness was halved from 18.3 to 9.4 species/225 m^2 when pristine grasslands were invaded by *P. juliflora* (Mbaabu *et al.*, 2020). In Ethiopia, invasion by *P. juliflora* reduced understorey basal cover for perennial grasses from 68 to 2% and reduced the number of grass species from seven to two on survey plots in the Allideghi Wildlife Reserve (Kebede and Coppock, 2015). In the Shilabo District, Ethiopia, survey plot studies revealed that areas invaded by *Prosopis* trees had significantly lower species diversity than uninvaded areas, and that the number of plant species (including trees, shrubs and grasses), density, abundance, important value index, species evenness and richness were all significantly lower in invaded areas than uninvaded areas (Bezaredie *et al.*, 2023). The cover of native perennial grasses and herbaceous plants declined from 15–20% where the basal area of *Prosopis* trees was <2 m^2/ha to zero where the basal area of *Prosopis* trees was >4.5 m^2/ha in South Africa's Northern Cape Province (Shackleton *et al.*, 2015b). In South Africa's Western Cape Province, reductions in native plant species richness and basal cover declined markedly once the level of invasion exceeded a 'critical threshold' of 20% *Prosopis* canopy cover (Ndhlovu *et al.*, 2011, 2016).

5.4.2.3 Native mammals

Native mammals respond to changes in habitat brought about by *Prosopis* invasions, while grazing species are directly affected by the loss

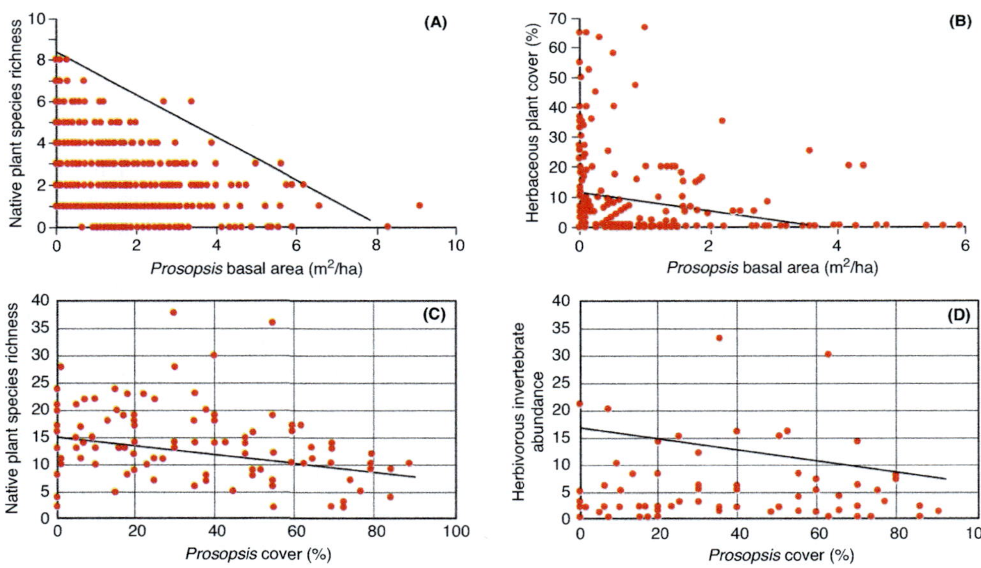

Fig. 5.4. Some examples of the significant effects of increasing cover of *Prosopis* trees on native biota. The measures are: (A) native plant species richness; (B) herbaceous plant cover; (C) native plant species richness; (D) abundance of herbivorous invertebrates. Panels A and B show data from the Northern Cape Province, South Africa, redrawn from Shackleton *et al.* (2015b). Panel C shows data from sites in Ethiopia and Kenya, and panel D shows data from sites in Kenya, redrawn from Linders *et al.* (2019).

of pasture species that accompanies invasions. In the Allideghi Wildlife Reserve in Ethiopia, serious negative implications of invasion by *P. juliflora* were reported for the survival of an isolated population of the endangered, grass-dependent Grevy's zebra (*Equus grevyi*) (Kebede and Coppock, 2015). A study in Namibia found that *Prosopis* invasions (including *P. glandulosa* var. *torreyana* and *P. velutina*) decreased the frequency of detection (using camera traps) of two antelope species (the steenbok *Raphicerus campestrus* and the springbok *Antidorcas marsupialis*) and one hare species (*Lepus capensis*) but increased the frequency of detection of African wild cats (*Felis silvestris lybica*) and one rodent species along the Swakop River (Williams *et al.*, 2014).

5.4.2.4 Native birds

As with mammal species, birds are affected by invasions due to changes in food resources and habitat structure. A study in South Africa noted that bird communities in native woodlands in the southern Kalahari were consistently more species rich and more diverse than sites invaded by *Prosopis* trees (Dean *et al.*, 2002). Additionally, invasion by *P. glandulosa* var. *torreyana* and *P. velutina* resulted in the elimination of raptors, and reduced the numbers of frugivorous and insectivorous birds, but had no effect on nectarivores, mixed feeders (seeds and insects) and granivores. In Kenya, thousands of lesser flamingos (*Phoeniconaias minor*) have been trapped in stands of *P. juliflora* around the shores of Lake Bogoria (Box 5.2).

5.4.2.5 Native invertebrates

Invertebrates are by far the largest contributors to species richness and ecosystem functioning (Wilson, 1992) but only two studies have so far been published on the impact of *Prosopis* tree invasions on these taxa.

In Ethiopia's Afar region, decreases of 49 and 19% in the abundance of predatory and herbaceous invertebrates, respectively, were found after *P. juliflora* had invaded rangelands (Linders *et al.*, 2019) (Fig. 5.4D). In addition, decreases of 46 and 13% in the abundance of predatory invertebrates and parasitoids, respectively, were recorded after *P. juliflora* had invaded rangelands in Baringo County, Kenya (Linders *et al.*, 2019). In South Africa's Northern Cape Province, there were more dung beetle species in undisturbed savanna compared to areas invaded by *P. glandulosa* (41 vs 34 species; Steenkamp and Chown, 1996). Larger dung beetles (>20 mg dry weight), particularly telecoprids, showed a marked aversion to invaded areas, but comprised 60% of the assemblage in uninvaded areas. Rare dung beetle species in the assemblage all showed a marked decline in invaded areas.

In Mali, the population density of malaria-spreading *Anopheles* mosquitoes dropped by 69% following removal of *Prosopis* flowers (Muller *et al.*, 2017).

5.4.2.6 Native microbial organisms

Microbial organisms are very important to the functioning of ecosystems, but the effects of invasions generally on these communities have been poorly studied. In Ethiopia, Gemeda (2019) found that the abundance and composition of both *Rhizobium* and mycorrhizal soil microorganisms were greater in bulk soils under *P. juliflora* than under other native species. The ecological implications of these changes have not been quantified.

5.4.3 Impacts on the structure of ecosystems

Invasion of African arid savannas by *Prosopis* trees changes the structure of these ecosystems in several ways. The invasive trees rapidly come to dominate the sites where they establish, in many cases changing the cover of the tree layer substantially, often covering as much as 100% in invaded thickets. The movement of large mammals through these thickets is further exacerbated by the presence of numerous thorns (Fig. 5.5). For example, in the Lake Baringo district in Kenya, there were numerous injuries to wildlife (and livestock) from thorns (Mwangi and Swallow, 2005). Although the pods can be nutritious, these dense, impenetrable thickets prevent the utilization of pods by wild and domestic ungulates, and also hamper access to patches of grazing and water resources. The presence of large thorns also causes numerous injuries (see Chapter 7, this volume).

Box 5.2. Rescuing Lake Bogoria's lesser flamingos.

Lake Bogoria is a saline, alkaline lake that lies south of Lake Baringo, Kenya. Like several other lakes in the region, Lake Bogoria is periodically home to one of the world's largest populations of lesser flamingos (*Phoeniconaias minor*). Lesser flamingos are classified as near threatened due to their declining populations and the limited number of breeding sites, all of which are degraded to some degree by human activities (Birdlife International, 2004). Lake Bogoria is a Ramsar site and was also proclaimed as a National Reserve in 1973.

Lesser flamingos, a long-lived species with a life span of over 40 years in the wild, feed primarily on *Spirulina* algae which grow only in very alkaline lakes. More than 1 million lesser flamingos, close to 50% of the global population, gather at Lake Bogoria when conditions allow. However, over the past two decades, the water levels in Lake Bogoria have risen, so that the shallow areas and saline flats that are used by flamingos to take off and land have become deeper or inundated. The lake's fringing vegetation has also become heavily invaded by *Prosopis juliflora* trees which were introduced in the 1980s.

Due to rising water levels, many invading *P. juliflora* trees grow directly along the shores or have become partially submerged around the periphery of the lake. Lesser flamingos are trapped by these *P. juliflora* trees as they attempt to either land or take off. Most trapped birds die as they cannot disentangle themselves from the thorny branches. Mortality has been reported to be more than 20,000 birds over the past few years (Ndonye, 2022).

In 2023, a decision was taken to clear the invading trees inside the entire Lake Bogoria National Reserve, which includes the lake. This process, funded by Woody Weeds +, took less than 1 year and was achieved in June 2024. The task of maintaining the area in an uninvaded state, i.e. removing new *Prosopis* seedlings emerging from the soil seed bank or introduced by livestock or wildlife, will be overseen in future by members of the County Implementation Group, together with the rangers of the National Reserve, the county wardens and Kenya Wildlife Service. These actions will have removed a significant threat to an endangered and iconic bird species.

Box Fig. 5.1. Lesser flamingos gather in large numbers on Lake Bogoria (A), where many have in the past been trapped by invading *Prosopis* trees (B). Photos courtesy of (A) Brian van Wilgen and (B) Joseph Kipsang.

The invasions also reduce the cover and biomass of herbaceous plants, sometimes eliminating them altogether. Studies have recorded decreases in herbaceous biomass of 40% in Ethiopia and 29% in Kenya (Linders *et al.*, 2019), while Mbaabu *et al.* (2020) found that dry herbaceous plant biomass declined from 1300g m^{-2} to almost zero when native grasslands were invaded by *P. juliflora* in Baringo County, Kenya. Similarly, there was a decline in understorey basal cover for perennial grasses from 68 to 2% following invasion, and an increase in soil surface exposure from 30 to 80% in the Allideghi Wildlife Reserve, Ethiopia (Kebede and Coppock, 2015). In the Western Cape Province, South Africa, invasion by *Prosopis* trees markedly reduced the cover of native vegetation (especially grasses) once the level of invasion had exceeded a critical threshold of 20% *Prosopis* canopy cover (Ndhlovu *et al.*, 2011; 2016).

Fig. 5.5. Thorns on invasive *Prosopis* trees (left) combine with the formation of dense thickets (right) to increase the difficulty that herbivorous mammals experience in gaining access to grazing or water or moving between different areas. Photos courtesy of Theo Linders (A) and Jano Barnard (B).

5.4.4 Impacts on the functioning of ecosystems

Invasions by *Prosopis* species can impact on the functioning of ecosystems by changing their capacity to assimilate and store carbon, recycle nutrients, produce and decompose biomass, and efficiently use water. These changes can have implications for the delivery of ecosystem services in invaded areas. There have, however been very few studies covering these aspects. Most studies have examined the effects of *Prosopis* trees in terms of the hydrological cycle. These are reported briefly here for completeness and are covered in more detail in Chapter 6 (this volume).

5.4.4.1 Impacts on carbon storage

Converting grasslands to woodlands or forests is often naively put forward as a means to sequester carbon, with the goal of mitigating the drivers of climate change (Parr *et al.*, 2014; Bond *et al.*, 2019). Planting trees into degraded grasslands (or allowing grasslands to become invaded by alien trees) may increase above-ground and soil carbon stocks, but it ignores the potential that healthy perennial grasslands have for carbon storage (Ward *et al.*, 2016). Additionally, it will carry a high cost in terms of the loss of other ecosystems services, for example losses of biodiversity and grazing for livestock and wildlife. Despite this, *Prosopis* tree invasions have been promoted as a possible carbon stock (Birhane *et al.*, 2017).

Mbaabu *et al.* (2020) assessed how invasion by *P. juliflora* and restoration of degraded grasslands affected soil organic carbon, biodiversity and fodder availability in a semi-arid ecosystem in Kenya. They found that grassland restoration over 30 years replenished soil organic carbon to 1m depth at a rate of 1.4% per year. They concluded that grassland restoration would be at least as effective in replenishing soil organic carbon levels in degraded rangelands as encroachment by invasive *Prosopis* trees would be. Clearing invasive *Prosopis* trees and restoring the grasslands would thus be an equally effective measure for mitigating climate change, and it would have the additional advantage of restoring the vital ecosystem services that can support pastoralists and agropastoralist communities (see also Chapter 7, this volume).

5.4.4.2 Impacts on hydrology

Prosopis trees use excessive amounts of water and thus alter the hydrological functioning of the ecosystems they invade, with knock-on effects on biodiversity. The ability of invasive *Prosopis* trees to utilize both surface and deep groundwater alters the hydrological cycle, depriving native plants of water. Excessive water use by stands of invasive *Prosopis* trees has been demonstrated in South Africa, where transpiration on the invaded sites was more than five times greater (544 mm y^{-1}) than sites occupied by native trees (91 mm y^{-1}) (Dzikiti *et al.*, 2017). Similarly, in Ethiopia, stands of *P. juliflora* transpired almost three times as much water (242 mm yr^{-1}) as the stands of native trees that they replaced (87 mm yr^{-1}; Shiferaw *et al.*, 2023). In addition, replacing diverse natural ecosystems with dense stands of *Prosopis* resulted in the net loss of ecosystem services (mainly groundwater and grazing resources) valued at USD 90 ha^{-1} over 31 years in Ethiopia's Afar region (Shiferaw *et al.*, 2019b). See Chapter 6, this volume for further details.

5.4.4.3 *Other impacts on ecosystem functioning*

Further changes may be brought about through nitrogen fixation (Geesing *et al.*, 2000) and allelopathic effects. Invasive species that increase soil nitrogen levels can make the site less suitable for native plant species and prevent successful restoration of the original vegetation on cleared sites (Marchante *et al.*, 2009).

5.4.4.4 *Indirect impacts on ecosystem functioning*

Most studies on the impact of biological invasions on ecosystems measure the direct effects arising from the displacement of native species from the ecosystem. For example, the effects of invasion by *Prosopis* trees have been quantified in terms of reductions in native plant species richness or biomass. However, these reductions can indirectly lead to further reductions in a range of ecosystem functions.

Linders *et al.* (2019) used structural equation modelling to separate direct effects of invasion by *P. juliflora* from indirect effects mediated through changes in biodiversity and herbaceous biomass at sites in Kenya and Ethiopia. They found that reductions in species richness and herbaceous biomass indirectly affected soil organic carbon, available phosphorus, soil stability, and the composition and abundance of a range of invertebrate communities. They concluded that control measures that simply removed the invasive species would be insufficient to restore full functionality to the ecosystems involved, and that the restoration of diverse and productive herbaceous communities would be necessary to reestablish the ecosystem services that had been lost through invasion (see Chapter 12 for a brief review of restoration following control).

5.5 Formal Assessments of Impact

The Environmental Impact Classification for Alien Taxa (EICAT) system (Blackburn *et al.*, 2014) is increasingly being used to formally and consistently evaluate the impact of invasive alien species on biodiversity.

In the case of the genus *Prosopis*, only one such evaluation has been completed to date, for *P. juliflora* in Iran (Sohrabi *et al.*, 2021). These authors found, based on ten available studies, that *P. juliflora* had a 'moderate' environmental impact based on competition and chemical or toxic effects. Their assessment was made with medium confidence as they reported that there was no information on 50% of the aspects that need to be considered by EICAT assessments. They did, however, note that the species' environmental impacts substantially outweighed its socio-economic benefits, and that the species was regarded as one of the top eight that needed to be prioritized for management in Iran. A moderate impact is defined by Blackburn *et al.* (2014) as causing 'declines in the population densities of native species, but no changes to the structure of communities or to the abiotic or biotic composition of ecosystems'.

Clearly, invasions by *Prosopis* trees do reduce the population densities of native species, and they change the community structure and biotic composition of ecosystems (as described above), and by taking this into account, a revised assessment would almost certainly recommend a 'major' impact category for *P. juliflora* in terms of the EICAT scheme. Much still needs to be done to quantify the full range of impacts on biodiversity that are brought about by *Prosopis* invasions, especially in terms of the longer-term effects of changing site conditions through nitrogen enrichment, allelopathy and altered hydrological regimes. There is, however, ample evidence at this stage to suggest that *Prosopis* invasions have, and will continue to have, major impacts on biodiversity.

5.6 Perceptions of Impact Among Indigenous Peoples and Local Communities

Any given country's ability to control *Prosopis* invasions over large areas will require the support and collaboration of indigenous peoples and local communities. It is unlikely that such communities would place a value on how invasions change the composition, structure and functioning of natural ecosystems per se. Rather, they would see the benefits and values of a resource, and the threats that the resource faces, which is seen broadly as 'ecosystem services' or more recently as 'nature's contribution to people' (Pascual *et al.*, 2017). However, the studies reported below were based on a predetermined set of questions that asked local households whether and how 'biodiversity' impacted on a range of issues including rangeland productivity, livestock production and general wellbeing. In addition, the studies were based on hundreds of interviews with households in areas that were lightly, moderately or heavily invaded by *Prosopis* trees.

The studies found that the views of communities differ both within and between the communities in the region. One study investigated whether the views of rural communities aligned with the findings of parallel studies that quantified the ecological effects of *Prosopis* invasions in the same areas (Baringo County in Kenya and the Afar region in Ethiopia; Bekele *et al.*, 2022). In these two sites, most people believed that *Prosopis* invasions had a negative impact on biodiversity (83.4% in Afar and 94.8% in Baringo). However, 10.3 and 2.8% of the respondents from Afar and Baringo, respectively, held contrary views. Similarly, 82.5% of respondents from Afar agreed that *Prosopis* invasion causes biodiversity loss, while 17.5% of the respondents disagreed (Bekele *et al.*, 2022). Thus, while the majority of the respondents' views were in agreement with the findings of ecological studies, a proportion did not hold these views. Interestingly, awareness of the negative effects of *Prosopis* invasions on biodiversity and water availability increased with increasing cover of invasive plants in the immediate surroundings of the respondents' households (Bekele *et al.*, 2022).

In another study, where rural communities were heavily reliant on natural resources, there was a clear understanding that *Prosopis* invasions caused reductions in native species richness (e.g. that native grass and tree species were reduced or eliminated by invasion). This was understood to impact on the provision of multiple services, such as fodder for livestock and the availability of medicinal plant species (Bekele *et al.*, 2018). The study found that each household would lose between USD 22.67 and 29.52 per year on average in Kenya and Ethiopia, respectively (about 4–5% of average annual income) due to the impact of *Prosopis* invasions on biodiversity. Despite some positive benefits associated with the invading trees, households from both regions were willing to pay between 37–50 USD per household per year for efforts to bring *Prosopis* invasions under control, mainly due to perceived impacts on biodiversity and water supplies. The main contributing factor to this willingness to pay was the loss of biodiversity.

Besides the direct influence of increasing cover of *Prosopis* on people's perceptions of impact, perceptions were also affected by socio-economic and sociocultural attributes, education level, social participation, diversified income sources and demographic characteristics (for details, see Chapter 8, this volume). For example, Haregeweyn *et al.* (2013) reported that 100% of pastoralists and 96.7% of agropastoralists in the Afar region agreed that invasions by *P. juliflora* had negative impacts on biodiversity, while only 40% of the charcoal makers agreed with this view due to their personal evaluation from a different perspective.

There are several important stakeholder groups who have interests in *Prosopis* trees. The importance of various desired benefits of *Prosopis* trees (which include changes in tourism potential, crop production, livestock production, honey production, shade, water and woody biomass) differs between these groups (Linders *et al.*, 2021). Increasing *Prosopis* cover was found to change the multifunctional value (i.e. the combined value of all desired benefits) as cover increased, providing an increased benefit to only two stakeholder groups: charcoal producers and, marginally, developmental non-governmental organizations. However, multifunctionality decreased significantly for the government, education, and conservation and tourism groups, while there was no detectable change for farmers and pastoralists (Linders *et al.*, 2021) (Fig. 5.6). The negative costs arose primarily from impacts on biodiversity and water.

Finally, despite heterogeneous views among the local communities, most households from the communities surveyed in Ethiopia and Kenya were willing to contribute to the control of *Prosopis* invasions, primarily driven by the potential for biodiversity and water improvements (Bekele *et al.*, 2018). Two things are thus clear. First, the perceptions of impact increase as the invasions become more dominant (by which time the initiation of effective control will be disproportionately more difficult). Second, there is potential for conflict between interest groups. Charcoal producers (who benefit from the sale of charcoal) and 'developmental' non-governmental organizations (who often take a view that *Prosopis* trees represent a resource that must be utilized, see https://forestsnews.cifor.org/tag/mesquite/), on the one hand, benefit while on the other hand everyone else does not.

5.7 Potential for and Benefits of Restoring Biodiversity

The removal of invasive trees and the restoration of functional ecosystems can bring substantial benefits. Eschen *et al.* (2021) combined ecological, social and financial data from a site in Kenya to demonstrate that it would be viable for local households to remove *Prosopis* invasions and restore productive livestock farming (see Chapter 10, this volume). The benefits came about through diversified income streams arising from charcoal production from cleared *Prosopis* trees, increased cattle production and carbon credits. These authors concluded that detailed ecological, social and financial geodata could be integrated to develop realistic invasive alien species management scenarios, and that in this case it was clear that active management of the invasions would bring a suite of benefits.

Studies from South Africa have also shown that native vegetation can recover after *Prosopis* clearing (Ndhlovu *et al.*, 2011, 2016). These studies found that invasion (c. 15% *Prosopis* canopy cover) reduced grazing capacity by 34%. Clearing of invaded areas, on the other hand, restored native plant cover to preinvasion levels. Much of the loss in grazing capacity was due to the displacement of the annual grasses that dominated herbaceous forage production prior to invasion. After clearing, grass cover and richness on the site rose due to increases in abundance of several annual and perennial grass species, leading to a 39% improvement in carrying capacity in cleared rangeland when compared to uninvaded rangeland due to a higher abundance of the annual and perennial grasses. The authors concluded that clearing *Prosopis* trees from invaded rangeland could, even under heavy grazing, substantially improve grazing capacity within 4–6 years.

Finally, an economic evaluation of the costs and benefits of *Prosopis* invasions in the Northern Cape Province, South Africa, found that ongoing increases in invasive cover would result in the net benefits (i.e. benefits minus costs) of *Prosopis* trees becoming negative within 5 years of the study (Wise *et al.*, 2012). Although these authors did not have enough information to estimate the value of lost biodiversity per se on the economic outcome, the outcome was determined by ongoing (and increasing) losses in pasture and water benefits, both of which are dependent on the structure and composition of the invaded communities.

5.8 Implications for Management

The biodiversity of eastern Africa is an important economic resource that is threatened in many places by the invasion of alien species, of which *P. juliflora* is a leading example. Invasion

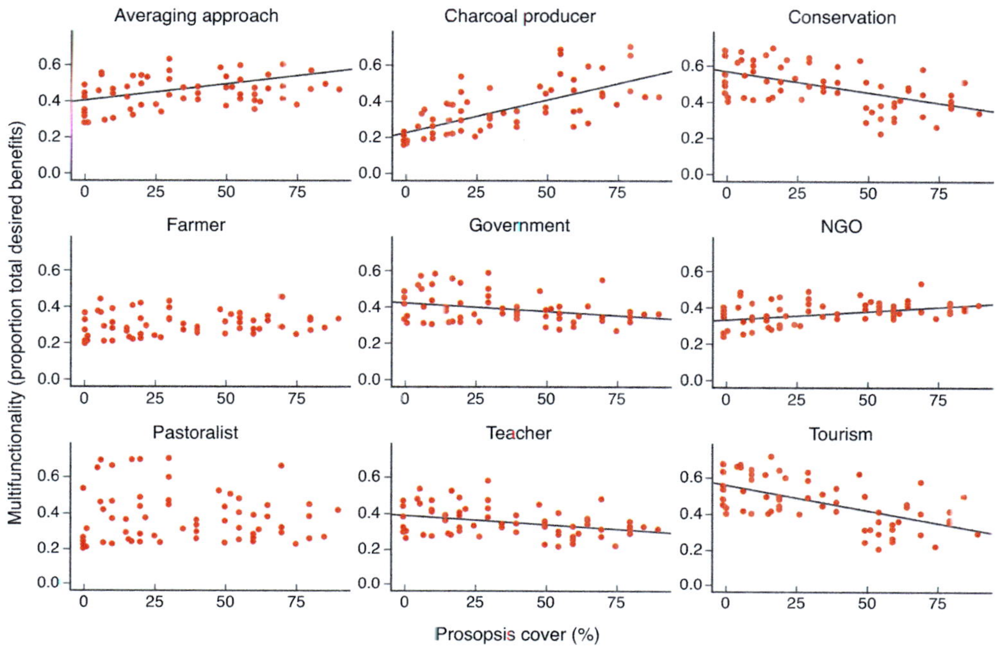

Fig. 5.6. Changes in the multifunctional value of net benefits (the sum of various desired ecosystem services) for different stakeholder groupings with an increase in *Prosopis* tree cover from 0 to 100%. Relationships were not significant for farmers and pastoralists. Data from Linders *et al.* (2021).

by *Prosopis* trees has been shown to reduce native biodiversity and alter ecosystem functioning. Studies have also indicated that control of invasions can restore the lost functions and that such action can be justified because of the positive returns on investment in control. Such control would achieve more if the diverse and productive herbaceous communities that are important for many ecosystem functions can also be restored.

The implications of these findings are that governments, aid agencies and communities would benefit from investing in control in order to protect biodiversity and ecosystem services by preventing further spread of *P. juliflora*, implementing targeted control efforts where the species has become established, and actively restoring biodiversity to areas where the invasive trees have been brought under control.

Acknowledgements

This work was supported by the Swiss Programme for Research on Global Issues for Development (r4d), funded by the Swiss National Science Foundation and the Swiss Agency for Development and Cooperation, through the project 'Woody invasive alien species in East Africa: Assessing and mitigating their negative impact on ecosystem services and rural livelihood' (Grant Number: 400440_152085). This work was also supported in part by the National Research Foundation of South Africa (Grant Numbers 150260 and 151653 to BWvW).

References

Ansley, R.J., Cooper, C. and Zhang, T. (2019) Invasion promotes invasion: Facilitation of C_3 perennial grass dominance in mixed C_3/C_4 grassland by an invasive C_3 woody sprouter (*Prosopis glandulosa*). *Ecology and Evolution* 9, 13438–13449.

Ayanu, Y., Jentsch, A., Müller-Mahn, D., Rettberg, S., Romankiewicz, C. *et al.* (2015) Ecosystem engineer unleashed: *Prosopis juliflora* threatening ecosystem services. *Regional Environmental Change* 15, 155–167.

Bekele, K., Haji, J., Legesse, B. and Schaffner, U. (2018) Economic impacts of *Prosopis* spp. invasions on dryland ecosystem services in Ethiopia and Kenya: Evidence from choice experimental data. *Journal of Arid Environments* 158, 9–18.

Bekele, K., Linders, T.E., Eschen, R., Shiferaw, H. and Haji, J. (2022) How well do local stakeholders' perceptions of environmental impacts of an invasive alien plant species relate to ecological data. *Ecological Indicators* 137, 108748.

Berhanu, A. and Tesfaye, G. (2006) The *Prosopis* dilemma, impacts on dryland biodiversity and some controlling methods. *Journal of the Drylands* 1, 158–164.

Bezaredie, D., Tadesse, Z. and Tadesse, Z. (2023) Effects of *Prosopis juliflora* on plant diversity on rangeland in Shilabo District, Somali Regional State, Ethiopia. *Heliyon* 9, e14049.

Bibi, S., Bibi, A., Al-Ghouti, M.A. and Abu-Dieyeh, M.H. (2023) Allelopathic effects of the invasive *Prosopis juliflora* (Sw.) DC. on native plants: Perspectives toward agrosystems. *Agronomy* 13, 590.

Birdlife International (2004) *Threatened Birds of the World*. Birdlife International, Cambridge, UK.

Birhane, E., Treydte, A.C. and Eshete, A. (2017) Can rangelands gain from bush encroachment? Carbon stocks of communal grazing lands invaded by *Prosopis juliflora*. *Journal of Arid Environments* 141, 60–67.

Blackburn, T.M., Essl, F., Evans, T., Hulme, P.E. and Jeschke, J.M. (2014) A unified classification of alien species based on the magnitude of their environmental impacts. *PLoS Biology* 12, e1001850.

Bond, W.J., Stevens, N., Midgley, G.F. and Lehmann, C.E. (2019) The trouble with trees: Afforestation plans for Africa. *Trends in Ecology & Evolution* 34, 963–965.

Dean, W.R.J., Anderson, M.D., Milton, S.J. and Anderson, T.A. (2002) Avian assemblages in native *Acacia* and alien *Prosopis* drainage line woodland in the Kalahari, South Africa. *Journal of Arid Environments* 51, 1–19.

Drake, J.A., Mooney, H.A., Castri, F., Grives, F., Kruger, F.J. *et al.* (1989) *Biological Invasions: A Global Perspective*. Wiley, Chichester, UK.

Dzikiti, S., Ntshidi, Z., Le Maitre, D.C., Bugan, R.D. and Mazvimavi, D. (2017) Assessing water use by *Prosopis* invasions and *Vachellia* karroo trees: Implications for groundwater recovery following alien plant removal in an arid catchment in South Africa. *Forest Ecology and Management* 398, 153–163.

Eschen, R., Bekele, K., Mbaabu, P.R., Kilawe, C.J. and Eckert, S. (2021) *Prosopis juliflora* management and grassland restoration in Baringo County, Kenya: Opportunities for soil carbon sequestration and local livelihoods. *Journal of Applied Ecology* 58, 1302–1313.

Geesing, D., Felker, P. and Bingham, R.L. (2000) Influence of mesquite (*Prosopis glandulosa*) on soil nitrogen and carbon development: Implications for global carbon sequestration. *Journal of Arid Environments* 46, 157–180.

Gemeda, W.S. (2019) Effects of *Prosopis juliflora* on soil microbial and other pathogenic activities: A review paper. *CPQ Microbiology* 3, 1–8.

Getachew, S., Demissew, S. and Woldemariam, T. (2012) Allelopathic effects of the invasive *Prosopis juliflora* (Sw.) DC. on selected native plant species in Middle Awash, Southern Afar rift of Ethiopia. *Management of Biological Invasions* 3, 105–114.

Haregeweyn, N., Tsunekawa, A., Tsubo, M., Meshesha, D. and Melkie, A. (2013) Analysis of the invasion rate, impacts and control measures of *Prosopis juliflora*: A case study of Amibara District, Eastern Ethiopia. *Environmental Monitoring and Assessment* 185, 7527–7542.

IPBES (Intergovernmental Science-Policy Platform on Biodiversity and Ecosystem Services) (2018) *Summary for Policymakers of the Regional Assessment Report on Biodiversity and Ecosystem Services for Africa of the Intergovernmental Science-Policy Platform an Biodiversity and Ecosystem Services*. IPBES Secretariat, Bonn, Germany.

Kaur, R., Gonzales, W.L., Llambi, L.D., Soriano, P.J., Callaway, R.M, et alE. (2012) Community impacts of *Prosopis juliflora* invasion: Biogeographic and congeneric comparisons. *PLoS One* 7(9), e44966.

Keane, R.M. and Crawley, M.J. (2002) Exotic plant invasions and the enemy release hypothesis. *Trends in Ecology & Evolution* 17, 164–170.

Kebede, A.T. and Coppock, D.L. (2015) Livestock-mediated dispersal of *Prosopis juliflora* imperils grasslands and the endangered Grevy's Zebra in Northeastern Ethiopia. *Rangeland Ecology & Management* 68, 402–407.

Kleinjan, C.A., Hoffmann, J.H., Heystek, F., Ivey, P. and Kistensamy, Y. (2021) Developments and prospects for biological control of *Prosopis* (Leguminosae) in South Africa. *African Entomology* 29, 859–874.

Linders, T.E.W., Schaffner, U., Eschen, R., Abebe, A., Choge, S.K. *et al.* (2019) Direct and indirect effects of invasive species: Biodiversity loss is a major mechanism by which an invasive tree affects ecosystem functioning. *Journal of Ecology* 107, 2660–2672.

Linders, T.E.W., Schaffner, U., Alamirew, T., Allan, E., Choge, S.K. *et al.* (2021) Stakeholder priorities determine the impact of an alien tree invasion on ecosystem multifunctionality. *People and Nature* 3, 658–672.

Marchante, E., Kjøller, A., Struwe, S. and Freitas, H. (2009) Soil recovery after removal of the N2-fixing invasive *Acacia longifolia*: Consequences for ecosystem restoration. *Biological Invasions* 11, 813–823.

Mbaabu, P.R., Ng, W.T., Schaffner, U., Gichaba, M., Olago, D. *et al.* (2019) Spatial evolution of *Prosopis* invasion and its effects on LULC and livelihoods in Baringo, Kenya. *Remote Sensing* 11, 1217.

Mbaabu, P.R., Olago, D., Gichaba, M., Eckert, S., Eschen, R. *et al.* (2020) Restoration of degraded grasslands, but not invasion by *Prosopis juliflora*, avoids trade-offs between climate change mitigation and other ecosystem services. *Scientific Reports* 10, 1–13.

Mittermeier, R.A., Gil, P.R. and Hoffmann, M. (2004) *Hotspots Revisited: Earth's Biologically Richest and Most Endangered Terrestrial Ecoregions*. CEMEX, Mexico City, Mexico.

Muller, G.C., Junnila, A., Traore, M.M., Traore, S.F., Doumbia, S. *et al.* (2017) The invasive shrub *Prosopis juliflora* enhances the malaria parasite transmission capacity of anopheles mosquitoes: A habitat manipulation experiment. *Malaria Journal* 16, 1–9.

Muturi, G.M., Poorter, L., Mohren, G.M.J. and Kigomo, B.N. (2013) Ecological impact of *Prosopis* species invasion in Turkwel riverine forest, Kenya. *Journal of Arid Environments* 92, 89–97.

Mwangi, E. and Swallow, B. (2005) *Invasion of* Prosopis juliflora *and Local Livelihoods: Case Study from the Lake Baringo Area of Kenya*, ICRAF Working Paper no. 3. World Agro-forestry Centre, Nairobi, Kenya.

Ndhlovu, T., Milton-Dean, S.J. and Esler, K.J. (2011) Impact of *Prosopis* (mesquite) invasion and clearing on the grazing capacity of semiarid Nama Karoo Rangeland, South Africa. *African Journal of Range and Forage Science* 28, 129–137.

Ndhlovu, T., Milton, S.J. and Esler, K.J. (2016) Effect of *Prosopis* (mesquite) invasion and clearing on vegetation cover in semi-arid Nama Karoo Rangeland, South Africa. *African Journal of Range and Forage Science* 33, 11–19.

Ndonye, M. (2022) Plan to stop mathenge weed spread. *The Standard*. Nairobi, Kenya.

Nikodinoska, N., Foxcroft, L.C., Rouget, M., Paletto, A. and Notaro, S. (2014) Tourists' perceptions and willingness to pay for the control of *Opuntia stricta* invasion in protected areas: A case study from South Africa. *Koedoe* 56, 1–8.

Noha, A., Nigatu, L. and Manikandan, R. (2021) Allelopathy effect of *Prosopis juliflora* on selected grass species (*Cenchrus ciliaris Paspalidium desertorum* and *Lintonia nutans*). *Journal of Phytology* 13, 75–78.

Noss, R.F. (1990) Indicators for monitoring biodiversity: A hierarchical approach. *Conservation Biology* 4, 355–364.

Parker, I.M., Simberloff, D., Lonsdale, W.M., Goodell, K., Wonham, M. *et al.* (1999) Impact: Towards a framework for understanding the ecological effects of invaders. *Biological Invasions* 1, 3–9.

Parr, C.L., Lehmann, C.E., Bond, W.J., Hoffmann, W.A. and Andersen, A.N. (2014) Tropical grassy biomes: Misunderstood, neglected, and under threat. *Trends in Ecology & Evolution* 29, 205–213.

Pascual, U., Balvanera, P., Díaz, S., Pataki, G., Roth, E. *et al.* (2017) Valuing nature's contributions to people: The IPBES approach. *Current Opinion in Environmental Sustainability* 26, 7–16.

Pasiecznik, N.M., Felker, P., Harris, P.J., Harsh, L., Cruz, G. *et al.* (2001) *The* Prosopis juliflora–Prosopis pallida *Complex: A Monograph*. HDRA Publishing, Coventry, UK.

Poynton, R.J. (1990) The genus *Prosopis* in Southern Africa. *South African Forestry Journal* 152, 62–66.

Pyšek, P. and Richardson, D.M. (2010) Invasive species, environmental change and management, and health. *Annual Review of Environment and Resources* 35, 25–55.

Pyšek, P., Hulme, P.E., Simberloff, D., Bacher, S., Blackburn, T.M. *et al.* (2020) Scientists' warning on invasive alien species. *Biological Reviews of the Cambridge Philosophical Society* 95, 1511–1534.

Rija, A., Kideghesho, J.R., Mwamende, K. and Selemani, I. (2013) Emerging issues and challenges in conservation of biodiversity in the rangelands of Tanzania. *Nature Conservation* 6, 1–29.

Sala, O.E., Stuart Chapin, F.I.I.I., Armesto, J.J., Berlow, E., Bloomfield, J. *et al.* (2000) Global biodiversity scenarios for the year 2100. *Science* 287(5459), 1770–1774.

Sawal, R.K., Ratan, R. and Yadav, S.B.S. (2004) Mesquite (*Prosopis juliflora*) pods as a feed resource for livestock – A review. *Asian-Australasian Journal of Animal Sciences* 17, 719–725.

Schachtschneider, K. and February, E.C. (2013) Impact of *Prosopis* invasion on a keystone tree species in the Kalahari desert. *Plant Ecology* 214, 597–605.

Shackleton, R., Pizarro, C. and Rossi, S.D. (2022) Tourist's knowledge, perceptions and behaviours towards invasive species. In: Barros, A., Shackleton, R., Rew, L., Pizarro, C. and Pauchard, A. (eds) *Tourism, Recreation and Biological Invasions*. CABI, Wallingford, UK, pp. 120–129.

Shackleton, R.T., Le Maitre, D.C. and Richardson, D.M. (2015a) Prosopis invasions in South Africa: Population structures and impacts on native tree population stability. *Journal of Arid Environments* 114, 70–78.

Shackleton, R.T., Le Maitre, D.C., van Wilgen, B.W. and Richardson, D.M. (2015b) The impact of invasive alien *Prosopis* species (mesquite) on native plants in different environments in South Africa. *South African Journal of Botany* 97, 25–31.

Shiferaw, H., Bewket, W. and Eckert, S. (2019a) Performances of machine learning algorithms for mapping fractional cover of an invasive plant species in a dryland ecosystem. *Ecology and Evolution* 9, 2562–2574.

Shiferaw, H., Bewket, W., Alamirew, T., Zeleke, G., Teketay, D. *et al.* (2019b) Implications of land use/land cover dynamics and prosopis invasion on ecosystem service values in Afar Region, Ethiopia. *Science of the Total Environment* 675, 354–366.

Shiferaw, H., Alamirew, T., Dzikiti, S., Bewket, W., Zeleke, G. *et al.* (2023) Water abstraction of invasive *Prosopis juliflora* and native Senegalia Senegal trees: A comparative study in the Great Rift Valley Area, Ethiopia. *Science of the Total Environment* 862, 160833.

Sohrabi, S., Pergl, J., Pyšek, P., Foxcroft, L.C. and Gherekhloo, J. (2021) Quantifying the potential impact of alien plants of Iran using the Generic Impact Scoring System (GISS) and Environmental Impact Classification for Alien Taxa (EICAT). *Biological Invasions* 23, 2435–2449.

Spawls, S. and Mathews, G. (2012) *Kenya: A Natural History*. T. & A.D. Poyser, Berkhamsted, UK.

Steenkamp, H.E. and Chown, S.L. (1996) Influence of dense stands of an exotic tree, *Prosopis glandulosa* Benson, on a Savanna dung beetle (Coleoptera: Scarabaeinae) assemblage in southern Africa. *Biological Conservation* 78, 305–311.

Strohbach, B.J., Ntesa, C., Kabajani, M., Shidolo, E.K. and d'Alton, C.D. (2015) *Prosopis* encroachment along the Fish River at Gibeon, Namibia. Habitat preferences, population densities and the effect on the environment. *Dinteria* 35, 53–73.

Tiawoun, M.A.P., Malan, P.W., Comole, A.A. and Moshobane, M.C. (2023) Impact of *Prosopis velutina* Wooton on the composition and diversity of native woody species in a semi-arid zone along the Molopo River, South Africa. *Plants* 12, 1561.

Tsinda, A., Mugisha, R., Mutesi, L., Bizoza, A., Kayitesi, E. *et al.* (2016) Biodiversity informatics in Eastern Africa: Status, drivers and barriers. *Journal for Nature Conservation* 32, 67–80.

van den Berg, E.C., Kotze, I. and Beukes, H. (2013) Detection, quantification and monitoring of *Prosopis* in the Northern Cape province of South Africa using remote sensing and GIS. *South African Journal of Geomatics* 2, 68–81.

van Wilgen, B.W., Zengeya, T.A. and Richardson, D.M. (2022) A review of the impacts of biological invasions in South Africa. *Biological Invasions* 24, 27–50.

Villagra, P.E., Vilela, A., Giordano, C. and Alvarez, J.A. (2010) Ecophysiology of *Prosopis* species from the arid lands of Argentina: What do we know about adaptation to stressful environments. In: Ramawat, K.G. (ed.) *Desert Plants: Biology and Biotechnology*. Springer, Berlin, pp. 321–340.

Wakie, T.T., Laituri, M. and Evangelista, P.H. (2016) Assessing the distribution and impacts of *Prosopis juliflora* through participatory approaches. *Applied Geography* 66, 132–143.

Ward, S.E., Smart, S.M., Quirk, H., Tallowin, J.R., Mortimer, S.R. *et al*. (2016) Legacy effects of grassland management on soil carbon to depth. *Global Change Biology* 22, 2929–2938.
Williams, D., Pettorelli, N., Henschel, J., Cowlishaw, G. and Douglas, C.M. (2014) Impact of alien trees on mammal distributions along an ephemeral river in the Namib desert. *African Journal of Ecology* 52, 404–413.
Wilson, E.O. (1988) *Biodiversity*. National Academy Press, Washington, DC, USA.
Wilson, E.O. (1992) *The Diversity of Life*. Penguin Books, London, UK.
Wise, R.M., van Wilgen, B.W. and Le Maitre, D.C. (2012) Costs, benefits and management options for an invasive alien tree species: The case of mesquite in the Northern Cape, South Africa. *Journal of Arid Environments* 84, 80–90.
Witt, A.B.R., Beale, T. and van Wilgen, B.W. (2018) An assessment of the distribution and potential ecological impacts of invasive alien plant species in Eastern Africa. *Transactions of the Royal Society of South Africa* 73, 217–236.
Zimmermann, H.G. (1991) Biological control of *Prosopis* spp. (Fabaceae), in South Africa. *Agriculture, Ecosystems & Environment* 37, 175–186.

6 The Impacts of Invading *Prosopis* Trees on Water Resources

Hailu Shiferaw[1]*, Sebinasi Dzikiti[2], Urs Schaffner[3] and Brian W. van Wilgen[4]
[1]Water and Land Resource Centre, Addis Ababa University, Ethiopia; [2]Department of Horticultural Science, Stellenbosch University, South Africa; [3]CABI, Delémont, Switzerland; [4]Centre for Invasion Biology, Department of Botany and Zoology, Stellenbosch University, South Africa

Abstract
Excessive water use is one of the major impacts of invasive alien trees, and it is especially important in arid and semi-arid parts of the world. In eastern Africa, a large proportion of the land receives less than 600 mm mean annual rainfall, and it is in these water-scarce areas that invasive alien *Prosopis* trees will have the largest impacts on water resources. *Prosopis* trees are physiologically well adapted to survive and outcompete native vegetation, having deep taproots as well as an extensive network of shallow roots that allows them to effectively utilize both deep groundwater and water deposited on the soil surface by rain. *Prosopis* trees also use this root network to redistribute water in the soil profile, providing surface water to shallow roots during times of drought and moving water into deep storage in times of excess. A variety of methods have been used to estimate water use by *Prosopis* trees, including measuring sap flow in individual trees with heat pulse techniques and stand-level evapotranspiration with micrometeorological methods such as the eddy covariance method. Stable isotopes of hydrogen and oxygen in water have also been used to determine the source of water used by trees. These studies have been used to estimate that *Prosopis juliflora* tree invasions are using between 3.1 and 3.3 billion m^3/year in the Afar region of Ethiopia alone (around 7 litres of water per tree per day on average). This has significantly contributed to the estimated loss of ecosystem services valued at USD 602 million per year over the past three decades, about four times greater than the annual budget of Afar State. Climate change and an increasing frequency of droughts are set to exacerbate these impacts, due to an increase in evapotranspiration and reductions in groundwater recharge brought about by invading *Prosopis* trees.

6.1 Introduction

Water resources are under threat globally (Vörösmarty *et al.*, 2000) due to direct abstraction and anthropogenic changes to vegetation composition, including invasion by alien plant species. Freshwater scarcity is considered to be the main threat to the sustainable development of human societies, particularly in arid and semi-arid areas which are by definition water-limited ecosystems (Mekonnen and Hoekstra, 2016; Greve *et al.*, 2018). In those regions, meeting the rising demands for water and protecting or restoring ecosystems so that they can sustainably replenish reserves of water will be one of the most difficult and important challenges of this century.

*Corresponding author: hailu2nd@gmail.com

© CAB International 2025. *The Ecology and Management of Invasive Prosopis Trees in Eastern Africa* (eds U. Schaffner *et al.*)
DOI: 10.1079/9781800623644.0006

Trees in the genus *Prosopis* (mesquite; Fabaceae) originate from the Americas and have been introduced to 129 countries outside their natural distribution range (Shackleton *et al.*, 2014) to provide a variety of benefits (Chapter 1, this volume). Several of these species have become invasive (Castillo *et al.*, 2021), spreading rapidly across the landscape and resulting in many negative impacts. One of the major impacts of these trees has been on water resources (Le Maitre *et al.*, 2015; Howari *et al.*, 2022).

The impacts of invading alien plants on water resources have not been widely studied. *Prosopis* species qualify as exceptions, as their water use has been studied both in their native range in the Americas (Snyder *et al.*, 1998; Hultine *et al.*, 2004; Scott *et al.*, 2006) and in the invaded range where most of the literature on this topic originates from South Africa (Le Maitre, 1999; Le Maitre *et al.*, 2015). More recently, though, a few additional studies have been conducted in eastern Africa (e.g. Shiferaw *et al.*, 2021, 2023), allowing for the development of general principles with regard to the impacts of *Prosopis* invasions on water resources across Africa. Understanding the level of water use by *Prosopis* trees is important because they can reduce water resources to below critical levels in semi-arid ecosystems (Le Maitre *et al.*, 1999; Ndhlovu *et al.*, 2016). Numerous reports have noted a decline in both groundwater (Fourie *et al.*, 2007) and stream flow (Le Maitre *et al.*, 2015) in areas that are heavily invaded by *Prosopis* trees.

In this chapter, we briefly review the status and importance of water resources in eastern Africa and cover how these resources are affected by alien *Prosopis* tree invasions. We also examine these consequences in economic terms, and how the predicted changes in climate are likely to affect the consequences.

6.2 Water Resources and Their Relevance in Eastern Africa

Rainfall over eastern Africa varies considerably, ranging from <200 to >1500 mm mean annual rainfall (Fig. 6.1). The rainfall regime is bimodal, with 'long rains' falling between March and May and 'short rains' occurring between October and December (Palmer *et al.*, 2023). Rainfall is lowest in the north-east of the region, i.e. in Somalia, Somaliland, eastern Ethiopia, northern Kenya, Djibouti and Eritrea.

The occurrence of invasive *Prosopis juliflora* trees is most marked in zones of transition to arid areas, noting that there have been no surveys in Somalia where additional invasions occur (Fig. 6.1) (Meroni *et al.*, 2017; Wudad and Abdulahi, 2021). The occurrence of drought is a common feature throughout the region, with periodic droughts being severe enough to incapacitate the society (Nicholson, 2017; Chapter 7, this volume). Most rural populations rely on groundwater for their water supply, through shallow wells, deep boreholes and springs, which are sources most likely to be affected by *Prosopis* tree invasions. Kenya is also a water-scarce country, where (as elsewhere in the region) droughts have increased in frequency (Nicholson, 2017), impacting negatively on the quality of life of pastoralists. In Somalia, 60% of the population live in rural areas as nomadic or semi-nomadic pastoralists, where droughts have resulted in declining groundwater levels, leading to reductions in income and to conflict (Chapter 7, this volume).

6.3 Structure and Ecophysiology of *Prosopis* Plants

6.3.1 Ability to spread

Prosopis trees can grow to a height of 12 m, with a trunk diameter of up to 1.2 m, although most invasive individuals are usually somewhat smaller, forming multiple branches that spread close to the ground. They produce copious amounts of seeds which are consumed by livestock and wildlife and spread over considerable distances in dung or are carried by episodic floods along ephemeral rivers (Shiferaw *et al.*, 2004), resulting in new foci for invasion. In particular, invasive *Prosopis* trees can form dense impenetrable thickets along river banks, flood plains and seasonal water courses (Dzikiti *et al.*, 2017; Shiferaw *et al.*, 2019a) (Fig. 6.2).

Prosopis trees have several additional features that allow them to grow and spread in

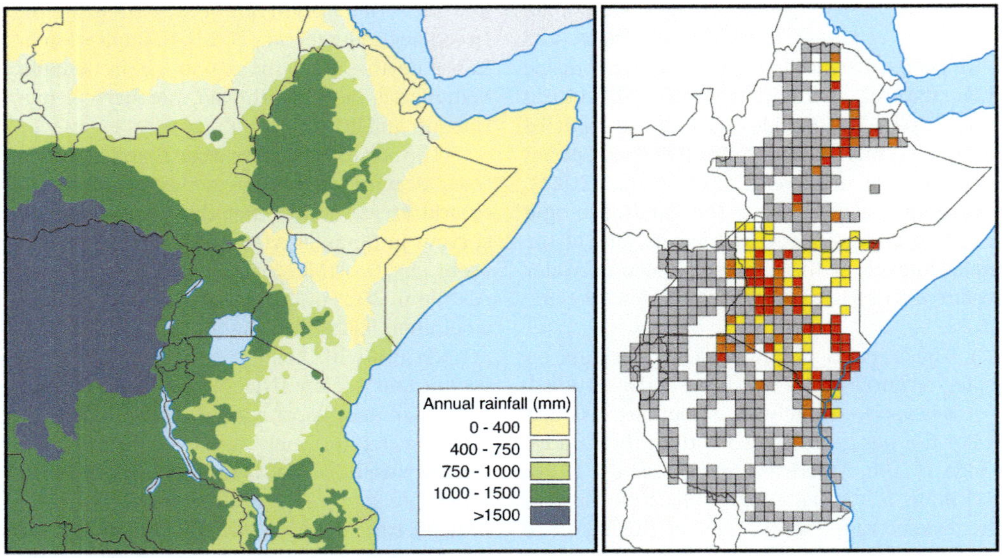

Fig. 6.1. Left: Mean annual rainfall across eastern Africa (redrawn from United States Geological Survey data). Right: Distribution of *Prosopis juliflora* (redrawn from Witt *et al*., 2018). Grey grid squares represent areas that were surveyed but where no *Prosopis* trees were seen; orange grid squares are areas where individuals were present and/or naturalized; red grid squares are where individuals were considered to be invasive; and yellow grid squares represent locality data obtained from other sources. Note that the surveys did not cover eastern Ethiopia or Somalia, where *Prosopis* trees are also present.

Fig. 6.2. *Prosopis juliflora* tree invasions along a river in the Bogoria District, Kenya. Photo courtesy of Urs Schaffner.

arid areas. These include their dimorphic root system, adaptations in leaf morphology and ecophysiology, and tolerance of salinity (Villagra *et al.*, 2010; López *et al.*, 2012; Garrido *et al.*, 2018; Zarzosa *et al.*, 2021). *Prosopis* species are generally deciduous, possibly due to frost tolerance (Pasiecznik *et al.*, 2001). However, *P. juliflora* is evergreen and abstracts groundwater even during peak dry seasons (Shiferaw *et al.*, 2021).

6.3.2 Root systems

Prosopis species develop extensive and deep taproots, enabling them to access groundwater at depths of 10–15m or more. In one extreme case, a taproot that reached 53m below the surface was recorded (Phillips, 1963; Stromberg, 2013). The surface roots can also extend laterally beyond the canopy area, which facilitates access to surface soil water after light rains or mist.

Prosopis trees are thus very adaptable and can survive and grow in arid regions by using deep soil water which can be reached by taproots and surface water obtained through lateral roots. For example, Snyder *et al.* (1998) used stable isotopes to show that *P. velutina* primarily used groundwater along a perennial river compared to predominantly using rain-derived water along rivers with ephemeral flows, particularly during the rainy season. Gile *et al.* (1997) studied the root systems of *P. glandulosa* and found that this species was able to establish itself over large areas due to several features, including the root system's ability to take advantage of sparse precipitation, to spread laterally over long distances, to grow upward and take advantage of water from occasional light rain, and to descend to considerable depths along cracks and other openings in the soil.

6.3.3 Leaf structure and physiology

Prosopis trees have small, thick leaves covered with a waxy cuticle which buffers the leaves from extremely high temperatures. *Prosopis* trees are known to survive air temperatures exceeding 50°C and soil temperatures up to 70°C, although the trees are susceptible to frost damage (Pasiecznik *et al.*, 2001). The leaflets of *Prosopis* trees also possess specialized adaptations that minimize water losses (e.g. sunken stomata, more stomata on adaxial than abaxial surfaces, and the presence of mucilaginous cells; Dzikiti *et al.*, 2017).

Metabolic changes can also occur within the leaf and the whole plant during droughts. For example, seasonal variations in leaf concentrations of proline, sugars and proteins have been observed and these promote processes such as osmotic adjustment that allow the leaves to maintain a positive turgor even when water is severely limited (Hailu Shiferaw, personal observation). The green stems of some *Prosopis* species are also an adaptation strategy to drought, allowing the trees to maintain some level of photosynthesis even when they shed their leaves.

Diurnal changes in photosynthesis and stomatal conductance show a characteristic midday depression when the atmospheric evaporative demand is greatest (Dellatore *et al.*, 2008). The leaves of *Prosopis* trees can also fold to reduce the transpiring leaf area when the atmospheric evapotranspiration rates are very high.

6.3.4 Tolerance of soil conditions

Prosopis trees can grow in a wide range of soils, from nutrient-poor sandy dune soils to cracking clayey soils. They can also grow in saline soils and soils with high pH (Shackleton *et al.*, 2014). However, the trees prefer deeper soils, and soil depth is a critical factor limiting growth. They can also tolerate seasonal waterlogging (Orwa *et al.*, 2009).

6.3.5 Nutritional and chemical advantages

Prosopis trees are nitrogen-fixing legumes and can moderately enrich the soil, more so when they form mycorrhizal associations (Orwa *et al.*, 2009). *Prosopis* trees also possess allelopathic and allelochemical properties (Schachtschneider,

2010), giving them a competitive advantage against other species by exclusion.

6.3.6 Frost as a limiting factor

The occurrence of frost and very cold weather is a major factor limiting the distribution of *Prosopis* trees globally. Infrequent frosts of short duration can cause branches to die back while severe frosts over extended periods can cause the death of entire trees. Frost damage is more severe in seedlings and younger plants (Muthana, 1974), and deciduous *Prosopis* species tolerate frost better than evergreen species (Pasiecznik *et al.*, 2001). Optimal conditions for the growth of *Prosopis* trees are mean annual temperatures in the range of 14–34°C and average annual rainfall in the range of 50–1200 mm, but post-introduction rapid evolution may ease these limitations (Chapter 3, this volume).

6.4 How Do *Prosopis* Trees Use Water?

The high water use by *Prosopis* trees arises from their ability to manipulate the environment much more efficiently than indigenous plants can, either by directly accessing both surface and groundwater at volumes that other species are unable to match or through processes such as hydraulic redistribution, allowing them to rapidly invade surrounding areas (Hultine *et al.*, 2004; Scott *et al.*, 2006).

Hydraulic redistribution occurs when the xylem water potential of the roots is intermediate between that of roots in the dry and wet portions of the soil when canopy transpiration is very low. Under these circumstances, the roots act as conduits that transfer water between the different soil layers. For instance, when light rain falls and wets the top soil layers such that the xylem water potential of the lateral roots becomes higher than that of the taproot, water can be siphoned from the upper soil layers and deposited deeper in the soil profile via the taproot through a process called hydraulic descent. Hultine *et al.* (2004) illustrated this process for *P. velutina* (Fig. 6.3) by monitoring the sap flow dynamics of tap and lateral roots following precipitation pulses. This process (which is akin to 'banking the water') is a drought survival strategy where rain or fog water stored near the soil surface is siphoned deeper into the soil. This not only minimizes losses through evaporation near the soil surface but also removes the water from shallow-rooted competitors.

The converse process occurs when the taproot has access to water in the deeper soil profiles such that the water potential of the roots near the surface is lower than that of the taproot. Under these circumstances, water is extracted via the taproot and deposited in the shallow soil layers via the lateral roots through a process called hydraulic lift. This process maintains soil moisture in the shallow soil layers during dry periods, thereby providing water to seedlings and younger trees that cannot access groundwater. The moist soil environment also promotes nutrient cycling, thus enhancing the growth, survival and rapid spread of the invasions even under hostile conditions (Dzikiti *et al.*, 2017).

According to Scott *et al.* (2006), significant volumes of water can be transferred between soil layers by hydraulic redistribution and this process has ecosystem-wide implications. For example, Kurz-Besson *et al.* (2006) observed that the roots of *Quercus suber* (cork oak) trees lifted 17–81% of the water that they used the following day for transpiration (Kurz-Besson *et al.*, 2006). In another study, Emerman and Dawson (1996) noted that *Acer saccharum* (sugar maple) hydraulically lifted 102 (±54) litres of water daily, about 25% of the transpiration amount.

The dual strategy of *Prosopis* trees to obtain water from the unsaturated zone and, depending upon moisture regimes, access groundwater using the long taproot may provide a competitive advantage over native African tree species, even if the latter are adapted to dry conditions. In areas invaded by *Prosopis*, native tree species experience increased mortality and are rapidly replaced by *Prosopis* (Robertson and Woodborne, 2001; Schachtschneider and February, 2013; Shackleton *et al.*, 2015). Using stable isotope analysis, Robertson and Woodborne (2001) confirmed that *Vachellia erioloba* trees competing with *Prosopis* trees died in times of water stress. They proposed that in areas where water is a limiting factor, *Prosopis* trees maintain a competitive advantage over *V. erioloba* by

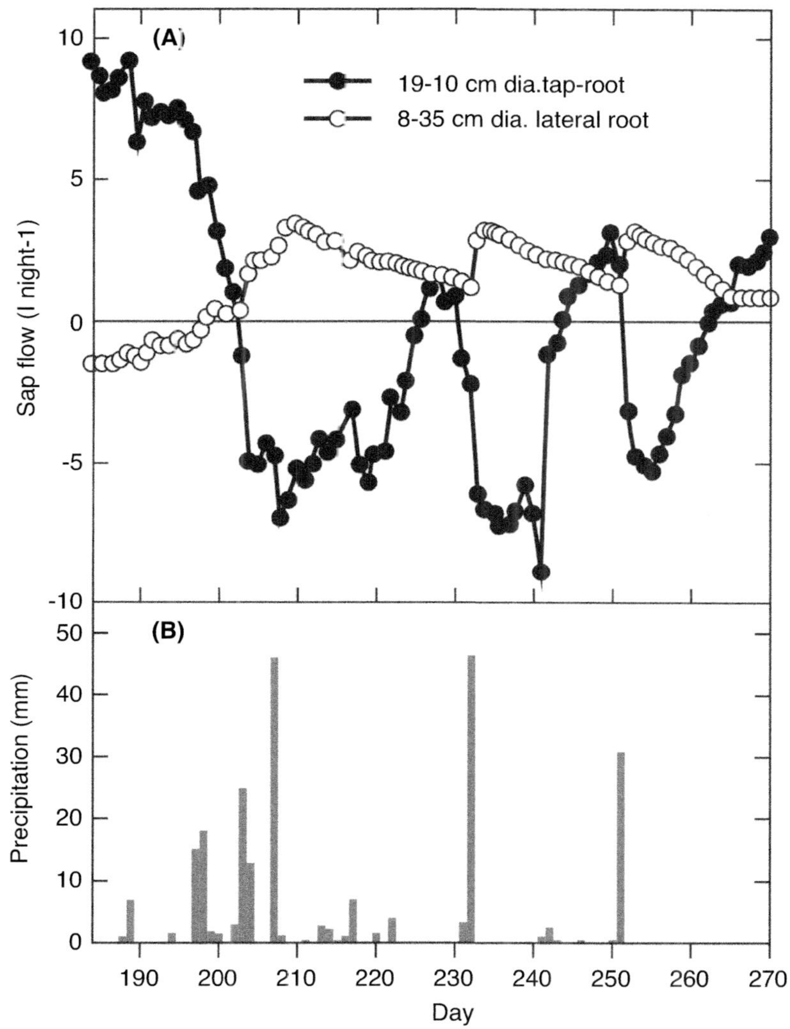

Fig. 6.3. (A) Total daily nocturnal sap flow of the taproot and lateral roots of a *Prosopis* tree calculated from half-hourly measurements from 20.00 h to 05.30 h during the 2002 growing season. Positive and negative values show upward (from deep soil layers to the upper roots) and downward (from upper soil to lower soil) flow respectively. (B) Daily precipitation totals (mm day^{-1}) (redrawn from Hultine *et al.*, 2004).

tracking changes in the level of the water table or extracting deeper groundwater. For instance, after a rainfall event, *Prosopis* trees initially use their lateral roots to extract water from the unsaturated zone until the water reserves are exhausted and then switch to the dormant taproot. In contrast, *V. erioloba* is almost entirely dependent upon groundwater (Robertson and Woodborne, 2001).

Studies by Dzikiti *et al.* (2013, 2017) and Shiferaw *et al.* (2021, 2023) using the heat ratio method of monitoring sap flow (Box 6.1) have shown that daily total water consumption by *Prosopis* trees depends on the availability of water in the root zone. For trees growing in a very dry environment with a very deep water table at Kenhardt in the arid Northern Cape province of South Africa, the trees maintained very low transpiration rates with a peak of <1 mm d^{-1} (Dzikiti *et al.*, 2013). In addition, the physiology of the trees was geared towards the arid conditions with very thin sapwood and thin leaves. In Dzikiti *et al.*'s (2013) study, a very strong correlation between the cumulative rise in water uptake by the trees and the cumulative decline in the groundwater levels was observed. Much higher transpiration rates were observed by Shiferaw *et al.* (2021) on larger trees growing at a relatively drier site in the Afar region in Ethiopia; the authors attributed this to the superior ability of *Prosopis* trees to access groundwater.

Box 6.1. Examples of how water use by trees can be determined

Several techniques have been developed to estimate water use by trees at various scales, and the most important ones that have been used on *Prosopis* invasions are outlined here.

The compensation heat pulse velocity method and heat ratio method (HRM) of monitoring sap flow are most widely used for quantifying the rate at which water flows up the stems of trees. These measurements are useful for comparing water use by different tree species under similar conditions (Dye *et al.*, 1996; Forster, 2017). Both methods involve the injection of heat as a tracer for sap movement up the tree stems. Because xylem sap is more than 99% water, measuring sap flow rates gives good approximations of tree transpiration rates. In both methods, a pulse of heat lasting 1 s or less is injected through a line heater installed between two temperature sensors aligned along the vertical axis of the tree stem. The temperature sensors measure the sapwood temperature before and after the release of the heat pulse.

In the *compensation heat pulse velocity method*, the heater probe is situated closer to the lower than the upper temperature probe. The time taken for the temperature readings of the two probes to equilibrate after release of the heat pulse is used to derive the sap velocity. The product of the sap velocity and sapwood area gives the sap flow volume.

In the *HRM method*, the upper and lower temperature probes are installed at equal distances (5 mm) from the central heater probe. Sap velocity is calculated from the ratio of the preheating temperatures for the upper and lower probes using an empirical equation (Burgess *et al.*, 2001). The HRM technique has been widely used in studies (Dzikiti *et al.*, 2013; Shiferaw *et al.*, 2021), given that it is suitable for measuring both low and reverse sap flow rates. This technique has been central to studying hydrological processes by which *Prosopis* trees manipulate the soil moisture regime in the root zone, such as hydraulic redistribution (Hultine *et al.*, 2004; Dzikiti *et al.*, 2017). Sap flow is limited in that it only gives an understanding of water use at the scale of the plant (transpiration) and upscaling to ecosystem-level evapotranspiration presents challenges.

Eddy covariance is a micro-meteorological method that is the gold standard in directly observing the exchanges of water, gases, energy and momentum between ecosystems and the atmosphere. This method obtains turbulent flux data by calculating the covariance of high-frequency fluctuations in the vertical wind velocity and in atmospheric water content. It is widely used to determine evapotranspiration and ecosystem net carbon exchange at the scale of an experimental site (100–2000 m), and to compare overall water use from, for example, invaded and uninvaded sites (e.g. Ponton *et al.*, 2006).

Stable isotopes of hydrogen and oxygen in water can be used to determine the source from which a tree is obtaining water (e.g. from recent rain or from soil-stored water at different depths). The method relies on determining the occurrence of stable isotopes of the molecules in water [hydrogen (^{2}H) and oxygen (^{18}O)], which can indicate the age of water up to 3–5 years. 'Young' water would indicate the use of recent rainfall, while 'older' water would indicate sourcing water from deeper in the soil profile. This technique was used, for example, to compare the ability of *Prosopis* trees and native trees to utilize groundwater at varying depths (Schachtschneider and February, 2013).

Paired catchment experiments can be used to gauge stream flow in response to different treatments, for example comparing stream flow from catchments with natural vegetation and those afforested with alien trees. These paired catchment experiments have typically shown that conversion from grasslands or shrublands to a closed canopy of trees has significantly reduced surface water run-off (Bosch and Hewlett, 1982).

Continued

Box Fig. 1. (A) Set-up to measure ecosystem water exchange (evapotranspiration) using eddy covariance, (B) sap flow monitoring instrumentation for transpiration measurement using the heat pulse method, and (C) instrumentation for weather and related data-logging at Melka Werer, Amibara, Middle Awash. Photos courtesy of Hailu Shiferaw.

6.5 Upscaling Estimates of Water Use by *Prosopis* Trees

One of the challenges faced by scientists is how to translate and upscale measurements made at the scale of a single tree, or on an experimental plot, to the landscape level where the realized impact needs to be assessed. One way is to clear *Prosopis* trees from an invaded area and monitor changes in water availability. Dzikiti *et al.* (2013) quantified the effect of invasion by *Prosopis* trees on groundwater by comparing data from a *Prosopis*-invaded area in South Africa and an adjacent cleared area which was predominantly covered by grasses, with a few indigenous shrubs and trees. They concluded that up to 70 m^3 of groundwater could be saved per month by clearing one hectare of *Prosopis* tree invasions.

A review of two sites (in the USA and Australia) by Doody *et al.* (2011) concluded that the actual volumes of water that can be saved by clearing invasive alien plants depend on: (1) the eco-hydrological setting of the aquifer (i.e. the specific interactions between water and plant communities where the aquifer occurs); (2) the water use characteristics of the vegetation that replaces the invasions once they have been cleared; and (3) the hydrogeological properties of the aquifer (i.e. the specific interactions between water and the geological features of the aquifer). The last point would not be responsive to invasion or clearing efforts, and more recent research on potential water salvage in invaded ecosystems has therefore focused on quantifying the incremental water that invasions by *Prosopis* trees use over and above that consumed by the native vegetation that would be restored after the invasions have been cleared.

These studies have yielded insights into the water use characteristics of *Prosopis* trees in comparison to native species with similar growth forms (Dzikiti *et al.*, 2017; Shiferaw *et al.*, 2023). These studies found no significant differences in the volumes of water that individual trees of similar sizes consumed. However, the deleterious impact of the invasions by *Prosopis* trees on the water resources arose from their ability to form dense stands which far exceed the densities achieved by native species. At the stand scale *Prosopis* trees transpired more than five times more water (~544 mm y^{-1}) than indigenous *Vachellia karroo* trees (~91 mm y^{-1}) since there were six times more *Prosopis* trees than the indigenous species. The removal of *Prosopis* trees slowed down the rate of decline of the water table in the dry season from a preclearing peak of ~8.9 to 5.0 mm d^{-1} after alien plant removal (Dzikiti *et al.*, 2017). Similarly, Shiferaw *et al.* (2023) found that at the stand scale *P. juliflora* trees transpired approximately three times more water (~242 mm y^{-1}) than the indigenous *Senegalia senegal* trees (~87 mm y^{-1}), because *P. juliflora* stands are denser than *S. senegal* stands and, in contrast to *S. senegal*, *P. juliflora* is evergreen and uses water throughout the year. Therefore, management activities that reduce the density of the *Prosopis* tree invasions would be beneficial in terms of restoring or protecting water resources.

Another approach to upscale estimates of water use by *Prosopis* trees to the landscape or even regional level was adopted by Shiferaw *et al.* (2021). In a study conducted in the Afar region, Ethiopia, sap flow rates of individual *P. julifora* trees were measured using the heat pulse method and estimates of evapotranspiration (quantified using the eddy covariance method) were used to quantify landscape-scale water impacts. Average transpiration by individual trees was 7 l/day (range 1–36 l/day), and the average transpiration of a dense stand of *Prosopis* trees was about 3.7 mm rainfall equivalent per day. These authors used these estimates together with a remotely sensed map of the cover and density of *P. julifora* (over an area of 1.2 million ha; Shiferaw *et al.*, 2019b) to estimate the water use from *Prosopis* trees. Water use was approximately 3.1–3.3 billion m^3/year for the Afar region, which corresponds to approximately 50% of the total rainfall in the region. Thus, *Prosopis* invasion significantly exacerbates water shortage in this semi-arid and arid region.

6.6 Economic Valuation of Water Loss Due to *Prosopis* Invasions

As indicated above, a study in the Afar region, Ethiopia, found that *P. juliflora* had invaded over 1.2 million ha (12.3% of the region) in the 31 years between 1986 and 2017 (Shiferaw *et al.*, 2019b). These invasions, along with other changes in land use and land cover, had resulted in declines in ecosystem services such as water supply and grasses for livestock. Invasion by *Prosopis* trees was ranked as the third most important driver of these changes (after drought and climate change; Shiferaw *et al.*, 2019b).

To translate the water consumed by *Prosopis* in the Afar region into monetary values, Shiferaw *et al.* (2021) used two different approaches: first, they multiplied the amount of water lost with the regional water tariff, which was at that time USD 0.00015 per m^3 water. This produced an estimated annual economic impact of *Prosopis* trees in the Afar region of USD 465,000–500,000 due to water loss. An alternative way to assess the economic impact is the loss of production capacity due to water use by *Prosopis* trees. The value of water is then typically much higher than water tariffs or delivery costs (Le Maitre *et al.*, 2020). With that in mind, Shiferaw *et al.* (2021) estimated the market price and net benefits of cotton and sugarcane that could potentially be grown with the amount of water consumed by *Prosopis* trees. They found that the amount of water consumed would be sufficient to irrigate about 460,000 ha of cotton or 330,000 ha of sugarcane. These are the main crops grown in the area and could generate net monetary values of USD 320 million or USD 470 million annually from cotton and sugarcane production, respectively. Hence, *P. juliflora* invasion in the Afar region can, by reducing water resources, potentially have serious impacts on the provision of other ecosystem services and ultimately on rural livelihoods.

These results support Le Maitre *et al.*'s (2020) notion that the actual economic impact of *Prosopis* may be significantly higher than that quantified using the regional water tariffs. Investing money in clearing *Prosopis* to make space for agriculture can be considered as a payment for ecosystem services by investors (i.e. food production).

Using the ecosystem service value (ESV) approach developed by Costanza and others (e.g. Costanza *et al.*, 2014), Shiferaw *et al.* (2019b) attempted to assess the changes in overall ecosystem services value in the Afar region between 1986 and 2017 and the main drivers of changes in land use and land cover. All together, they estimated that annual ESVs in the Afar region dropped by USD 602 million (range USD 112–1092 million) over the 31 years of the study, which is equivalent to four times the Afar region's annual budget (Shiferaw *et al.*, 2019b). *Prosopis*-dominated vegetation was the land cover type with the highest amount of change over the study period. These findings suggest that invasions by *Prosopis* trees have significantly contributed to a loss of ESVs in Afar, and that these losses are in a similar range to the estimated economic losses due to water use by invasive *Prosopis* trees.

In an investigation of whether restoration projects pay or not, Crookes and Blignaut (2019) reviewed the benefits and costs associated with alien tree clearing projects in South Africa, and the restoration and conservation of natural capital. They concluded that such projects were 'a step towards viewing the investment in restoration not merely as an expenditure item to be minimised, but as a truly worthwhile investment in the future wellbeing of both people and the planet – an investment in the national security of the country'.

6.7 Implications of Climate Change and Management Recommendations

As is the case elsewhere in the world, anthropogenic climate change is set to have significant impacts on eastern African ecosystems. The projected changes in precipitation across the Greater Horn of Africa are not that clear at this stage, but the length of dry periods is projected to increase and that of wet periods to decrease (Osima *et al.*, 2018). Droughts have already become longer and more intense, and they tend to continue into rainy seasons, although their causes are not adequately understood (Nicholson, 2017). In a retrospective analysis covering the mid-1980s to 2010, Palmer *et al.* (2023) reported that the long rains had delivered less rain (−0.65 to −2.95 mm season^{-1} a^{-1}), and that the short rains had become wetter (1.44 to 2.36 mm season^{-1} a^{-1}). These trends are expected to affect the severity and frequency of extreme droughts, and there is some consensus that the short rains will deliver more rainfall than the long rains by 2030–2040 (Osima *et al.*, 2018; Palmer *et al.*, 2023).

It has been shown that invasive *Prosopis* trees can increase transpiration and evapotranspiration losses and thus reduce surface water or groundwater recharge (Le Maitre *et al.*, 2020). Given the already increasing frequency of droughts in dry regions of sub-Saharan Africa, and the predictions that they will increase, these findings are cause for concern. This is particularly worrying in the case of *Prosopis* trees because of their superior ability to utilize both surface water and deeply stored groundwater, and other adaptations that allow them to survive in arid and semi-arid environments.

Studies of the impacts of *Prosopis* trees on water resources in eastern and southern Africa have demonstrated just how important it will be to reduce the density of existing invasions as well as prevent further spread. In a region where droughts are frequent and increasing in length and severity, the additional strain on the region's people and economy cannot be afforded. The estimated costs of the impacts are substantial and are set to grow. This information provides ample justification for investing in effective management interventions, making use of all available control methods in appropriate combinations (integrated control; see Chapter 12, this volume). The returns on investment, if the control is effectively carried out, should compensate amply for control costs. Studies have demonstrated that it would be far more effective to focus on sparse invasions, and on the leading edges of invasion fronts (Higgins *et al.*, 2000; van Wilgen *et al.*, 2001), than on densely invaded areas, and this approach is to be encouraged. By focusing clearing efforts on lightly invaded areas and preventing further spread, the formation of new, dense stands, and consequent increases in water losses, can be prevented more effectively.

Acknowledgements

This work was supported by the Swiss Programme for Research on Global Issues for Development (r4d), funded by the Swiss National Science Foundation and the Swiss Agency for Development and Cooperation, through the project 'Woody invasive alien species in East Africa: Assessing and mitigating their negative impact on ecosystem services and rural livelihood' (Grant Number: 400440_152085). This work was also supported in part by the National Research Foundation of South Africa (Grant Numbers 150260 and 151653 to BWvW).

References

Bosch, J.M. and Hewlett, J.D. (1982) A review of catchment experiments to determine the effect of vegetation changes on water yield and evapotranspiration. *Journal of Hydrology* 55, 3–23.

Burgess, S.S., Adams, M.A., Turner, N.C., Beverly, C.R. and Ong, C.K. (2001) An improved heat pulse method to measure low and reverse rates of sap flow in woody plants. *Tree Physiology* 21, 589–598.

Castillo, M.L., Schaffner, U., van Wilgen, B.W., Montaño, N.M., Bustamante, R.O. *et al.* (2021) Genetic insights into the globally invasive and taxonomically problematic tree genus *Prosopis*. *AoB PLANTS* 13, laa069.

Costanza, R., de Groot, R., Sutton, P., van der Ploeg, S., Anderson, S.J. *et al.* (2014) Changes in the global value of ecosystem services. *Global Environmental Change* 26, 152–158.

Crookes, D.J. and Blignaut, J.N. (2019) Investing in natural capital and national security: A comparative review of restoration projects in South Africa. *Heliyon* 5(5), 01765.

Dellatore, S.M., Garcia, A.S. and Miller, W.M. (2008) Mimicking stem cell niches to increase stem cell expansion. *Current Opinion in Biotechnology* 19, 534–540.

Doody, T.M., Nagler, P.L., Glenn, E.P., Moore, G.W., Morino, K. *et al.* (2011) Potential for water salvage by removal of non-native woody vegetation from dryland river systems. *Hydrological Processes* 25, 4117–4131.

Dye, P.J., Soko, S. and Poulter, A.G. (1996) Evaluation of the heat pulse velocity method for measuring sap flow in *Pinus patula*. *Journal of Experimental Botany* 47, 975–981.
Dzikiti, S., Schachtschneider, K., Naiken, V., Gush, M., Moses, G. *et al.* (2013) Water relations and the effects of clearing invasive *Prosopis* trees on groundwater in an arid environment in the Northern Cape, South Africa. *Journal of Arid Environments* 90, 103–113.
Dzikiti, S., Ntshidi, Z., Le Maitre, D.C., Bugan, R.D., Mazvimavi, D. *et al.* (2017) Assessing water use by *Prosopis* invasions and *Vachellia karroo* trees: Implications for groundwater recovery following alien plant removal in an arid catchment in South Africa. *Forest Ecology and Management* 398, 153–163.
Emerman, S.H. and Dawson, T.E. (1996) Hydraulic lift and its influence on the water content of the rhizosphere: An example from sugar maple, *Acer saccharum*. *Oecologia* 108, 273–278.
Forster, M.A. (2017) How reliable are heat pulse velocity methods for estimating tree transpiration?. *Forests* 8, 350.
Fourie, F., Mbatha, K., Verster, H. and van Dyk, G. (2007) The effect of vegetation (*Prosopis* sp.) on groundwater levels in Rugseer river, Kenhardt, South Africa. In: *Groundwater and Ecosystems, XXXV IAH Congress*. pp. 17–21.
Garrido, M., Silva, H., Franck, N., Arenas, J. and Acevedo, E. (2018) Evaluation of morpho-physiological traits adjustment of *Prosopis tamarugo* under long-term groundwater depletion in the hyper-arid Atacama Desert. *Frontiers in Plant Science* 9, 453.
Gile, L.H., Gibbens, R.P. and Lenz, J.M. (1997) The near-ubiquitous pedogenic world of mesquite roots in an arid basin floor. *Journal of Arid Environments* 35, 39–58.
Greve, P., Kahil, T., Mochizuki, J., Schinko, T., Satoh, Y. *et al.* (2018) Global assessment of water challenges under uncertainty in water scarcity projections. *Nature Sustainability* 1, 486–494.
Higgins, S.I., Richardson, D.M. and Cowling, R.M. (2000) Using a dynamic landscape model for planning the management of alien plant invasions. *Ecological Applications* 10, 1833–1848.
Howari, F.M., Sharma, M., Nazzal, Y., El-Keblawy, A., Mir, S. *et al.* (2022) Changes in the invasion rate of *Prosopis juliflora* and its impact on depletion of groundwater in the northern part of the United Arab Emirates. *Plants* 11, 682.
Hultine, K.R., Scott, R.L., Cable, W.L., Goodrich, D.C. and Williams, D.G. (2004) Hydraulic redistribution by a dominant, warm-desert phreatophyte: Seasonal patterns and response to precipitation pulses. *Functional Ecology* 18, 530–538.
Kurz-Besson, C., Otieno, D., Lobo do Vale, R., Siegwolf, R., Schmidt, M. *et al.* (2006) Hydraulic lift in cork oak trees in a savannah-type Mediterranean ecosystem and its contribution to the local water balance. *Plant and Soil* 282, 361–378.
Le Maitre, D.C. (1999) *Prosopis and Groundwater: A Literature Review and Bibliography*. Council for Scientific and Industrial Research, Stellenbosch.
Le Maitre, D.C., Scott, D.F. and Colvin, C. (1999) A review of information on interactions between vegetation and groundwater. *Water SA* 25, 137–152.
Le Maitre, D.C., Gush, M.B. and Dzikiti, S. (2015) Impacts of invading alien plant species on water flows at stand and catchment scales. *AoB PLANTS* 7, lv043.
Le Maitre, D.C., Blignaut, J.N., Clulow, A., Dzikiti, S., Everson, C.S. *et al.* (2020) Impacts of plant invasions on terrestrial water flows in South Africa. In: van Wilgen, B.W., Measey, J., Richardson, D.M. and Wilson, J.R. (eds) *Biological Invasions in South Africa*. Springer, Cham, Switzerland, pp. 431–458.
López, D., Fernández, M. and Verga, A. (2012) Drought stress tolerance of *Prosopis chilensis* and *Prosopis flexuosa* species and their hybrids. *Trees* 27, 285–296.
Mekonnen, M.M. and Hoekstra, A.Y. (2016) Sustainability: Four billion people facing severe water scarcity. *Science Advances* 2, 1–7.
Meroni, M., Ng, W.T., Rembold, F., Leonardi, U. and Atzberger, C. (2017) Mapping *Prosopis juliflora* in west Somaliland with Landsat 8 satellite imagery and ground information. *Land Degradation & Development* 28, 494–506.
Muthana, K. (1974) A note on frost susceptibility in arid zone trees. *Annals of Arid Zone* 13, 370–373.
Ndhlovu, T., Milton, S.J. and Esler, K.J. (2016) Impact of *Prosopis* (mesquite) invasion and clearing on vegetation species composition and diversity in semi-arid Nama-Karoo rangeland, South Africa. *African Journal of Range & Forage Science* 33, 101–110.
Nicholson, S.E. (2017) Climate and climatic variability of rainfall over Eastern Africa. *Reviews of Geophysics* 55, 590–635.

Orwa, C., Mutua, A., Kindt, R., Jamnadass, R. and Simons, A. (2009) Agroforestree database: A tree reference and selection guide version 4.0. Available at: https://www.cifor-icraf.org/knowledge/publication/33826/ (accessed 30 July 2024).

Osima, S., Indasi, V.S., Zaroug, M., Endris, H.S., Gudoshava, M. *et al.* (2018) Projected climate over the greater horn of Africa under 1.5 C and 2 C global warming. *Environmental Research Letters* 13, 065004.

Palmer, P.I., Wainwright, C.M., Dong, B., Maidment, R.I., Wheeler, K.G. *et al.* (2023) Drivers and impacts of Eastern African rainfall variability. *Nature Reviews Earth & Environment* 4, 254–270.

Pasiecznik, N.M., Felker, P., Harris, P.J., Harsh, L., Cruz, G. *et al.* (2001) *The Prosopis juliflora – Prosopis pallida Complex: A Monograph*. HDRA, Coventry, UK.

Phillips, W.S. (1963) Depth of roots in soil. *Ecology* 44, 424.

Ponton, S., Flanagan, L.B., Alstad, K.P., Johnson, B.G., Morgenstern, K.A.I. *et al.* (2006) Comparison of ecosystem water-use efficiency among Douglas-fir forest, aspen forest and grassland using eddy covariance and carbon isotope techniques. *Global Change Biology* 12, 294–310.

Robertson, I. and Woodborne, S. (2001) Carbon isotopes confirm the competitive advantages of *Prosopis* over *Acacia erioloba*. In: *Conference on Study of Environmental Change Using Isotope Techniques, Vienna, April 2001*, International Atomic Energy Board, Vienna, Austria.

Schachtschneider, K. (2010) Water sourcing by riparian trees along ephemeral riverbeds. PhD thesis, Department of Botany, University of Cape Town, South Africa.

Schachtschneider, K. and February, E.C. (2013) Impact of *Prosopis* invasion on a keystone tree species in the Kalahari Desert. *Plant Ecology* 214, 597–605.

Scott, R.L., Huxman, T.E. and Williams, D.G. (2006) Ecohydrological impacts of woody-plant encroachment: Seasonal patterns of water and carbon dioxide exchange within a semiarid riparian environment. *Global Change Biology* 12, 311–324.

Shackleton, R.T., Le Maitre, D.C., Pasiecznik, N.M. and Richardson, D.M. (2014) *Prosopis*: A global assessment of the biogeography, benefits, impacts and management of one of the world's worst woody invasive plant taxa. *AoB PLANTS* 6, plu027.

Shackleton, R.T., Le Maitre, D.C., van Wilgen, B.W. and Richardson, D.M. (2015) The impact of invasive alien *Prosopis* species (mesquite) on native plants in different environments in South Africa. *South African Journal of Botany* 97, 25–31.

Shiferaw, H., Teketay, D., Nemomissa, S. and Assefa, F. (2004) Some biological characteristics that foster the invasion of *Prosopis juliflora* in middle Awash Rift Valley area. *Journal of Arid Environments* 58, 134–154.

Shiferaw, H., Schaffner, U., Bewket, W., Alamirew, T., Zeleke, G. *et al.* (2019a) Modelling the current fractional cover of an invasive alien plant and drivers of its invasion in a dryland ecosystem. *Scientific Reports* 9, 1–12.

Shiferaw, H., Bewket, W., Alamirew, T., Zeleke, G., Teketay, D. *et al.* (2019b) Implications of land use/land cover dynamics and *Prosopis* invasion on ecosystem service values in Afar Region, Ethiopia. *Science of the Total Environment* 675, 354–366.

Shiferaw, H., Alamirew, T., Dzikiti, S., Bewket, W., Zeleke, G. *et al.* (2021) Water use of *Prosopis juliflora* and its impacts on catchment water budget and rural livelihoods in Afar region, Ethiopia. *Scientific Reports* 11, 1–14.

Shiferaw, H., Alamirew, T., Dzikiti, S., Bewket, W., Zeleke, G. *et al.* (2023) Water abstraction of invasive *Prosopis juliflora* and native *Senegalia senegal* trees: A comparative study in the Great Rift Valley area, Ethiopia. *Science of the Total Environment* 862, 160833.

Snyder, K.A., Williams, D.G. and Gempko, V.L. (1998) Water source determination in cottonwood/willow and mesquite forests on the San Pedro river in Arizona. In: *Proceedings of the 78th Annual Meeting of the American Meteorological Society Special Symposium on Hydrology*, Phoenix, Arizona, pp. 185–188.

Stromberg, J.C. (2013) Root patterns and hydrogeomorphic niches of riparian plants in the American Southwest. *Journal of Arid Environments* 94, 1–9.

van Wilgen, B., Richardson, D. and Higgins, S.I. (2001) Integrated control of invasive alien plants in terrestrial ecosystems. *Land Use and Water Resources Research* 1, 1732–2016.

Villagra, P.E., Vilela, A., Giordano, C. and Alvarez, J.A. (2010) Ecophysiology of *Prosopis* species from the arid lands of Argentina: What do we know about adaptation to stressful environments?. In: Ramawat, K.G. (ed.) *Desert Plants*. Springer, Berlin, pp. 321–340.

Vörösmarty, C.J., Green, P. and Salisbury, J. (2000) Global water resources: Vulnerability from climate change and population growth. *Science* 289(5477), 284–288.

Witt, A.B.R., Beale, T. and van Wilgen, B.W. (2018) An assessment of the distribution and potential ecological impacts of invasive alien plant species in eastern Africa. *Transactions of the Royal Society of South Africa* 73, 217–236.

Wudad, A. and Abdulahi, A. (2021) Expansion of *Prosopis juliflora* and land use land cover change in Korahey zone of somali regional state, eastern Ethiopia. *Journal of Materials and Environmental Science* 12, 728–737.

Zarzosa, P.S., Mendieta-Leiva, G., Navarro-Cerrillo, R.M., Cruz, G., Grados, N. *et al*. (2021) An ecological overview of *Prosopis pallida*, one of the most adapted dryland species to extreme climate events. *Journal of Arid Environments* 193, 104576.

7 *Prosopis* Invasions in Eastern Africa's Rangelands: Impacts and Management Challenges

Staline Kibet[1]* and Brian W. van Wilgen[2]

[1]Department of Resource Management and Agricultural Technology, University of Nairobi, Kenya; [2]Centre for Invasion Biology, Department of Botany and Zoology, Stellenbosch University, South Africa

Abstract

This chapter reviews information on the impacts of invasive alien *Prosopis* trees on rangelands in eastern Africa. Natural rangelands cover 3.3 million km^2 (about 47%) of eastern Africa and support a variety of activities, including livestock production and wildlife-based tourism. In the region, pastoral practices contribute between 10% and 44% and tourism between 9% and 13% of the gross domestic product of eastern African countries respectively. Alien *Prosopis* trees were introduced into the region in the early 20th century with the aim of improving rangeland productivity by providing an additional source of fodder and shade, and to combat erosion. However, these potential benefits have largely been overshadowed by negative impacts associated with invasion, resulting in declines in rangeland carrying capacity rather than enhancing it. Invasion has been shown to reduce palatable grasses through competition, sometimes totally. Domestic and wild mammalian grazers therefore have to consume excessive amounts of *Prosopis* pods, leading to direct injury and toxic effects, and unhealthy imbalances in their diet. As a result, there have been significant declines in livestock and wildlife populations in invaded areas in eastern Africa, leading in turn to substantial negative impacts on the quality of life of nomadic pastoralists and the sustainability of their livelihoods. Invasion of rangelands by *Prosopis juliflora* in eastern Africa is one of several threats to the continued practice of rangeland-based pastoralism in the region, and it will be important to contain further spread to prevent substantial economic losses.

7.1 Introduction

Rangelands are extensive areas of land supporting native herbaceous or shrubby plants that are grazed by domestic or wild mammalian herbivores. They have many economic, ecological, social and cultural values, and a wealth of biodiversity that supports ecosystem health (International Livestock Research Institute, 2021). In eastern Africa, rangelands cover substantial areas (e.g. 44% of Uganda, 65% of Ethiopia, 74% of Tanzania and over 80% of Kenya; Bolo *et al.*, 2019). Rangelands provide much needed ecosystem services and support to local livelihoods, and contribute to national economies through tourism and employment. Pastoralism, which is the economic mainstay in African rangelands, is estimated to contribute 10–44% of the gross domestic products (GDP) of eastern African countries, affecting the lives

*Corresponding author: staline@uonbi.ac.ke

© CAB International 2025. *The Ecology and Management of Invasive Prosopis Trees in Eastern Africa* (eds U. Schaffner *et al.*)

DOI: 10.1079/9781800623644.0007

of millions of people living on over 40% of the continent's total land mass (African Union, 2013). In Kenya, tourism, most of which takes place in rangelands, accounts for 13% of GDP and the pastoral sector is estimated to be worth USD 1.13 billion, with livestock providing 92% and non-livestock the remaining 8% (Nyariki and Amwata, 2019). National meat consumption in Kenya is estimated at 553,200 tons, of which about 28% comes from pastoral practices in arid and semi-arid lands (the rest coming from ranches and smallholdings; Nyariki and Amwata, 2019). In Tanzania, where tourism is also centred on rangelands, tourism contributed 9% of the total GDP, attracted 26% of total foreign income, provided 8.2% of total employment and 3.3% of total investment in 2017 (World Travel and Tourism Council, 2018). Similarly, Ethiopian rangelands contribute substantially toward the country's economy. The country's agropastoral and pastoral production systems jointly supported 28, 100, 70 and 50% of the national cattle, camel, goat and sheep populations, respectively (Centre for Pastoral Areas and Livestock Development, 2020).

Africa's rangelands, and the pastoralists and tour operators who depend on them, face many challenges. These include bush encroachment, overgrazing, harvesting of trees for wood and charcoal, climate change and conversion to crop agriculture. In addition to these threats, invasion by alien plant species contributes substantially to rangeland degradation (Witt *et al.*, 2017; O'Connor and van Wilgen 2020). In this regard, one of the greatest threats comes from trees in the genus *Prosopis*, which were first introduced from central America into Ethiopia and Kenya during the second half of the 20th century (Chapter 2, this volume). The intention of these introductions was to provide an additional source of fodder for livestock, curb desertification and provide a source of firewood (Chapter 2, this volume), but the trees rapidly became invasive and now threaten rangelands across the region.

In this chapter, we review what is known about the value of rangelands in eastern Africa and the impact that invasion by *Prosopis* trees is having on their productivity. We review briefly the intended benefits for livestock that the introduction of these trees was meant to bring and why they did not materialize. We assess the impacts of invasions on rangeland productivity and on the social systems that depend on rangelands. The impacts of invasive *Prosopis* trees on rangelands are significant but they must be viewed in the context of other social, biophysical and political trends. These factors, in combination with invasion by *Prosopis* trees, threaten the potential for rangeland ecosystems to continue to make valuable and sustainable contributions to the economy and human wellbeing in the region.

7.2 Importance of Rangelands in Eastern Africa

7.2.1 Current and future uses of rangelands

In eastern Africa, rangelands cover approximately 3.3 million km^2 (about 47% of the region) with major land cover types including grassland (25%), savanna (44%), crops (6%) and degraded land (12%) (Global Center on Adaptation, 2022). These rangelands contribute significantly to regional economies by providing the basis for livestock production, tourism, agriculture and renewable energy, among others. There is further potential for rangelands to contribute to the region's economy through emerging opportunities such as carbon credits, game ranching, non-timber forest products (e.g. gums, resins, honey and herbal products) and as a source of biotechnology materials.

Despite the potential value of pastoralism, many African governments currently pursue a policy of modernizing agriculture by displacing nomadic or transhumance livestock production with irrigated crop agriculture. Pastoral communities are therefore under pressure as they become marginalized and excluded from the wetter and more fertile parts of territories previously available to them (Notenbaert *et al.*, 2012). The most important of these activities (livestock production and tourism) are discussed in the sections below.

7.2.2 Value for livestock production

The mobile livestock production system (i.e. production based on seasonal movements between wet and dry season grazing areas), practised by nomadic pastoralists in African rangelands, is recognized by several scholars as the most productive and preferred form of land use (Nori, 2007; Krätli and Schareika, 2010; Khazanov and Schlee, 2012; Oba, 2012). For example, state-owned, large-scale irrigated cotton or sugarcane production in the Awash Valley, Ethiopia, not only yields lower returns than those from pastoral animal husbandry (live animals, meat and milk) but also causes environmental harm in terms of soil salinity and loss of biodiversity, both of which are important for food security (Notenbaert *et al.*, 2012; Behnke and Kerven, 2013).

Pastoral and agropastoral communities contribute over 70% of the livestock numbers in eastern African rangelands (East African Community, 2016). In this review, pastoralism has been used to mean extensive livestock production systems normally practised in dryland areas. It involves the tracking and use of grazing and water across rangelands. On the other hand, agropastoralism supports livelihoods that integrate crop production and livestock production, practised by settled, nomadic or transhumance communities.

It is estimated that the pastoral sector in Kenya has an overall value of EUR 750 million and an annual marketed value of EUR 50–80 million (Nyariki and Amwata, 2019). In Uganda, 4% of total GDP and 16% of agricultural GDP come from the livestock sector (Economic Policy Research Centre, 2021). Similarly, in Ethiopia, over 20% of national exports (90% of which comprise live animals from pastoralists) and 80% of milk supply are derived from rangelands (Abduletif, 2019). Over 70% of milk produced in Tanzania comes from pastoral and agropastoral systems. In addition, this sector provides goods and raw materials to other economic sectors in both formal and informal supply chains, such as butchers, restaurant operators, traders and transporters.

7.2.3 Value for tourism

Tourism contributes significantly to the economy of eastern Africa. In 2019, prior to the COVID-19 pandemic, the tourism industry contributed 7.7, 6.3, 10.6 and 5.9% of GDP for Kenya, Ethiopia, Tanzania and Uganda respectively (World Travel and Tourism Council, 2023). The sector depends predominantly on nature-based visits to wildlife destinations such as national parks, game reserves and wildlife conservancies, the majority of which are situated in rangelands. Protected areas are extensive in eastern Africa, and cover 11.1, 11.4, 48.2 and 17.8% of total land area for Ethiopia, Kenya, Uganda and Tanzania respectively (Bacsi *et al.*, 2023).

Community-based natural resource management has gained prominence amongst pastoral communities in eastern Africa in recent decades as an innovative approach that allows them to benefit from the natural resources under their jurisdiction (Kenya Wildlife Conservancies Association, 2019; Lee, 2021). In Kenya, for example, there are over 160 wildlife conservancies covering 11% of the country, with over 700,000 households deriving benefits by generating income from tourism and employment. Some of these conservancies host rare or critically endangered species such as black and white rhinoceros (*Diceros bicornis* and *Ceratotherium simum*), Grévy's zebra (*Equus grevyi*) and hirola (*Beatragus hunteri*) (Kenya Wildlife Conservancies Association, 2019; Government of Kenya, 2022).

7.3 The Benefits and Drawbacks of *Prosopis* Trees for Free-ranging Livestock

In their native environment, *Prosopis* trees are adapted to survive and thrive in very harsh arid environments. Trees in this genus are thus regarded as having substantial potential to provide much needed fodder and shade for livestock and people (Bekele *et al.*, 2018) and to reduce soil erosion, among other things. Bekele *et al.* (2018) estimated the monetary value of microclimate regulation due to *Prosopis* trees to be USD 5.36 and USD 3.89 per year per household in Baringo County, Kenya, and Afar region, Ethiopia, respectively. These attributes led to their introduction in the 19th and 20th centuries in many parts of the world where the supposed benefits were needed, including eastern Africa (Chapter 2, this volume).

The pods of *Prosopis* trees are a source of protein and carbohydrates, providing 9 and 56%

by weight respectively (Patnaik *et al.*, 2017) As feed supplement for lambs or goats, 200–250 g/day of *Prosopis* pods resulted in weight gains without any adverse effects (Obeidat *et al.*, 2008; Ali *et al.*, 2012; Obeidat and Shdaifat, 2013; Koech *et al.*, 2014). *Prosopis* pods can also be substituted for grass or corn by up to 30–47.5% in the diet of sheep, goats and cattle without negative effects (Girma *et al.*, 2011; Medeiros *et al.*, 2012; Chaturvedi and Sahoo, 2013; Pereira *et al.*, 2013; Hintsa *et al.*, 2015). The suitability of *Prosopis* pods as feed for horses, poultry and fishes has also been assessed with mixed results (Girma *et al.*, 2011; Silva *et al.*, 2015; Odero-Waitituh *et al.*, 2016). However, it is not possible to regulate the proportions of feed ingested by free-ranging livestock, and excessive intakes of *Prosopis* pods can cause neurological disorders (Silva *et al.*, 2015), being toxic to neurons of cranial nerve nuclei (Tabosa *et al.*, 2000; Tabosa *et al.*, 2006), and weaken and break teeth (Mutavi, 2020). Further, as the pods alone cannot satisfy the nutritional needs of grazing livestock or wildlife, access to natural grasses is still needed (Patnaik *et al.*, 2017). Such access is substantially reduced following invasion (Chapter 5, this volume), resulting in livestock having to rely on an excessive intake of *Prosopis* pods, with detrimental effects.

One of the other desired benefits of introducing *Prosopis* trees was to provide shade for livestock to cope with the very high temperatures experienced in African rangelands. However, because *Prosopis* trees rapidly form dense, impenetrable stands, their ability to provide shade is lost. On balance, therefore, the potential benefits of *Prosopis* trees have largely been overshadowed by their negative impacts, and the carrying capacity of rangelands is actually substantially reduced rather than enhanced by their presence.

7.4 Impact of *Prosopis* Invasions on Rangeland Carrying Capacity

Alien *Prosopis* trees are highly invasive, turning productive rangelands into impenetrable thickets with low carrying capacities (Fig. 7.1). Studies in eastern and southern Africa have quantified many of these changes, in terms of both their ecological and social consequences. These invasions are a widespread problem. For example, *Prosopis* trees are threatening 11 of the 32 districts of Afar region in Ethiopia, most of which are largely pastoral rangelands (Shiferaw *et al.*, 2022), and invasion by *Prosopis* trees reduced the ability of pastoralists to keep large herds of livestock in affected areas in Kenya (Maundu *et al.*, 2009). *Prosopis* trees also negatively affect water resources, reducing groundwater supplies to below the critical levels needed to sustain a pastoral lifestyle (Chapter 6, this volume). As a result of these changes, there have been significant declines in livestock and

Fig. 7.1. (A) Uninvaded rangeland in the Ngorongoro Conservation Area, Tanzania, supporting both domestic livestock and wild ungulates. (B) Rangeland in the Moshi District, Tanzania, where important pasture grasses have been eliminated by invading *Prosopis juliflora* trees. Photos courtesy of Brian van Wilgen (A) and CABI (B).

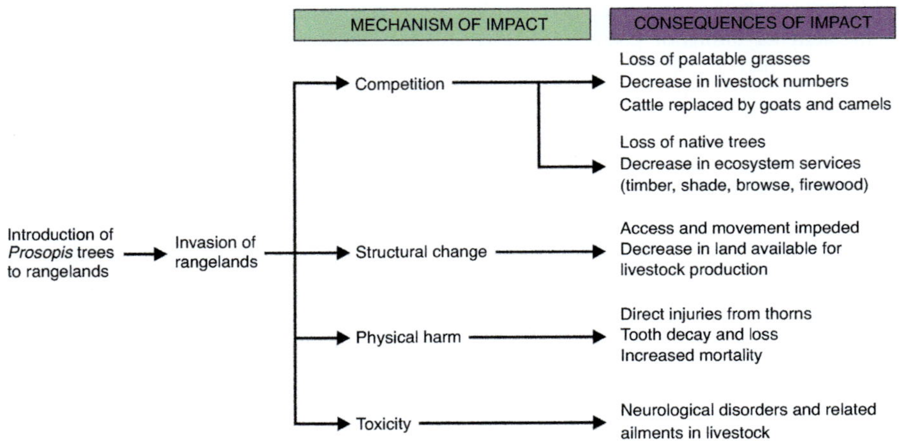

Fig. 7.2. Mechanisms by which invasive *Prosopis* trees generate impact in African rangelands, and the consequences for livestock production in invaded rangelands. See Section 7.4 for details.

wildlife populations, reducing livestock production and productivity in invaded areas in eastern Africa. Ultimately, these changes have had substantial negative impacts on the quality of life and sustainability of pastoralist's livelihoods (Chapter 8, this volume). The mechanisms that generate impacts and their consequences for livestock production are shown in Fig. 7.2 and discussed in the sections below.

7.4.1 Competition with herbaceous vegetation

One of the most noticeable effects is the displacement, sometimes total, of palatable grasses through competition (Fig. 7.1; Chapter 5, this volume). In the Afar region, Mekuyie *et al.* (2018) found that invasion by *Prosopis* trees prevented recruitment of the palatable grasses *Cenchrus ciliaris*, *Chrysopogon plumulous* and *Setaria acromelaena*. In Baringo, understorey grass cover was found to be five times lower under the canopy of *Prosopis juliflora* trees than in open areas (Kahi *et al.*, 2009). In the Allideghi Wildlife Reserve, Ethiopia, invasion by *Prosopis* trees reduced the understorey basal cover of perennial grasses from 68 to 2%, increased soil surface exposure from 30 to 80%, and reduced the number of native grass species from seven to two (Kebede and Coppock, 2015). In the Shibalo District, Ethiopia, Bezaredie *et al.* (2023) found that the number of native plant species, and their density, abundance and richness were significantly higher in uninvaded areas than invaded areas. In the Northern Cape, South Africa, Shackleton *et al.* (2015) recorded a decline in the cover of native perennial grasses and herbaceous plants from 15–20% where the basal area of *Prosopis* trees was <2 m^2/ha to zero where the basal area of *Prosopis* trees was >4.5 m^2/ha. In the Western Cape, South Africa, invasion (c.15% *Prosopis* canopy cover) reduced grazing capacity by 34% (Ndhlovu *et al.*, 2011).

The reduction in herbaceous cover and diversity has led to negative impacts on livestock farming (Fig. 7.3). Both Haregeweyn *et al.* (2013) and Ayanu *et al.* (2015) noted that invasion by *Prosopis* trees has caused a considerable decline in livestock production and productivity due to loss of dry season grazing areas in Ethiopia. In Afar, Mehari (2015) found that 84% of pastoral households considered invasion by *P. juliflora* to be responsible for reductions in their ability to raise livestock. The invasions caused about 80% livestock loss and each household, on average, lost 6.5 small stock and seven cattle over ten years. Linders *et al.* (2020) found that both in Afar and Baringo, the supply of herbaceous biomass decreased by over 90% with increasing *Prosopis* tree cover. In the most heavily invaded areas, livestock consumed all the available herbaceous biomass within a few days.

Fig. 7.3. Undernourished cattle in a rangeland invaded by *Prosopis juliflora* in Kenya. Note the total absence of grasses and other herbaceous plants. Photo courtesy of Urs Schaffner.

Several other studies noted that rangeland degradation, to which invasion by *P. juliflora* was a major contributor, resulted in a 65% loss of grassland cover and a 114% increase in bush and shrub cover between 1985 and 2015. As a result, pastoralists' access to rangelands became increasingly restricted (Ayanu *et al.*, 2015; Bekele *et al.*, 2018; Mekuyie *et al.*, 2018). Cattle and camel populations in the Amibara zone of Ethiopia declined by 36 and 20% respectively between 1997 and 2011 due to *Prosopis* invasions (Yosef *et al.*, 2013). In the Somali region of the Rift Valley, Ethiopia, Kassahun *et al.* (2008) noted that vegetation changes, including invasion by *Prosopis* trees, resulted in a decrease in cattle and an increase in camels and small ruminants in the livestock mix owned by nomadic pastoralists, leading to increased poverty.

7.4.2 Competition with trees and shrubs

Invasive *Prosopis* trees also displace native trees, with negative effects. In the Afar region, Mekuyie *et al.* (2018) found that invasion by *Prosopis* trees prevented the recruitment of native shrubs that are important as browse species, including *Vachellia tortilis*, *Senegalia senegal* and *Vachellia nilotica*. In the same area, Mitiku (2010) noted that *Prosopis* trees severely invaded dense acacia woodlands, riverine forest and agricultural lands, where they displaced 9.91 km^2 of acacia woodlands between 1986 and 2001. As a consequence, many ecosystem services associated with savanna trees, including shade and browse for domestic and wild ungulates, timber and firewood, are reduced.

7.4.3 Structural changes to the vegetation

Invasion of rangelands by *P. juliflora* in eastern Africa changes open savannas to dense, impenetrable thickets which hamper the movement of livestock (Senayit *et al.*, 2004; Taye *et al.*, 2004; Shiferaw *et al.*, 2022). As a result, access to water and remaining (but surrounded) patches of rangeland is prevented, and herding becomes more difficult. In addition, herders often complain that dense *Prosopis* thickets provide cover for predators (Lekenit *et al.*, 2023).

7.4.4 Features that cause physical harm

Prosopis trees carry numerous sharp and sturdy thorns, which inflict wounds on livestock, wildlife and people. Livestock losses can be high due to such injuries (Mwangi and Swallow, 2005; Obiri, 2011). In addition, the jaws of livestock are disfigured by chewing hard *Prosopis* pods too frequently (Mwangi and Swallow, 2005, 2008; Maundu *et al.*, 2009). The high sugar content in the pods causes tooth decay and loss, which in some cases leads to animals starving to death. Similarly, the heavy consumption of pods by camels results in the loss of teeth, affecting their overall feeding capacity and health.

7.4.5 Toxic effects

Due to the reductions in herbaceous vegetation brought about by competition, domestic and wild mammalian grazers are forced to consume excessive amounts of *Prosopis* pods in the absence of suitable grazing grasses, leading to toxic effects and unhealthy imbalances in their diet. Consumption of *P. juliflora* pods has been found to cause neurological disorders and related ailments in livestock, including partial anorexia, salivation, twitching, dehydration, bloody diarrhea, bile duct hyperplasia, degeneration of renal tubules, chronic and progressive injury to neurons, and denervation atrophy (Patnaik *et al.*, 2017).

7.5 Social Impacts of *Prosopis* Invasions

7.5.1 Threats to pastoralism

About 25% of Africa's population are pastoralists and agropastoralists. About 50 million pastoralists and up to 200 million agropastoralists live in the arid and semi-arid lands of Africa, especially in Somalia, Mauritania, Ethiopia, Sudan and Kenya (Leal Filho *et al.*, 2020). Pastoralists are resourceful, innovative and entrepreneurial people, but ongoing uncertainty about both climate and emerging infectious diseases (e.g. Rift Valley fever and Middle East respiratory syndrome coronavirus) (Hassell *et al.*, 2020) within pastoral rangelands necessitates flexibility and adaptability (Lind *et al.*, 2020). Key resources such as water and pastures are unevenly distributed both in space and time, due to erratic and unreliable rainfall in much of eastern Africa's drylands. The distances that livestock have to travel between these key resource areas could vary from tens to hundreds of kilometres.

The issue of how to deal with pastoralism as a land use practice and traditional way of life is complex. The key question, politically, is whether transhumance (i.e. the practice of moving livestock from one grazing area to another in a seasonal cycle) should be supported or whether pastoralists should be encouraged and supported to become sedentary (Behnke and Kerven, 2013; Haji and Legesse, 2013). On the one hand, pastoralism is seen as a livelihood system in which the dual objectives of development and environmental conservation can be achieved in traditional ways that have existed for centuries. On the other hand, there is a view that although transhumance has historically been sustainable, modern stresses and changes in policies and practices (e.g. restricting access to land and water) have increased the environmental impacts of pastoralism.

Pastoralism in eastern Africa's rangelands faces many threats. In Ethiopia, nearly 40% of valuable dry season grazing pastures in Afar have been converted to state-run or commercial sugar plantations and as a result, pastoralists have lost significant access to the Awash River, the only supply of fresh water (Abbink *et al.*, 2014; Taye, 2022). Fencing has resulted in disruption of the traditional management of rangelands as common property in Borana, Ethiopia, and elsewhere (Letai and Lind, 2012). Additional threats to the nomadic pastoral way of life arise because of land alienation largely by government for other 'productive' land uses, such as large-scale irrigation schemes in Ethiopia and Sudan (Abbink *et al.*, 2014; Shiferaw *et al.*, 2019), the establishment of protected areas (e.g. national parks and game reserves) or land fragmentation due to infrastructure development. Tree-planting schemes, intended to offset carbon emissions (Box 2.2, Chapter 2, this volume) are targeting Africa's rangelands, which will further reduce the extent of these important ecosystems and have far-reaching negative impacts (Bond *et al.*, 2019). Importantly, these schemes do not clearly consider the risks of introducing invasive trees, which would ultimately increase the extent of their negative impacts.

Haji and Legesse (2013) argued that, due to increased environmental stresses, the loss of dry season grazing areas and changes in policies and practices, pastoralism has become a livelihood option that is more detrimental to the environment than sedentary farming. The loss of access to dry season grazing areas, and of the development of common water points, has led to overgrazing in the remaining areas and a shift from grazers (cattle) to browsers (goats and camels), which in turn are more likely to further degrade the environment. Other studies show that despite considerable investment by governments in supporting more sedentary livelihoods, pastoralism has proved to be consistently more profitable than either cotton or sugarcane farming, while avoiding many of the environmental costs associated with large-scale irrigation projects (Behnke and Kerven, 2013). Importantly, from the perspective of this review, Haji and Legesse (2013) found evidence that transhumance increased the expansion of *P. juliflora*, which they considered to be another factor that needs to be borne in mind when assessing the future sustainability of pastoralism.

7.5.2 The impact of *Prosopis juliflora* invasions on pastoral livelihoods

Invasion of rangelands by *P. juliflora* has had complex social and cultural implications beyond the environmental and economic costs (Rogers *et al.*, 2017). Common access to rangelands has traditionally been managed through a set of rules and regulations enforced by traditional institutions. Unfortunately, many of these customs and their effective enforcement have broken down, leading to land degradation in most of the traditional rangelands (Shiferaw *et al.*, 2019). During prolonged droughts such as the one experienced between 2020 and 2023 in the Horn of Africa, pastoralists were forced to move over long distances (sometimes more than 500 km). In such cases, pastoralists are not aware of the relevant customs for managing grazing in the relevant territory, or they choose to ignore them (Kibet, personal observation). In some cases, livestock owners are elites (powerful entrepreneurs, power brokers and/or individuals in government) and can use their positions to access resources without following any regulations (Hesse and Thébaud, 2006; Kibet and Oyieke, unpublished data).

The complete breakdown of traditional land management systems has not only led to land degradation due to overgrazing but has increased uncontrolled invasion by alien species such as *P. juliflora*. The Tana delta in Kenya is a case in point. This area is a critical dry season grazing area that is now densely invaded (80% cover) by *P. juliflora*. Local people have lost any motivation to manage these invasions because they now must share the area with people from other areas.

The introduction of *P. juliflora* into an already stressed social-ecological system has thus exacerbated the situation substantially. Even if the social problems associated with transhumance (e.g. regulatory restrictions imposed on a growing human population against a background of shrinking resources) could be solved, the additional threats posed by *P. juliflora* invasions may be the final factor preventing such solutions from being successfully implemented. Finding ways in which to sustainably control *P. juliflora* invasions, and to prevent their further spread, would then be essential if a return to what some regard as the most sustainable form of land use is to be realized.

7.6 Potential for Controlling *Prosopis* Invasions in Rangelands

The capacity for rangeland users to control invasions by *P. juliflora* in eastern African rangelands is limited. The average labour cost of clearing a *Prosopis* thicket 3–4 years old in a 10 × 10m plot was found to range from KES 1000 (USD

13) in Loiyangalani, to KES 1200 in Baringo, to KES 2800 in Garissa, Kenya, which cannot be afforded by most rangeland users (Maundu *et al.*, 2009). Mbaabu *et al.* (2019) have demonstrated the obvious benefits of restoration on small plots, but scaling this up to a landscape level would probably not be easily achieved.

Studies elsewhere in Africa have shown that removing invasive alien *Prosopis* trees and also invasive Australian *Acacia* species has allowed for some recovery of native grasses (Ndhlovu *et al.*, 2016; Yapi *et al.*, 2018). However, the capacity of ecosystems to recover from invasion by alien trees declines with both the density and the duration of the invasions (Holmes *et al.*, 2020), meaning that a strategy of relying on passive restoration could be compromised in areas that had been densely invaded for a relatively long time. In such cases, active restoration, involving the replanting of grasses and other plants, would be needed; such undertakings would add substantially to the costs of removing the trees and preventing reinvasion.

Finding ways to restore rangelands that have been cleared of invasive *Prosopis* trees, and the scale at which such restoration would be practically possible and economically justifiable, remains a priority area for further research. Experience elsewhere has demonstrated that it is far more effective to focus control efforts on areas that are still relatively lightly invaded, rather than attempting to clear dense invasions (Higgins *et al.*, 2000). This suggests that preventing the spread of *Prosopis* trees to areas where they are not yet present or concentrating on the leading edges of invasion fronts would be the best approaches to adopt (Chapter 11, this volume). Second, it is clear from experience in Australia that biological control of invasive alien *Prosopis* trees is probably the only way that will enable other control methods, such as mechanical clearing, to be effective (Chapter 10, this volume).

7.7 Conclusions

It is evident from this review that rangelands in eastern Africa are important ecosystems, making significant contributions to the economy by providing the basis for livestock production and tourism. Invasions by *P. juliflora* will substantially reduce the ability of these ecosystems to provide these contributions. Although the trees were originally introduced to provide a range of benefits to pastoralists, these benefits have proved to be short-lived. Invasive *P. juliflora* trees eliminate native pasture grasses, impact negatively on water resources (Chapter 6, this volume) and cause injuries and toxic reactions in livestock. As a result, rangeland carrying capacity is markedly reduced and the composition of livestock herds changes by reducing the proportion of cattle to goats and camels. Besides the environmental impacts, the negative social consequences for people living in, or benefiting from, rangelands are serious.

Effective steps will need to be taken to reduce the rate of spread of these invasions (Chapters 10 and 11, this volume) and prevent their establishment in important rangelands that are not yet invaded.

Acknowledgements

This work was supported by the Swiss Programme for Research on Global Issues for Development (r4d), funded by the Swiss National Science Foundation and the Swiss Agency for Development and Cooperation, through the project 'Woody invasive alien species in East Africa: Assessing and mitigating their negative impact on ecosystem services and rural livelihood' (Grant Number: 400440_152085). BWvW thanks the National Research Foundation of South Africa for funding (Grant Numbers 150260 and 151653).

References

Abbink, J., Askew, K., Feyissa, D., Fratkin, E., Gabbert, E.C. *et al.* (2014) Lands of the Future: Transforming Pastoral Lands and Livelihoods in Eastern Africa. Working paper no. 154, Max Planck Institute for Social Anthropology, Halle, Germany.

Abduletif, A.A. (2019) Benefits and challenges of pastoralism system in Ethiopia. *Studia Mundi Economica* 6, 56–67.

African Union (2013) *Policy Framework for Pastoralism in Africa: Securing, Protecting and Improving the Lives, Livelihoods and Rights of Pastoralist Communities*. Department of Rural Economy and Agriculture, African Union, Addis Ababa, Ethiopia.

Ali, A.S., Tudsri, S., Rungmekarat, S. and Kaewtrakulpong, K. (2012) Effect of feeding *Prosopis juliflora* pods and leaves on performance and carcass characteristics of Afar sheep. *Agriculture and Natural Resources* 46, 871–881.

Ayanu, Y., Jentsch, A., Müller-Mahn, D., Rettberg, S., Romankiewicz, C. *et al.* (2015) Ecosystem engineer unleashed: *Prosopis juliflora* threatening ecosystem services. *Regional Environmental Change* 15, 155–167.

Bacsi, Z., Yasin, A.S. and Bánhegyi, G. (2023) Tourism competitiveness in Eastern Africa: RCA and TTCI. *Heritage* 6, 5997–6015.

Behnke, R. and Kerven, C. (2013) Counting the costs: Replacing pastoralism with irrigated agriculture in the Awash Valley. In: Catley, A., Lund, J. and Scoones, I. (eds) *Pastoralism and Development in Africa*. Routledge, Oxford, UK, pp. 57–70.

Bekele, K., Haji, J., Legesse, B. and Schaffner, U. (2018) Economic impacts of *Prosopis* spp. invasions on dryland ecosystem services in Ethiopia and Kenya: Evidence from choice experimental data. *Journal of Arid Environments* 158, 9–18.

Bezaredie, D., Tadesse, Z. and Tadesse, Z. (2023) Effects of *Prosopis juliflora* on plant diversity on rangeland in Shilabo District, Somali Regional State, Ethiopia. *Heliyon* 9(3), e14049.

Bolo, P.O., Sommer, R., Kihara, J.M., Kinyua, M., Nyawira, S. *et al.* (2019) *Rangeland Degradation: Causes, Consequences, Monitoring Techniques and Remedies*. International Center for Tropical Agriculture, Nairobi, Kenya.

Bond, W.J., Stevens, N., Midgley, G.F. and Lehmann, C.E. (2019) The trouble with trees: Afforestation plans for Africa. *Trends in Ecology and Evolution* 34, 963–965.

Centre for Pastoral Areas and Livestock Development (2020) *Total Economic Valuation of Pastoralism in Ethiopia Study Report*. IGAD Centre for Pastoral Areas and Livestock Development, Nairobi, Kenya.

Chaturvedi, O.H. and Sahoo, A. (2013) Nutrient utilization and rumen metabolism in sheep fed *Prosopis juliflora* pods and Cenchrus grass. *SpringerPlus* 2, 1–7.

East African Community (2016) East African Community livestock policy. Available at: https://www.eac.int/component/documentmananger/?task=download.document&file=bWFpbl9kb2N1bWVudHNfcGRmX1JuSmZ0dmJ5bW9YemZoR1pRZWRJVkRNRUFDIExJVkVTVE9DSyBQT0xJQ1kg&counter=671 (accessed 22 May 2023).

Economic Policy Research Centre (2021) *How Gaps in Traceability Systems Affect Uganda's Livestock Sector Transformation*. Ministry of Finance, Planning and Economic Development, Kampala, Uganda.

Girma, M., Urge, M. and Animut, G. (2011) Ground *Prosopis juliflora* pods as feed ingredient in poultry diet: Effects on growth and carcass characteristics of broilers. *International Journal of Poultry Science* 10, 970–976.

Global Center on Adaptation (2022) *State and Trends in Adaptation Reports 2021 and 2022: Executive Summaries and Syntheses*. Global Center on Adaptation, Rotterdam, The Netherlands and Abidjan, Côte d'Ivoire.

Government of Kenya (2022) *Impact of COVID-19 on Tourism in Kenya. The Measures Taken and the Recovery Pathways*. Ministry of Tourism and Wildlife, Nairobi, Kenya.

Haji, J. and Legesse, B. (2013) *Environmental Consequences of Pastoral and Sedentary Production Systems: Local Perceptions from Oromia and Afar*. Drylands Coordination Group, Addis Ababa, Ethiopia.

Haregeweyn, N., Tsunekawa, A., Tsubo, M., Meshesha, D. and Melkie, A. (2013) Analysis of the invasion rate, impacts and control measures of *Prosopis juliflora*: A case study of Amibara District, Eastern Ethiopia. *Environmental Monitoring and Assessment* 185, 7527–7542.

Hassell, J.M., Zimmerman, D., Fèvre, E.M., Zinsstag, J., Bukachi, S. *et al.* (2020) Africa's nomadic pastoralists and their animals are an invisible frontier in pandemic surveillance. *American Journal of Tropical Medicine and Hygiene* 103, 1777.

Hesse, C. and Thébaud, B. (2006) Will pastoral legislation disempower pastoralists in the Sahel? *Indigenous Affairs* 6, 4–23.

Higgins, S.I., Richardson, D.M. and Cowling, R.M. (2000) Using a dynamic landscape model for planning the management of alien plant invasions. *Ecological Applications* 10, 1833–1848.

Hintsa, K., Balehegn, M. and Birhane, E. (2015) Utilization of pods of *Prosopis juliflora*, an invasive tree, as a replacement to concentrate feed for goats in Ethiopia. *Livestock Research for Rural Development* 27, 9.

Holmes, P.M., Esler, K.J., van Wilgen, B.W. and Richardson, D.M. (2020) Ecological restoration of ecosystems degraded by invasive alien plants in South African Fynbos: Is spontaneous succession a viable strategy? *Transactions of the Royal Society of South Africa* 75, 111–139.

International Livestock Research Institute (2021) *Rangelands Atlas*. International Livestock Research Institute, Nairobi, Kenya.

Kahi, C.H., Ngugi, R.K., Mureithi, S.M. and Ng'ethe, J.C. (2009) The canopy effects of *Prosopis juliflora* (DC.) and *Acacia tortilis* (Hayne) trees on herbaceous plant species and soil physico-chemical properties in Njemps flats, Kenya. *Tropical and Subtropical Agroecosystems* 10, 441–449.

Kassahun, A., Snyman, A. and Smit, N. (2008) Impact of rangeland degradation on the pastoral production systems, livelihoods and perceptions of the Somali pastoralists in Eastern Ethiopia. *Journal of Arid Environments* 72, 1265–1281.

Kebede, T. and Coppock, L. (2015) Livestock-mediated dispersal of *Prosopis juliflora* imperils grasslands and the endangered Grevy's zebra in northeastern Ethiopia. *Rangeland Ecology and Management* 68, 402–407.

Kenya Wildlife Conservancies Association (2019) Fourth Annual National Conservancies Leaders Conference 2019 report. Available at: https://kwcakenya.com/wp-content/uploads/2019/11/Final-KWCA-Conservancy-Leaders-report-.pdf (accessed 21 May 2023).

Khazanov, A.M. and Schlee, G. (2012) *Who Owns the Stock? Collective and Multiple Property Rights in Animals*, Vol. 5. Berghahn Books, New York.

Koech, O.K., Kinuthia, R.N. and Wahome, R.G. (2014) Use of *Prosopis juliflora* seedpod as livestock feed supplement in the arid and semi-arid rangelands of Kenya. In: Behnassi, M., Shahid, S.A. and Mintz-Habib, N. (eds) *Science, Policy and Politics of Modern Agricultural System*. Springer, Dordrecht, pp. 197–209.

Krätli, S. and Schareika, N. (2010) Living off uncertainty: The intelligent animal production of dryland pastoralists. *European Journal of Development Research* 22, 605–622.

Leal Filho, W., Taddese, H., Balehegn, M., Nzengya, D., Debela, N. *et al.* (2020) Introducing experiences from African pastoralist communities to cope with climate change risks, hazards and extremes: Fostering poverty reduction. *International Journal of Disaster Risk Reduction* 50, 101738.

Lee, D. (2021) Tanzanian community wildlife management areas are successfully conserving wildlife. Available at: https://www.oneearth.org/tanzanian-community-wildlife-management-areas-are-successfully-conserving-wildlife/ (accessed 9 February 2023).

Lekenit, P.P., Mabwoga, S.O. and Omollo, M.O.J. (2023) Impacts of *Vachelia reficiens* and *Prosopis juliflora* on the environment of Samburu East Sub-County, Kenya. *International Journal of Natural Resource Ecology and Management* 8, 95–108.

Letai, J. and Lind, J. (2012) Squeezed from all sides: Changing resource tenure and pastoralist innovation on the Laikipia Plateau, Kenya. In: Catley, A., Lund, J. and Scoones, I. (eds) *Pastoralism and Development in Africa*. Routledge, Oxford, UK, pp. 164–176.

Lind, J., Sabates-Wheeler, R., Caravani, M., Kuol, L.B.D. and Nightingale, D.M. (2020) Newly evolving pastoral and post-pastoral rangelands of Eastern Africa. *Pastoralism: Research, Policy and Practice* 10, 1–14.

Linders, T.E., Bekele, K., Schaffner, U., Allan, E., Alamirew, T. *et al.* (2020) The impact of invasive species on social-ecological systems: Relating supply and use of selected provisioning ecosystem services. *Ecosystem Services* 41, 101055.

Maundu, P., Kibet, S., Morimoto, Y., Imbumi, M. and Adeka, R. (2009) Impact of *Prosopis juliflora* on Kenya's semi-arid and arid ecosystems and local livelihoods. *Biodiversity* 10, 33–50.

Mbaabu, P.R., Ng, W.T., Schaffner, U., Gichaba, M., Olago, D. *et al.* (2019) Spatial evolution of *Prosopis* invasion and its effects on LULC and livelihoods in Baringo, Kenya. *Remote Sensing* 11, 1217.

Medeiros, M.A., Riet-Correa, F., Pessoa, A.F., Pessoa, C.R., Batista, J.A. *et al.* (2012) Use of *Prosopis juliflora* pods as food for cattle and horses. *Pesquisa Veterinária Brasileira* 32, 1014–1016.

Mehari, Z.H. (2015) The invasion of *Prosopis juliflora* and Afar pastoral livelihoods in the Middle Awash area of Ethiopia. *Ecological Processes* 4, 1–9.

Mekuyie, M., Jordaan, A. and Melka, Y. (2018) Land-use and land-cover changes and their drivers in rangeland-dependent pastoral communities in the southern Afar Region of Ethiopia. *African Journal of Range and Forage Science* 35, 33–43.

Mitiku, G.S. (2010) *The Ecological and Socio-Economic Impacts of Invasive Plant Species: The Ecological Distribution and Socio-Economic Impacts of* Prosopis juliflora *in the Afar Regional State of Ethiopia*. VDM Verlag Dr Müller, Riga, Latvia.

Mutavi, S.K. (2020) Effects of *Prosopis juliflora* pod and leaf meal on physical characteristics of teeth and bones of goats in Kitui County, Kenya. *Livestock Research for Rural Development* 32, Article #163.

Mwangi, E. and Swallow, B. (2005) *Invasion of Prosopis juliflora and Local Livelihoods: Case Study from the Lake Baringo Area of Kenya*. Working paper #3. World Agroforestry Centre, Nairobi, Kenya.

Mwangi, E. and Swallow, B. (2008) Invasion and rural livelihoods in the Lake Baringo Area of Kenya. *Conservation and Society* 6, 130–140.

Ndhlovu, T., Milton-Dean, S.J. and Esler, K.J. (2011) Impact of *Prosopis* (mesquite) invasion and clearing on the grazing capacity of semiarid Nama Karoo Rangeland, South Africa. *African Journal of Range and Forage Science* 28, 129–137.

Ndhlovu, T., Milton, S.J. and Esler, K.J. (2016) Effect of *Prosopis* (mesquite) invasion and clearing on vegetation cover in semi-arid Nama Karoo Rangeland, South Africa. *African Journal of Range and Forage Science* 33, 11–19.

Nori, M. (2007) *Mobile Livelihoods, Patchy Resources and Shifting Rights: Approaching Pastoral Territories*. International Land Coalition, Rome, Italy.

Notenbaert, A.M., Davies, J., De Leeuw, J., Said, M., Herrero, M. *et al.* (2012) Policies in support of pastoralism and biodiversity in the heterogeneous drylands of East Africa. *Pastoralism: Research, Policy and Practice* 2, 1–17.

Nyariki, D.M. and Amwata, D.A. (2019) The value of pastoralism in Kenya: Application of total economic value approach. *Pastoralism: Research, Policy and Practice* 9, 1–13.

Oba, G. (2012) Harnessing pastoralists' indigenous knowledge for rangeland management: Three African case studies. *Pastoralism: Research, Policy and Practice* 2, 1–25.

Obeidat, B.S. and Shdaifat, M.M. (2013) Partial substitution of barley grain with *Prosopis juliflora* pods in lactating Awassi ewe's diets: Effect on intake, digestibility, and nursing performance. *Small Ruminant Research* 111, 50–55.

Obeidat, B.S., Abdullah, A.Y. and Al-Lataifeh, F.A. (2008) The effect of partial replacement of barley grains by *Prosopis juliflora* pods on growth performance, nutrient intake, digestibility, and carcass characteristics of Awassi lambs fed finishing diets. *Animal Feed Science and Technology* 146, 42–54.

Obiri, J.F. (2011) Invasive plant species and their disaster-effects in dry tropical forests and rangelands of Kenya and Tanzania. *Journal of Disaster Risk Studies* 3, 417–428.

O'Connor, T. and van Wilgen, B.W. (2020) The impact of alien plants on rangelands in South Africa. In: van Wilgen, B.W., Measey, G.J., Richardson, D.M., Wilson, J.R. and Zengeya, T. (eds) *Biological Invasions in South Africa*. Springer, Cham, Switzerland, pp. 445–473.

Odero-Waitituh, J.A., King'ori, A.M. and Guliye, A.Y. (2016) Effect of replacing maize with milled mature pods of *Prosopis juliflora* on performance of finishing broiler chicken. *Livestock Research for Rural Development* 28, Article #22.

Patnaik, P., Abbasi, T. and Abbasi, S.A. (2017) *Prosopis* (*Prosopis juliflora*): Blessing and bane. *Tropical Ecology* 58, 455–483.

Pereira, T.C.D.J., Pereira, M.L.A., Oliveira, C.A.S.D., Argôlo, L.D.S., Silva, H.G.D.O. *et al.* (2013) Mesquite pod meal in diets for lactating goats. *Revista Brasileira de Zootecnia* 42, 102–108.

Rogers, P., Nunan, F. and Fentie, A.A. (2017) Reimagining invasions – the social and cultural impacts of *Prosopis* on pastoralists in southern Afar, Ethiopia. *Pastoralism: Research, Policy and Practice* 7, 22.

Senayit, R., Agajie, T., Taye, T., Adefires, W. and Getu, E. (2004) *Invasive Alien Plant Control and Prevention in Ethiopia. Pilot Surveys andControl Baseline Conditions*. Report submitted to EARO, Ethiopia and CABIunder the PDF B phase of the UNEP GEF Project – Removing Barriers to Invasive Plant Management in Africa. Addis Ababa, Ethiopia: East Africa Regional Office.

Shackleton, R.T., Le Maitre, D.C., van Wilgen, B.W. and Richardson, D.M. (2015) The impact of invasive alien *Prosopis* species (mesquite) on native plants in different environments in South Africa. *South African Journal of Botany* 97, 25–31.

Shiferaw, H., Bewket, W., Alamirew, T., Zeleke, G., Teketay, D. *et al.* (2019) Implications of land use/ land cover dynamics and *Prosopis* invasion on ecosystem service values in Afar Region, Ethiopia. *Science of the Total Environment* 675, 354–366.

Shiferaw, W., Demissew, S., Bekele, T. and Aynekulu, E. (2022) Community perceptions towards invasion of *juliflora*, utilization, and its control options in Afar Region, Northeast Ethiopia. *PLoS One* 17(1), e0261838.

Silva, T.R.M., Chung, S., de Araújo, T.A.T., de Azevedo, K.S.P., dos Santos, M.C. *et al.* (2015) Replacing corn by mesquite meal (*Prosopis juliflora*) in diets for juvenile Nile tilapia reared in low temperature. *Revista Brasileira de Ciências Agrárias (Agrária)* 10, 460–465.

Tabosa, I.M., Souza, J.C., Graca, D.L., Barbosa-Filho, J.M., Almeida, R.N. *et al.* (2000) Neuronal vacuolation of the trigeminal nuclei in goats caused by ingestion of *Prosopis juliflora* pods (mesquite beans). *Veterinary and Human Toxicology* 42, 155–158.

Tabosa, I.M., Riet-Correa, F., Barros, S.S., Summers, B.A., Simões, S.V.D. *et al.* (2006) Neurohistologic and ultrastructural lesions in cattle experimentally intoxicated with the plant *Prosopis juliflora*. *Veterinary Pathology* 43, 695–701.

Taye, M. (2022) Opinion: Drought in pastoral areas of Ethiopia: The missing link between problem and solution? Available at: https://addisstandard.com/opinion-drought-in-pastoral-areas-of-ethiopia-the-missing-link-between-problem-and-solution/ (accessed 15 February 2023).

Taye, T., Ameha, T., Adefiris, W. and Getu, E. (2004) *Biological Impact Assessment on Selected IAS Plants on Native Species Biodiversity*. Report submitted to EARO, Ethiopia and CABI under the PDF-B phase of the UNEPGEF Project – Removing Barriers to Invasive Plant Management in Africa. Addis Ababa, Ethiopia: East African Regional Office.

Witt, A.B.R., Kiambi, S., Beale, T. and van Wilgen, B.W. (2017) A preliminary assessment of the extent and potential impacts of alien plant invasions in the Serengeti-Mara ecosystem, East Africa. *Koedoe* 59, a1426.

World Travel and Tourism Council (2018) Travel and Tourism Economic Impact 2018 – Tanzania. Available at: https://s3-eu-west-1.amazonaws.com/s3.sourceafrica.net/documents/119574/Tanzania2018-WORLD-TOURISM-COUNCIL.pdf (accessed 18 December 2023).

World Travel and Tourism Council (2023) Travel and Tourism Global Economic Impact and Trends. Available at: https://wttc.org/research/economic-impact (accessed 12 May 2023).

Yapi, T.S., O'Farrell, P.J., Dziba, L.E. and Esler, K.J. (2018) Alien tree invasion into a South African montane grassland ecosystem: Impact of *Acacia* species on rangeland condition and livestock carrying capacity. *International Journal of Biodiversity Science, Ecosystem Services and Management* 14, 105–116.

Yosef, T., Mengistu, U., Solomon, A., Mohammed, Y.K. and Kefelegn, K. (2013) Camel and cattle population dynamics and livelihood diversification as a response to climate change in pastoral areas of Ethiopia. *Livestock Research for Rural Development* 25, 1–10.

8 Impacts of Invasions by *Prosopis* Species on Rural Livelihoods: The Human Dimension

Ketema Bekele[1]*, Beatrice Adoyo[2], Charles J. Kilawe[3] and Urs Schaffner[4]
[1]School of Agricultural Economics, Haramaya University, Dire Dawa, Ethiopia; [2]Department of Geography, Population, and Environmental Studies, University of Nairobi, Kenya; [3]Department of Ecosystems and Conservation, Sokoine University of Agriculture, Morogoro, Tanzania; [4]CABI, Delémont, Switzerland

Abstract

Prosopis juliflora and other *Prosopis* species were introduced into eastern Africa for their presumed benefits to people and the environment. However, *P. juliflora* became invasive, spreading across valuable ecosystems on which rural communities rely for their livelihoods. This chapter reviews the impacts of *P. juliflora* on rural livelihoods, with a focus on people's perceptions, costs and benefits to households, and willingness to pay for managing *P. juliflora* invasions. About 70% of the population of eastern Africa are rural, and more than half (>60%) of them are directly dependent on ecosystem services. *Prosopis* trees generate benefits as well as negative impacts, and people's perceptions of the net value of these trees vary considerably with the regional context and stage of invasion. Importantly, the net value (benefits minus costs) does not change in a linear fashion, but starts off being positive, then becomes increasingly negative at an exponential rate as the cover of invasive trees increases. People's perceptions of the potential or actual impact of these invasions are not always aligned with ecological evidence and change relatively slowly in comparison with the growth of impacts. Despite this, 98% of 503 surveyed households were willing to pay between USD 38 and 50 per year towards managing invasions. This would amount to annual investments of USD 4.2 million and USD 6.3 million in Baringo County, Kenya and Afar region, Ethiopia, respectively. The current estimated economic costs of *Prosopis* invasions could grow to USD 375 million/yr in Ethiopia, Kenya and Tanzania jointly if all suitable habitat becomes invaded, suggesting a net present cost of USD 7 billion over the next 30 years. By recognizing the potentially significant future costs and allocating appropriate amounts to control and containment efforts now, it would be possible to avoid much greater social costs in the future.

8.1 Introduction

Over the last 10,000 years, nomadic pastoralism has been practised along the Great Rift Valley, effectively making use of the rich grasslands in a semi-arid environment that experienced alternating wet and dry seasons (Galaty, 2021). However, since the 20th century, pastoralist communities have faced serious challenges caused by anthropogenic and natural developments (Chapter 7, this volume). They have lost land to expanding crop

*Corresponding author: ketmimi@yahoo.com

© CAB International 2025. *The Ecology and Management of Invasive Prosopis Trees in Eastern Africa* (eds U. Schaffner *et al.*)
DOI: 10.1079/9781800623644.0008

production in former dry-season grazing areas along rivers or the shoreline of lakes. National and regional governments have increasingly privatized communal land to encourage the production of crops for domestic use or for export (Rettberg, 2010). Moreover, over the last 30 years, droughts, which are often coupled with famine and ethnic conflicts, have become more frequent and severe in eastern Africa (Haile *et al.*, 2019).

Besides land-use change, climate change and an ongoing marginalization of pastoralism, pastoralist communities in eastern Africa attribute their increased vulnerability to the invasion by *Prosopis juliflora* (Rettberg, 2010; Mehari, 2015). *P. juliflora* was introduced together with other *Prosopis* tree species from the Americas to eastern Africa in the late 1970s and early 1980s for the provision of various ecological and livelihood benefits such as firewood, a feedstock for livestock or for the stabilization of degraded areas (Elfadl and Luukkanen, 2003; Shackleton *et al.*, 2016; Chapter 2, this volume). About two to three decades after its introduction, *P. juliflora* became invasive, with negative consequences for the composition, structure and functioning of the natural and semi-natural ecosystems in the region (Mwangi and Swallow, 2008; Ayanu *et al.*, 2015).

Currently, in spite of its intended benefits, *P. juliflora* has become a serious threat by invading grasslands, croplands, wetlands and other ecosystems, resulting in conflicting views on its relative value. The invasion resulted in both positive and negative effects on the supply of ecosystem services on which people's livelihoods in the region depend, and the net effect varies with the level of invasion cover (Figure 15.1, Chapter 15, this volume). In order to gain a meaningful understanding of how communities are affected by *P. juliflora* invasions, it would thus be necessary to sample perceptions and impacts along a gradient of *P. juliflora* cover from sparse to dense invasions.

An understanding of people's perceptions about the relative benefits and impacts of an invasive species is needed to develop effective strategies for their management. Despite this need, most research has focused on ecological effects, without considering diverse local stakeholder perceptions. This gap in our understanding probably arises because of difficulties in measuring perceptions about impacts and the presence of differing perceptions among different stakeholders (García-Llorente *et al.*, 2008).

Further, most studies in the region analysed the negative and positive effects of *P. juliflora* on rural livelihoods by assuming that the impacts were similar regardless of the level of invasion (e.g. Mwangi and Swallow, 2008; Maundu *et al.*, 2009; Haregeweyn *et al.*, 2013; Jema and Abdu, 2013; Mehari, 2015; Zeray *et al.*, 2017). These and other similar studies may ultimately result in the proposal of 'one-size-fits-all' solutions. As a result, such uniform policies may not be sustainable, and investigating impacts across a continuum of invasion cover would provide a more accurate picture of how *P. juliflora* impacts on rural households' livelihoods (Shackleton *et al.*, 2007; Vilà *et al.*, 2011; Bekele *et al.*, 2018b).

Currently, the net benefits of *P. juliflora* would change from positive to negative as cover increases (Wise *et al.*, 2012; van Wilgen and Richardson, 2014), and it would be useful to know at which point this occurs (Mwangi and Swallow, 2008). The onset of invasion marks the point at which negative impacts begin to emerge, but it takes time before the cost of these impacts exceeds the value of benefits. Thus, during the initial stages of invasion, the net value of benefits, and people's perceptions, will remain positive. By the time that perceptions switch from predominantly positive to predominantly negative, the invasions are already at an advanced stage and thus more challenging to control. An understanding of these trends in real and perceived costs will be important for designing effective interventions and can be achieved in part by establishing the net value to rural households comprehensively in monetary terms. This is usually done in the framework of total economic valuation of local households' preferences or willingness to participate in managing the invasion.

This chapter reviews recent research on the impacts of invasion by *P. juliflora* on rural livelihoods by investigating non-linear responses to net value across a continuum from low to high cover of invasion and identifying factors influencing local perceptions. Local households' willingness to pay for managing the invasion is also reviewed and scaled up to measure economic welfare lost both at national and regional levels.

8.2 Dependence of Rural Communities on Ecosystem Services in Eastern Africa

People in low-income and middle-income countries are often highly reliant on services derived from natural ecosystems (Sangha *et al.*, 2022), and are more vulnerable to ecosystem shocks than communities in developed countries. Most rural communities' livelihoods in African countries are derived from traditional farming, pastoralism and agropastoralism. As such, they depend heavily on ecosystem services, either through the collection of natural resources for direct use or for generating income or benefiting indirectly from a range of regulating ecosystem services (Egoh *et al.*, 2012). In Africa, more than 62% of the population in rural areas depends directly on ecosystem services, and a significant amount of urban and periurban people also supplement their incomes, energy, medicinal needs and other essentials from ecosystem-based resources (Schmeller *et al.*, 2017).

In eastern Africa, most (70%) of the population is rural (about 330 million people; United Nations Department of Economic and Social Affairs, 2021), and the region's arid and semi-arid rangelands provide numerous vital goods and services to millions of pastoralists and agropastoralists (Davies and Hatfield, 2007). This contributes significantly to national and regional economies through the supply of livestock and their products (Schmeller *et al.*, 2017). Specifically, the contribution of pastoralism to agricultural GDP in Ethiopia and Kenya was estimated to be 31% and 41%, respectively (Bacsi *et al.*, 2023). However, the arid and semi-arid ecosystems of the region are under broad and dynamic trajectories of ecological change, induced by an increasing human population and livestock density, changes in climate and land use, and by invasive alien plant species such as *P. juliflora* (Lind *et al.*, 2016).

8.3 Perceptions of Rural Communities About the Impacts of *Prosopis juliflora*

Information on local stakeholders' perceptions about the impacts of invasive species can play a fundamental role in improving context-specific management aimed at maximizing benefits and minimizing harm (Urgenson *et al.*, 2013; Shackleton *et al.*, 2019). The way in which local communities perceive the benefits and negative impacts of an invasion can implicitly affect the success of management options (García-Llorente *et al.*, 2008). This is within the theory of planned behaviour which argues that an individual's perceptions influence their intention to act in a certain way (Ajzen, 1991). Explicitly, understanding local individual's perceptions provides information about their potential responses in protecting their resources and the level of commitment and support they will have for public policy management options (Vanslembrouck *et al.*, 2002). For instance, if local stakeholders perceive that an invasion results in a net negative outcome, they would be motivated to plan and implement management options that reduce the adverse effects.

Decisions by rural people regarding the ecosystem services they rely on are influenced by their perception of the current status of these services and whether they will meet their current as well as future needs. Hence, a critical aspect of environmental decision-making in socio-ecological systems is whether people's perceptions are accurate or not.

Based on evidence from arid and semi-arid rangelands of Ethiopia and Kenya, Bekele *et al.* (2022) compared local perceptions and ecological data on the effects of invasions by *P. juliflora* on biodiversity and three ecosystem services (water, soil fertility and soil stability). They found that while *P. juliflora* had significant and changing ecological effects along a gradient of invasion cover, local householders' perceptions were not aligned with this understanding. Comparing the two study areas, positive effects (benefits) were less acknowledged in Ethiopia than in Kenya, which may be at least partly explained by different environmental contexts. At the time *P. juliflora* was introduced to Ethiopia, some areas still had extensive and relatively healthy grasslands (Shiferaw *et al.*, 2019), while in contrast, some lowland areas in Kenya were already heavily degraded at that time (Elfadl and Luukkanen, 2003). Further, there were marked differences in perception between individual households due to different socio-economic contexts (Linders *et al.*, 2020; Bekele *et al.*, 2022; Malila *et al.*, 2023). For instance, households that generate income from charcoal sales had positive perceptions about the trees, while pastoralist households' perceptions were negative (Tabe-Ojong, 2023).

The perceived net impact of *P. juliflora* invasion is a key determinant that shapes the management decisions of land users (Bekele *et al.*, 2018b, 2022; Bagavathiannan *et al.*, 2019). Generally, successful management of invasive alien species requires incentives for stakeholders to engage in collective management actions (Malila *et al.*, 2023). However, preventative management (stopping impacts from occurring) is far less easy to promote than the initiation of control measures once the invasion becomes obviously detrimental. Consequently, invasive species are allowed to establish and spread, and control measures are only initiated after the point at which their spread becomes very difficult to effectively control. In this regard, it is imperative to evaluate the factors that influence community perceptions about the impacts of invasive species, and how those perceptions may be influenced, so that effective early management can be implemented.

The variation of perceptions about the net impact of *P. juliflora* invasion among stakeholders was explained by several factors, including the cover of invasive *P. juliflora*, land tenure rights, participation in invasive species management programmes, and demographic characteristics such as education level and diversified income sources (Adoyo and Choge, 2018; Bekele *et al.*, 2022; Malila *et al.*, 2023). Overall, land users perceived that *P. juliflora* had a net negative impact on ecosystem services and livelihoods (Adoyo and Choge, 2018; Bekele *et al.*, 2022). However, the level of invasion may define their preferred management options. Early detection and rapid removal of *P. juliflora* was preferred at the early stage of invasion, i.e. when *P. juliflora* densities were still low and the ecosystems provided at least some of the original ecosystem services, but utilization of *P. juliflora* became preferred as the cover increased and the invaded ecosystem became increasingly degraded (Adoyo and Choge, 2018).

Importantly, respondents from both highly and sparsely invaded areas were uncomfortable with any further spread of *P. juliflora*. There was a perceived need for more sustainable management measures that would both halve the cover of *P. juliflora* in heavily invaded areas and prevent it from spreading into uninvaded areas (Adoyo and Choge, 2018). Importantly, though, people's perceptions were based on the abundance and impacts at a very localized scale. For example, stakeholders' perceptions of the ecological impacts of *P. juliflora* were significantly influenced by invasion cover 100m around individual households, rather than at the larger (village) level (Bekele *et al.*, 2022). Yet, while perceptions are influenced by impacts at a localized scale, sustainable management can only be achieved through collective action at a landscape level, which provides a further challenge to effective control.

The above findings suggest that local stakeholders manage according to their perceptions, which are in turn dependent on context. According to Wiesmann *et al.* (2011), local actors' management actions are defined by, among other things, their knowledge of the implications of their actions. Such information may be gained through education and training, expert advice or social participation and experience (Bekele *et al.*, 2022). Respondents in Baringo, Kenya, with prior involvement in attempts to control invasions through utilization, were more likely to perceive net impact as positive (Adoyo and Choge, 2018). This reflected the focus of initial extension services and management interventions, which one-sidedly promoted utilization as a form of managing invasions. The focus on potential benefits, while playing down the adverse impacts of *P. juliflora*, has thus made it difficult to implement effective control as utilization creates an overdependence on the species without recognizing its adverse impacts. This is supported by Tabe-Ojong (2023) who asserted that the positive perception of *P. juliflora* results in engagement in charcoal production but limits community willingness to participate in control efforts.

In essence, utilization is a form of livelihood enhancement that benefits a few people, while many more suffer the negative consequences, and people need to be made aware of this distinction. In addition, utilization is not an effective form of control, as was revealed in a study by Mbaabu *et al.* (2019) who showed that the invasion of *P. juliflora* in Baringo continued to increase despite the widespread implementation of a strategy to achieve control through utilization. This suggests that stakeholders' perceptions were influenced by the fact that *P. juliflora* can be a source of income, reducing their willingness to control spread. Unlike Adoyo and Choge (2018), Bekele *et al.* (2022) found that respondents who had engaged in social participation had more access to information on impacts in Afar region, where utilization of *P. juliflora* has been repeatedly banned because it led to the simultaneous felling of native trees.

The above observations imply that providing communities with relevant information can influence the way in which they make decisions. This emphasizes the importance of disseminating credible information at the grassroots level to promote sustainable actions in managing invasions.

A major challenge to the management of *P. juliflora* and other invasive alien tree species is that the perceptions of local stakeholders about their net impacts are likely to change over time. This is particularly so if the trees were initially planted in degraded areas but subsequently invaded healthy and important habitats, such as dry season grazing areas, wetlands, croplands or settlements. In Afar region, most stakeholders considered *P. juliflora* as a beneficial species when it was first introduced; today, only 1% still favour it while 90% strongly disfavour it (Seid *et al.*, 2020).

The initially positive stakeholder perceptions of a deliberately introduced tree species may be a major obstacle when attempting to implement control efforts at the early stage of invasion when these control efforts may still be effective. Adoyo and Choge (2018) found that stakeholders in Baringo prefer complete removal of *P. juliflora* in sparsely invaded areas, unlike the situation in Afar where at least some people preferred retaining the trees (Seid *et al.*, 2020). However, stakeholders in Baringo inhabit sparsely invaded regions next to areas with high *P. juliflora* densities, giving them an example of what might happen if the invasive species is not controlled in the early stages of invasion. In contrast, at the time when *P. juliflora* was introduced into Afar, nobody in the region had these kinds of insights about the potential long-term impacts of this tree species.

8.4 Impacts of Invasion on Rural Households' Livelihoods

8.4.1 Comparative beneficial and negative consequences

While *P. juliflora* was introduced intentionally for its various ecological and livelihood benefits, it has become an aggressive invader with both positive and negative effects on the livelihoods of local communities in eastern Africa (Table 8.1). The positive and negative impacts of *P. juliflora* on rural communities' livelihoods in arid and semi-arid rangelands have been reported in numerous studies based on interviews with stakeholders (Table 8.1; Maundu *et al.*, 2009; Ilukor *et al.*, 2016; Bekele *et al.*, 2018b; Linders *et al.*, 2020).

8.4.2 Benefits

Prosopis juliflora has the potential to provide a wide range of economic and other benefits, which was the original reason for introducing the species to Africa and elsewhere (Table 8.1). In heavily degraded areas, *P. juliflora* offers opportunities for rural households to generate new income sources from charcoal and fuel wood production for sales in arid and semi-arid rural areas (Maundu *et al.*, 2009; Haregeweyn *et al.*, 2013; Ilukor *et al.*, 2016) (Fig. 8.1A). While the large stems of the trees are used for making charcoal, the smaller branches are used for firewood.

However, the income from charcoal production is highly dependent on local communities' sociocultural contexts. For instance, a study by Bekele *et al.* (2018b) showed that while *P. juliflora* invasion had a positive impact on income from charcoal and fuel wood sales in Baringo County, Kenya, income from charcoal was not widely realized in Afar region, Ethiopia. One

Table 8.1. Benefits and negative impacts of invasion by alien *Prosopis juliflora* trees on rural livelihoods in eastern Africa.

Type of impact	Nature of impact or benefit	Description of impact	Sources
Beneficial	Wood	Sale of charcoal and firewood provides additional income to some households; reduces the burdens on women to collect fuel wood for home consumption	Mwangi and Swallow (2008); Maundu *et al.* (2009); Ilukor *et al.* (2016); Linders *et al.* (2020); Tabe-Ojong (2023);
	Nutrition of livestock	The pods provide an additional source of food for cattle, sheep and goats, provided that they are given with other fodder	Berhanu and Tesfaye (2006); Maundu *et al.* (2009)
	Windbreaks	Tall trees can provide a windbreak for planted crops	Berhanu and Tesfaye (2006); Argaw (2015)
	Shade	Tall trees can provide shade for people and livestock, but this advantage is lost when impenetrable thickets are formed	Bekele *et al.* (2018a)
	Fencing	Branches and stems can be used for fencing poles	Berhanu and Tesfaye (2006); Ilukor *et al.* (2016)
	Construction material	Wood used in the construction of houses or barns, or to make beds	Ayanu *et al.* (2015); Ilukor *et al.* (2016)
	Nectar	Flowers provide pollen for bees, providing a source of honey	Tabe-Ojong (2023)
	Soil protection	Dense cover can reduce soil erosion, relative to bare ground	Haregeweyn *et al.* (2013)
	Soil fertility	*P. juliflora* trees are able to fix nitrogen	Berhanu and Tesfaye (2006)
Negative	Forage production	Invasion of rangelands displaces grasses and forbs and substantially reduces carrying capacity for livestock	Linders *et al.* (2020); Chapter 7, this volume
	Access and movement	Invasions by *P. juliflora* trees form dense, impenetrable thickets. Livestock cannot access grazing and water, and movement is impeded for both people and livestock	Rettberg (2010); Ayanu *et al.* (2015); Bekele *et al.* (2018a)
	Physical injury	Thorns of *P. juliflora* trees cause injury to people and livestock	Ayanu *et al.* (2015); Chapter 9, this volume
	Disease	*P. juliflora* trees increase populations of malaria-transmitting mosquitoes	Muller *et al.* (2017)
	Nutrition	Feeding on pods and leaves is dangerous to livestock health	Ayanu *et al.* (2015); Chapter 9, this volume
	Biodiversity	Displaces native trees, grasses and forbs	Bekele *et al.* (2022); Chapter 5, this volume
	Water	*P. juliflora* trees use large amounts of water, reducing the levels of groundwater and causing wells to dry up	Shiferaw *et al.* (2023); Chapter 6, this volume

Continued

Table 8.1. Continued

Type of impact	Nature of impact or benefit	Description of impact	Sources
	Landscape processes	Deviation of rivers, flooding of settlements	Nzumira (2014)
	Conflict	As access is impeded and natural grazing is reduced or eliminated, conflicts between people, livestock and wildlife increase	Mwangi and Swallow (2008)

reason could be the culture of pastoralist communities, i.e. pastoralists have no experience and knowledge about charcoal production and only a few people may benefit from charcoal production as a source of income (Datona, 2014). In pastoralist and agropastoralist communities, where livestock production is the main base of livelihood, any increased income from charcoal and fuel wood is more than offset by reductions in rangeland carrying capacity. In those cases, income from charcoal and fuel wood cannot really be considered as additional income, rather as a way to offset the losses caused by reduced income from livestock production (Linders *et al.*, 2020).

As *P. juliflora* is a drought-resistant and fast-growing tree, it is sometimes the only cheap and easily available energy source for home consumption in water-scarce and degraded drylands of the region. Thus, while Seid *et al.* (2020) reported that 90% of stakeholders in Afar disliked *P. juliflora*, Ilukor *et al.* (2016), found that in the same region 96% of their sampled households agreed that *P. juliflora* was useful to them as fuel wood for their home consumption. In the same vein, Mwangi and Swallow (2008) argued that *P. juliflora* reduces the burdens on women, for instance, by reducing travel time for collecting fuel wood for home consumption. Their findings from Baringo County revealed that women received as much as 17 times more aggregate benefits than men in one area, while in another area no gender-specific benefits were found. In the rural communities of eastern Africa, poles are harvested from *P. juliflora* for use in house construction, repairing and fencing (Mwangi and Swallow, 2008), thus significantly reducing local households' costs for construction materials. For instance, Mwangi and Swallow (2008) estimated that the average annual per household economic benefit of *P. juliflora* from construction and fencing poles was Kenyan shilling 8600 in Baringo. The figure was estimated to be Ethiopian birr 1538 in Ethiopia (Ilukor *et al.*, 2016).

Prosopis juliflora is also acknowledged for indirectly contributing to rural communities (who are highly dependent on natural resources) by delivering ecological benefits which included microclimate regulation, serving as shade for humans and livestock and acting as windbreaks (Fig. 8.1C), improving soil fertility, controlling soil erosion and desalinating soils (Berhanu and Tesfaye, 2006). Maundu *et al.* (2009) concluded that these ecological benefits improved land productivity and ultimately local livelihoods. Considering microclimate regulation benefits of *P. juliflora*, Bekele *et al.* (2018a) estimated that the average per household monetary values in Afar and Baringo County were USD 5.36 and USD 3.89 per year, respectively. It should be noted, though, that the benefits of microclimate regulation, shade and controlling soil erosion only manifest themselves in degraded areas because the native vegetation cannot play this role any more (Elfadl and Luukkanen, 2003).

8.4.3 Negative impacts

Despite their potential and realized benefits, *P. juliflora* trees are disliked for their negative impacts on livelihoods in the region. All the negative impacts of *P. juliflora* trees (Table 8.1) substantially reduce incomes from different agricultural production systems, while simultaneously increasing household expenditure on health issues. These two outcomes ultimately reduce households' consumption expenditures,

Fig. 8.1. (A) Charcoal produced from *Prosopis juliflora* trees is sold on the main road to Djibouti in Afar, Ethiopia, while standing trees provide shade to the charcoal sellers. (B) *P. juliflora* trees planted as a fence. (C) Sedentary pastoralists in Afar use *P. juliflora* as a shade tree for their livestock. Photos courtesy of Ketema Bekele.

a measure that can be used as an indicator of wellbeing (Meyer and Sullivan, 2003).

Prosopis juliflora rapidly invades rangelands and replaces beneficial native plant species, particularly grasses and forbs (Chapter 7, this volume), which in turn reduces rangeland carrying capacity. For instance, between 1986 and 2016, grassland cover in Afar decreased by 25% (599,000 ha loss), which was mainly attributed to invasion by *P. juliflora* (Shiferaw *et al.*, 2019). The negative impacts of *Prosopis* invasion on rangelands are well understood by local communities in the Afar region (Shiferaw *et al.*, 2022).

In arid and semi-arid rangelands in eastern Africa, riparian habitats and wetlands, which are the main areas supporting livestock during the dry season, have now been severely invaded or are under threat of invasion by *P. juliflora*. Pastoralists in Baringo County claim that, as a result of these invasions, they have to move 40–50 km away from where they reside in search of grazing land (Mwangi and Swallow, 2008). In these rural communities that are highly dependent on natural resources, diminishing access to good grazing causes deterioration of livelihoods by reducing incomes from livestock and crops.

Ten years of evidence from Baringo and Afar indicate that invasion by *P. juliflora* has significantly reduced income from livestock (Linders *et al.*, 2020). Specifically, these authors found that *P. juliflora* reduced average income from livestock by 75% (from USD 243 to USD 63 per household per year) in Afar and by 98% (from USD 169 to USD four per household) in Baringo County. Similarly, an empirical study in Dire Dawa, Ethiopia, found that *P. juliflora* had decreased the annual income of agropastoral households from the sale of livestock and other products by 29% (Jema and Abdu, 2013). Bekele *et al.* (2018b) also showed that invasion by *P. juliflora* had a negative impact on household income from livestock production in Afar.

The invasion also directly caused an increase in household health expenditure because households had significant treatment costs for injuries (Bekele *et al.*, 2018b; Fig. 8.2) that resulted from the thorns of *P. juliflora*. In addition, *P. juliflora* has been linked to increases in malaria-carrying mosquitoes (Muller *et al.*, 2017), making further health costs possible.

Prosopis juliflora also invades croplands and increases farm labour costs. The area under crop production in the Gash delta in Sudan decreased by 27% (from 32,125 to 23,538 ha) between 1979 and 2013 due to the invasion by *P. juliflora* trees (Shiferaw *et al.*, 2022). The same study also showed that the invasion disrupted water flow and decreased canal discharge capacity in the Gash delta, further reducing crop yields. Due to their high rate of water consumption (Chapter 6, this volume), *P. juliflora* trees can aggravate water scarcity in the arid and semi-arid rangelands of eastern Africa where most livelihoods are dependent on groundwater supplies.

The rapid encroachment by *P. juliflora* trees into paths, villages, water sources, crop fields and rangelands, and irrigation schemes, coupled with their thorny nature, limits mobility of animals and people (Fig. 8.3). Secured grazing lands and mobility are important cultural identity markers (Rettberg, 2010), and mobility is an important customary strategy for managing natural resources which enables eastern African pastoralists to survive repeated droughts.

Further, in pastoralist and agropastoralist communities in eastern Africa, the reduction in rangelands and crops due to invasion generates social crises by causing intra- and intertribe conflicts (Rogers *et al.*, 2017) which in turn disturbs traditional social orders and local livelihoods. In one case, Mwangi and Swallow (2008) reported that crop farmers from Chemonke village, Kenya, have had to settle elsewhere because they had lost their land to invasions by *P. juliflora*, resulting in conflict with the adjacent communities.

8.5 Net Impacts of *Prosopis juliflora* Invasion on Rural Livelihoods

8.5.1 Changes in impact with increasing cover

The social impacts of invasion by *Prosopis* species change over time in a non-linear fashion as the density of the invasion changes. A simple comparison of the positive and negative impacts of *P. juliflora* trees is maybe a first step in an overall assessment of the perceived or actual impacts that they may have on rural livelihoods. Such a comparison tends to be misleading, though, because the change in the level of impacts is not linear and is also dependent on the context in

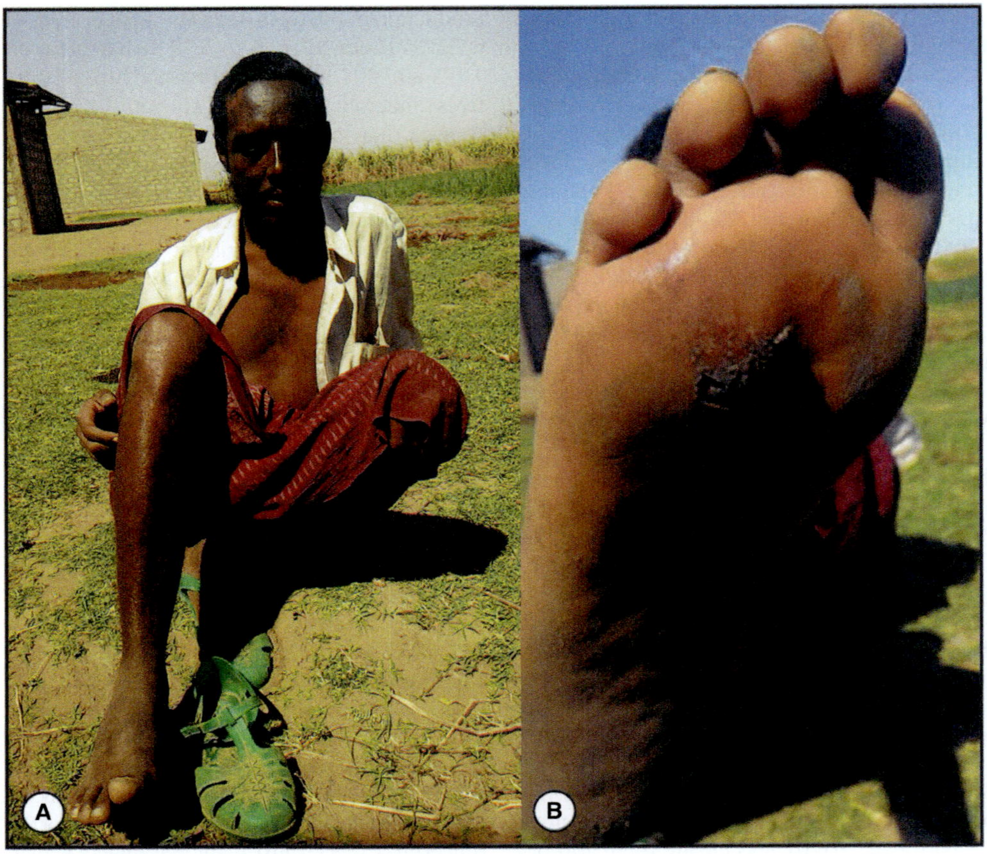

Fig. 8.2. *Prosopis juliflora* can inflict physical injuries on humans and animals, ultimately resulting in increased household health expenditure. Photos courtesy of Ketema Bekele.

which the invasion takes place (Vilà *et al.*, 2011). It is therefore plausible to assume that both the positive and negative impacts of *P. juliflora* invasions on different livelihood indicators will vary with invasion cover. Accordingly, it is necessary to analyse impacts by taking their density-dependent non-linear nature into account.

An additional aspect about the impact of *P. juliflora* tree invasion is to give due attention to the invasion's indirect (but not insubstantial) contribution to the exacerbation of rural poverty via its adverse effects on non-cognitive abilities of the local communities. The following subsections of this chapter discuss a more integrative and holistic approach to assess the non-linear impact of the invasion and its indirect impacts on rural poverty via causing deterioration of non-cognitive abilities of local communities.

8.5.2 Non-linear impacts of invasion on livelihoods

Per-capita household consumption expenditure can be a useful indicator of wellbeing (Meyer and Sullivan, 2003). Bekele *et al.* (2018b) argued that invasion by *P. juliflora* will indirectly affect per-capita consumption expenditure through its causal effects on other rural livelihood indicators. Household consumption expenditure is affected by a range of positive and negative indicators of household wellbeing, many of which are in turn related to the cover of invasive *P. juliflora* trees. Bekele *et al.* (2018b) used household income from livestock, charcoal and fuel wood sales, and household health expenditure across a

Fig. 8.3. In the Amibara district of the Afar region, Ethiopia, *Prosopis juliflora* has formed impenetrable thickets that block human and livestock mobility and hinder the traditional nomadic lifestyle. Photo courtesy of Ketema Bekele.

gradient of *P. juliflora* cover to determine the consumption expenditure of a rural household. The presumption was that increasing levels of invasion by *P. juliflora* trees decrease household income from livestock and increase household health expenditures, but simultaneously increase income from charcoal and fuel wood (Linders *et al.*, 2020).

Studies have shown that household annual consumption expenditure initially increases to an optimum value and then decreases again as the cover of invasive trees increases (Bekele *et al.*, 2018b). The invasion cover that delivered the highest annual household per capita consumption expenditure (USD 198) in Afar region was 22% (Fig. 8.4). At low invasion cover, the adverse effect of *P. juliflora* on income from livestock was less than its positive contribution in charcoal production (Shackleton *et al.*, 2007). Similarly, low cover of *P. juliflora* had no significant impact on annual household health expenditure so that its adverse effect on a given household's money to be spent on consumption was minimal.

However, expenditure declines after a certain level of *P. juliflora* cover is reached, especially in pastoralist-dominated rural households where livestock production is a mainstay of household livelihoods. Due to the aggressively invasive nature of *P. juliflora* and the resultant shrinkage of rangeland, encroachment of water points and injuries due to thorns, the total costs increased after this threshold of invasion cover was reached, thus correspondingly decreasing the net expenditure as the invasion proceeded (Bekele *et al.*, 2018b). Although this may be read as suggesting that containing *P. juliflora* invasions at an optimal level of 22% would benefit rural households' livelihoods in pastoralist-dominated areas, such a proposal would be naïve. The cost of maintaining the area covered by an aggressively invasive alien tree at a desired optimal level would simply be unaffordable.

8.5.3 Impact of *Prosopis juliflora* invasion on non-cognitive abilities

The adverse impacts of invasion by *P. juliflora* are not limited to the well-understood and direct impacts on rural livelihoods such as reduced

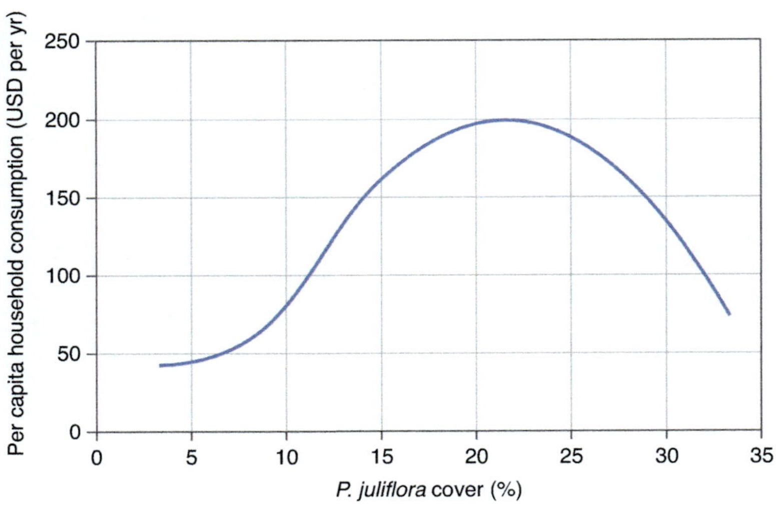

Fig. 8.4. Non-linear impacts of *Prosopis juliflora* on per capita household consumption expenditure in Afar, Ethiopia (for the detailed impact methodology, see Bekele *et al*., 2018b).

household income. The invasion exacerbates rural poverty indirectly through its negative effect on individuals' non-cognitive skills such as locus of control and self-efficacy. Using evidence from Baringo County, Kenya, Tabe-Ojong (2022) found that invasion by *P. juliflora* constitutes an environmental shock and as such causes adverse effects on locus of control and self-efficacy in which these non-cognitive mindsets, in turn, can affect welfare and poverty (Heckman *et al*., 2006).

When faced with the adverse impacts of invasion, individuals are less motivated to think they have any control over how their lives will turn out. Citing Eker (2005), Tabe-Ojong (2022) put the issue succinctly – 'Rich people believe 'I create my life' while poor people believe 'life happens to me''. This means that via non-cognitive skills, invasion by *P. juliflora* influences the economic decision-making behaviour of affected people negatively. The highly affected people are demotivated and lose commitment to sustainably manage invasions – they surrender to the problem by believing that they cannot stop the invasion and so prevent its adverse impacts. The indirect impacts of invasion by *P. juliflora* on rural poverty in eastern Africa should be taken seriously and further studied in order to provide improved support for those affected.

8.5.4 Impact on local household's willingness to pay for management

Ilukor *et al*. (2016) conducted an analysis of *P. juliflora* invasion by considering both benefits and costs of management and concluded that, in the short term, *P. juliflora* trees had a positive net benefit in the Afar region of Ethiopia. However, these authors admitted that their economic valuation was incomplete as it ignored values of some non-marketed ecosystem services such as water use and microclimate regulation. They therefore undervalued local communities' willingness to pay by estimating it to be USD 13 and 38 labour days per household per year.

To comprehensively assess the value of ecosystem services affected by the invasion of *P. juliflora*, or by their management, both marketable and non-marketable ecosystem services need to be considered. Bekele *et al*. (2018a) conducted an economic valuation by applying choice experimental data from Afar region, Ethiopia, and Baringo County, Kenya, in the framework of total economic valuation, and estimated mean local households' willingness to pay for generic management options that would reduce the adverse impacts of the invasion. In applying choice experiments, the study area's ecosystem services affected by invasion were identified (a total of 18 services were identified – ten were negatively affected and eight were affected positively; see Appendix A in Bekele *et al.*, 2018a). These were prioritized and ranked by local communities through a participatory approach. The impacts of invasion by *P. juliflora* were well understood by most rural communities in both study regions, and the majority were willing to pay for management to maximize the benefits and minimize the negative impacts. Almost all (98.4% and 99.2%) of the sample respondents from Afar (253 samples) and Baringo County (250 samples) preferred to participate in managing *P. juliflora* by contributing money and/or labour, believing that reducing the negative impacts of the invasion was desirable (Bekele *et al.*, 2018a).

Most respondents were willing to participate in management options that improve biodiversity, water availability and mobility in their localities. As the communities in both study areas were predominantly pastoralists, these communities would prefer management options that improved aspects that are important to them. These included water availability for home and livestock consumption, plant species diversity that improves the carrying capacity of rangelands, and improved seasonal mobility for themselves and their herds. Despite the adverse impacts of *P. juliflora* on biodiversity and water availability, the local communities in both study areas acknowledged that *P. juliflora* does improve microclimate regulation in these arid and semi-arid areas.

Finally, the mean annual household willingness to pay for *P. juliflora* management to improve the prioritized ecosystem services was estimated to be USD 50 and USD 38 for Afar and Baringo, respectively (Bekele *et al.*, 2018a). By upscaling and aggregating these estimated mean household values for willingness to pay, annual total economic welfare lost due to the invasion was estimated to be USD 6.1 and 4.2 million in Afar and Baringo County, respectively (Bekele *et al.*, 2018a). This implies that, for instance, it would be economically justifiable to annually invest USD 4.2 million to manage *P. juliflora* invasions in Baringo County, assuming that such management would be effective in preventing the estimated losses.

8.6 Upscaling Economic Welfare

The marginal values estimated above and the total area that could become invaded by *P. juliflora* trees in Ethiopia, Kenya and Tanzania (estimated by Eckert *et al.*, 2020) were used to determine the aggregated annual monetary value of impacts of the invasion (Table 8.2). Further, the net present value of these impacts was calculated over 30 years considering different discount rates. Assuming that the invasion is not controlled, the total areas to be invaded were assumed by Eckert *et al.* (2020) to reach the predicted total area suitable for *P. juliflora* invasion within 30 years. The household annual willingness to pay values per year were converted to values per km^2 per year for Afar and Baringo (Table 8.2) and the average of the two was applied in the case of Tanzania. This allowed for the estimation of mean annual economic

Table 8.2. Values used to convert mean willingness to pay estimates per household per year to mean willingness to pay per km^2 per year for the Afar study region in Ethiopia. Note that the same technique was used for Baringo.

Description of unit	Quantity
Total population of Afar region	2 million
Total area of Afar region (km^2)	72,053
Population density (people/km^2)	27.76
Average number of persons per household	6
Average number of persons per household/km^2	4.63
Marginal willingness to pay per household/yr (USD)	50.42
Marginal willingness to pay per km^2/yr (USD)	233.25

welfare for Ethiopia, Kenya and Tanzania (Table 8.3).

Accordingly, upscaling the economic welfare loss due to invasion by *P. juliflora* at a country or regional level may help to inform those who develop cross-border policies to manage invasions (Sagoff, 1998; Freeman III *et al.*, 2014). For instance, if invasion by *P. juliflora* trees is not controlled, the national aggregate economic welfare loss due to *P. juliflora* invasions would increase from USD 6 million to more than USD 75 million per year in Ethiopia (Table 8.3). In other words, if the Ethiopian government is not currently willing to annually spend USD 6 million on the management of invasions by *P. juliflora*, it would ultimately stand to lose more than USD 75 million per year in future.

This analysis suggests that if the current invasions by *P. juliflora* are not contained, it is highly likely that the annual economic costs of the invasions would reach USD 75 million, 175 million and 107 million in Ethiopia, Kenya and Tanzania, respectively, in the future. Further, considering only Ethiopia, Kenya and Tanzania, the annual aggregated economic cost of invasions by *P. juliflora* in these three countries is likely to be more than USD 375 million/yr. Expressed as a net present value over the next 30 years, the cost would amount to between USD 4.4 and 7 billion at discount rates between 7% and 3%, respectively. Further, studies show that future climate favours *P. juliflora* over native plants (e.g. Heshmati *et al.*, 2019; Sintayehu *et al.*, 2020), which indicates an increase in the invasion impacts on rural livelihoods by further escalating the impact and management costs.

8.7 Management Implications

Land users and land managers in eastern Africa face stark choices regarding the management of *P. juliflora* invasions. Clearly, the perceived and realized negative impacts of *Prosopis* trees will come to substantially outweigh any benefits that they deliver to rural communities if they are not brought under control. In many places, this is already the case. Most people realize this and are willing to contribute to such control,

Table 8.3. Estimation of mean annual economic welfare for three eastern African countries, and for the three together. Aggregated mean annual economic welfare was estimated at three discount rates (3%, 5% and 7%) over 30 years.

	Geographical scale			
	Ethiopia	Kenya	Tanzania	All three countries
Marginal willingness to pay per km^2 per year (USD)	233.25	487.77	360.51[b]	–
Predicted total suitable area for *P. juliflora* invasion (in km^2)[a]	322,599	359,280	295,843	977,723
Aggregated mean annual economic welfare (millions of USD)	75.2	175.2	106.7	357.1
Aggregated mean discounted economic welfare at 3% (millions of USD over 30 years)	1474.9	3434.9	2090.5	7000.3
Aggregated mean discounted economic welfare at 5% (millions of USD over 30 years)	1156.7	2694.0	1639.5	5490.2
Aggregated mean discounted economic welfare at 7% (millions of USD over 30 years)	933.8	2174.6	1323.5	4431.9

[a]Prediction based on Eckert *et al.* (2020).

[b]As marginal willingness to pay for Tanzania was not estimated, the average of the two other countries was used.

but the arguments for retaining some of the benefits continue to bedevil the formulation of clear and achievable goals that would come from a focus on sustainable management. Given the species' preadaptation to the eastern African environment and the possible effects of climate change, it is likely that costs will grow. By recognizing the potentially significant future costs and allocating appropriate amounts to control and containment efforts now, it would theoretically be possible to avoid much greater costs in the future. Communicating the urgency of this issue to land users, land owners, governments and development organizations should receive high priority.

Finally, the use of biological control as an essential component of integrated control (Chapter 10, this volume) could reduce the costs of control considerably, and this offers some hope that the predicted impacts can be substantially reduced if the right decisions are made.

Acknowledgements

This work was supported by the Swiss Programme for Research on Global Issues for Development (r4d), funded by the Swiss National Science Foundation and the Swiss Agency for Development and Cooperation, through the project 'Woody invasive alien species in East Africa: Assessing and mitigating their negative impact on ecosystem services and rural livelihood' (Grant Number: 400440_152085).

References

Adoyo, O.B. and Choge, S. (2018) Local actor's perceptions about *P. juliflora's* impacts on ecosystem services and human livelihoods: The case of Baringo county, Kenya. In: *Proceedings of 5th KEFRI Scientific Conference 17-19 April 2018*, KEFRI Headquarters Muguga, pp. 795–811.

Ajzen, I. (1991) The theory of planned behavior. *Organizational Behavior and Human Decision Processes* 50, 179–211.

Argaw, T. (2015) Impacts of utilizing invasive *Prosopis juliflora* (SWARTZ) DC. on rural household economy at Gewane District, Afar regional state, North-Eastern Ethiopia. *Journal of Economics and Sustainable Development* 6, 81–97.

Ayanu, Y., Jentsch, A., Müller-Mahn, D., Rettberg, S., Romankiewicz, C. *et al.* (2015) Ecosystem engineer unleashed: *Prosopis juliflora* threatening ecosystem services?. *Regional Environmental Change* 15, 155–167.

Bacsi, Z., Gebbisa, M.B. and Dávid, L.D. (2023) Pastoralism and tourism in Eastern Africa – quantitative analysis from 2004 to 2018. *Sustainability* 15, 9723.

Bagavathiannan, M.V., Graham, S., Ma, Z., Barney, J.N., Coutts, S.R. *et al.* (2019) Considering weed management as a social dilemma bridges individual and collective interests. *Nature Plants* 5, 343–351.

Bekele, K., Haji, J., Legesse, B. and Schaffner, U. (2018a) Economic impacts of *Prosopis* spp. invasions on dryland ecosystem services in Ethiopia and Kenya: Evidence from choice experimental data. *Journal of Arid Environments* 158, 9–18.

Bekele, K., Haji, J., Legesse, B., Shiferaw, H. and Schaffner, U. (2018b) Impacts of woody invasive alien plant species on rural livelihood: Generalized propensity score evidence from *Prosopis* spp. invasion in afar region in Ethiopia. *Pastoralism* 8, 1–17.

Bekele, K., Linders, T.E.W., Eschen, R., Shiferaw, H., Haji, J. *et al.* (2022) How well do local stakeholders' perceptions of environmental impacts of an invasive alien plant species relate to ecological data?. *Ecological Indicators* 137, 108748.

Berhanu, A. and Tesfaye, G. (2006) The *Prosopis* dilemma: Impacts on dryland biodiversity and some controlling methods. *Journal of the Drylands* 1, 158–164.

Datona, M. (2014) Socio-economic impacts of *Prosopis juliflora*-related charcoal trade in Gewane Woreda, Afar region. Managing *Prosopis juliflora* for better (agro-) pastoral livelihoods in the Horn of Africa.

Available at: https://wocatpedia.net/images/4/46/Proceedings_regional_conference_on_prosopis_juliflora%2C_Addis_Abeba.pdf (accessed 30 July 2024).

Davies, J. and Hatfield, R. (2007) The economics of mobile pastoralism: A global summary. *Nomadic Peoples* 11, 91–116.

Eckert, S., Hamad, A., Kilawe, C.J., Linders, T.E., Ng, W.T. *et al.* (2020) Niche change analysis as a tool to inform management of two invasive species in Eastern Africa. *Ecosphere* 11, e02987.

Egoh, B.N., O'Farrell, P.J., Charef, A., Gurney, L.J., Koellner, T. *et al.* (2012) An African account of ecosystem service provision: Use, threats and policy options for sustainable livelihoods. *Ecosystem Services* 2, 71–81.

Eker, T.H. (2005) *Secrets of the Millionaire Mind: Mastering the Inner Game of Wealth*. Harper Business, London, UK.

Elfadl, M.A. and Luukkanen, O. (2003) Effect of pruning on *Prosopis juliflora*: Considerations for tropical dryland agroforestry. *Journal of Arid Environments* 53, 441–455.

Freeman III, A.M., Herriges, J.A. and Kling, C.L. (2014) *The Measurement of Environmental and Resource Values: Theory and Methods*. Resources for the Future Press, Routledge, New York.

Galaty, J. (2021) Pastoralism in Eastern Africa. In: *Oxford Research Encyclopedia of African History*. Available at: https://oxfordre.com/africanhistory (accessed July 2024).

García-Llorente, M., Martín-López, B. and González, J.A. (2008) Social perceptions of the impacts and benefits of invasive alien species: Implications for management. *Biological Conservation* 141, 2969–2983.

Haile, G.G., Tang, Q., Sun, S., Huang, Z., Zhang, X. *et al.* (2019) Droughts in East Africa: Causes, impacts and resilience. *Earth-Science Reviews* 193, 146–161.

Haregeweyn, N., Tsunekawa, A. and Tsubo, M. (2013) Analysis of the invasion rate, impacts and control measures of *Prosopis juliflora*: A case study of Amibara district, Eastern Ethiopia. *Environmental Monitoring and Assessment* 185, 7527–7542.

Heckman, J.J., Stixrud, J. and Urzua, S. (2006) The effects of cognitive and noncognitive abilities on labor market outcomes and social behavior. *Journal of Labor Economics* 24(3), 411–482.

Heshmati, I., Khorasani, N., Shams-Esfandabad, B. and Riazi, B. (2019) Forthcoming risk of *Prosopis juliflora* global invasion triggered by climate change: Implications for environmental monitoring and risk assessment. *Environmental Monitoring and Assessment* 191, 1–12.

Ilukor, J., Rettberg, S., Treydte, A. and Birner, R. (2016) To eradicate or not to eradicate? Recommendations on *Prosopis juliflora* management in Afar, Ethiopia, from an interdisciplinary perspective. *Pastoralism* 6, 14.

Jema, H. and Abdu, M. (2013) Economic impact of *Prosopis juliflora* on agro-pastoral households of Dire Dawa administration, Ethiopia. *African Journal of Agricultural Research* 8, 768–779.

Lind, J., Sabates-Wheeler, R. and Kohnstamm, S. (2016) Changes in the drylands of Eastern Africa: A review of evidence and data and their implications for efforts to strengthening resilience. *Institute of Development Studies* 12413, 12082. Available at: https://opendocs.ids.ac.uk/opendocs/handle/20.500 (accessed 30 July 2024).

Linders, T.E., Bekele, K., Schaffner, U., Allan, E., Alamirew, T. *et al.* (2020) The impact of invasive species on social-ecological systems: Relating supply and use of selected provisioning ecosystem services. *Ecosystem Services* 41, 101055.

Malila, B.P., Kaaya, O.E. and Lusambo, L.P. (2023) Factors influencing smallholder farmers' willingness to adopt sustainable land management practices to control invasive plants in northern Tanzania. *Environmental and Sustainability Indicators* 19, 100284.

Maundu, P., Kibet, S. and Morimoto, Y. (2009) Impact of *Prosopis juliflora* on Kenya's semi-arid and arid ecosystems and local livelihoods. *Biodiversity* 10, 33–50.

Mbaabu, P.R., Ng, W.T., Schaffner, U., Gichaba, M., Olago, D. *et al.* (2019) Spatial evolution of *Prosopis* invasion and its effects on land use/land cover and livelihoods in Baringo, Kenya. *Remote Sensing* 11, 1217.

Mehari, Z.H. (2015) The invasion of *Prosopis juliflora* and Afar pastoral livelihoods in the Middle Awash area of Ethiopia. *Ecological Processes* 4, 1–9.

Meyer, B.D. and Sullivan, J.X. (2003) Measuring the well-being of the poor using income and consumption. *Journal of Human Resources* 38, 1180–1220.

Muller, G.C., Junnila, A., Traore, M.M., Traore, S.F., Doumbia, S. *et al.* (2017) The invasive shrub *Prosopis juliflora* enhances the malaria parasite transmission capacity of anopheles mosquitoes: A habitat manipulation experiment. *Malaria Journal* 16, 237.

Mwangi, E. and Swallow, B. (2008) Invasion and rural livelihoods in the lake Baringo area of Kenya. *Conservation and Society* 6, 130–140.

Nzumira, H. (2014) Mesquite trees infestation of the Gash spate irrigation system in Kassala state, Sudan: Impacts and remedial measures. MSc thesis, Institute for Water Education, Delft, The Netherlands.

Rettberg, S. (2010) Contested narratives of pastoral vulnerability and risk in Ethiopia's Afar region. *Pastoralism: Research, Policy and Practice* 1, 248–273.

Rogers, P., Nunan, F. and Fentie, A.A. (2017) Reimagining invasions: The social and cultural impacts of *Prosopis* on pastoralists in Southern Afar, Ethiopia. *Pastoralism* 7, 1–13.

Sagoff, M. (1998) Aggregation and deliberation in valuing environmental public goods: A look beyond contingent pricing. *Ecological Economics* 24, 213–230.

Sangha, K.K., Gordon, I.J. and Costanza, R. (2022) Ecosystem services and human wellbeing-based approaches can help transform our economies. *Frontiers in Ecology and Evolution* 10, 841215.

Schmeller, D.S., Niemelä, J. and Bridgewater, P. (2017) The intergovernmental science-policy platform on biodiversity and ecosystem services (IPBES): Getting involved. *Biodiversity and Conservation* 26, 2271–2275.

Seid, O., Haji, J. and Legesse, B. (2020) Rural households' perception on the effects of *Prosopis juliflora* invasion: The case of Amibara district of Afar National Regional State, Ethiopia. *Pastoralism* 10, 1–9.

Shackleton, C.M., McGarry, D., Fourie, S., Gambiza, J., Shackleton, S.E. *et al.* (2007) Assessing the effects of invasive alien species on rural livelihoods: Case examples and a framework from South Africa. *Human Ecology* 35, 113–127.

Shackleton, R.T., Le Maitre, D.C., van Wilgen, B.W. and Richardson, D.M. (2016) Identifying barriers to effective management of widespread invasive alien trees: *Prosopis* species (mesquite) in South Africa as a case study. *Global Environmental Change* 38, 183–194.

Shackleton, R.T., Richardson, D.M., Shackleton, C.M., Bennett, B., Crowley, S.L, *et al.* (2019) Explaining people's perceptions of invasive alien species: a conceptual framework. *Journal of Environmental Management* 229, 10–26.

Shiferaw, H., Bewket, W., Alamirew, T., Zeleke G., Teketay, D. *et al.* (2019) Implications of land use/land cover dynamics and *Prosopis* invasion on ecosystem service values in Afar region, Ethiopia. *Science of the Total Environment* 675, 354–366.

Shiferaw, H., Alamirew, T., Dzikiti, S., Bewket, W., Zeleke, G. *et al.* (2023) Water abstraction of invasive *Prosopis juliflora* and native Senegalia Senegal trees: A comparative study in the Great Rift Valley Area, Ethiopia. *Science of the Total Environment* 862, 160833.

Shiferaw, W., Demissew, S., Bekele, T. and Aynekulu, E. (2022) Community perceptions towards invasion of *Prosopis juliflora*, utilization, and its control options in Afar region Northeast Ethiopia. *PLoS One* 17, e0261838.

Sintayehu, D.W., Egeru, A., Ng, W.T. and Cherenet, E. (2020) Regional dynamics in distribution of *Prosopis juliflora* under predicted climate change in Africa. *Tropical Ecology* 61, 437–445.

Tabe-Ojong, M.P.J. (2022) Ecological shocks and non-cognitive skills: Evidence from Kenya. *Ecological Economics* 194, 107330.

Tabe-Ojong, M.P.J. (2023) Action against invasive species: Charcoal production, beekeeping, and *Prosopis* eradication in Kenya. *Ecological Economics* 203, 107614.

United Nations Department of Economic and Social Affairs (2021) *World Social Report 2021: Reconsidering Rural Development*. United Nations Department of Economic and Social Affairs, New York. Available at: https://desapublications.un.org/publications/world-social-report-2021-reconsidering-rural-development (accessed 23 October 2024).

Urgenson, L.S., Prozesky, H.E. and Esler, K.J. (2013) Stakeholder perceptions of an ecosystem services approach to clearing invasive alien plants on private land. *Ecology and Society* 18, 26.

Vanslembrouck, I., Huylenbroeck, G. and Verbeke, W. (2002) Determinants of the willingness of Belgian farmers to participate in agri-environmental measures. *Journal of Agricultural Economics* 53, 489–511.

van Wilgen, B.W. and Richardson, D.M. (2014) Challenges and trade-offs in the management of invasive alien trees. *Biological Invasions* 16, 721–734.

Vilà, M., Espinar, J.L., Hejda, M., Hulme, P.E., Jarošík, V. *et al.* (2011) Ecological impacts of invasive alien plants: A meta-analysis of their effects on species, communities and ecosystems. *Ecology Letters* 14, 702–708.

Wiesmann, U., Ott, C., Ifejika Speranza, S., Kiteme, B.P., Müller-Böker, U. *et al.* (2011) A human actor model as a conceptual orientation in interdisciplinary research for sustainable development. In:

Perspectives of the Swiss National Centre of Competence in Research (NCCR) North-South, Vol. 6. pp. 231–256.

Wise, R.M., van Wilgen, B.W. and Le Maitre, D.C. (2012) Costs, benefits and management options for an invasive alien tree species: The case of mesquite in the Northern Cape, South Africa. *Journal of Arid Environments* 84, 80–90.

Zeray, N., Legesse, B., Mohamed, J.H. and Aredo, M.K. (2017) Impacts of *Prosopis juliflora* invasion on livelihoods of pastoral and agro-pastoral households of Dire Dawa administration, Ethiopia. *Pastoralism: Research, Policy and Practice* 7, 1–14.

9 Using the One Health Concept to Assess the Multidimensional Impact of Invasion by *Prosopis* Trees

Urs Schaffner[1*], David M. Richardson[2,3] and Brian W. van Wilgen[3]
[1]CABI, Delémont, Switzerland; [2]Institute of Botany, Czech Academy of Sciences, Průhonice, Czech Republic; [3]Centre for Invasion Biology, Department of Botany and Zoology, Stellenbosch University, South Africa

Abstract

The One Health concept focuses on interactions between people, animals, plants and the environment, recognizing the interdependence of human, animal, plant and ecosystem health. This chapter applies the One Health approach to assess the effects of invasion by alien trees in the genus *Prosopis* on the four dimensions of health, drawing on research findings from eastern Africa and including, where appropriate, results from other parts of Africa. A review of the literature of *Prosopis* invasions shows that these invasions, while bestowing some benefits, have negative consequences on all dimensions of health, and that they are increasing in magnitude as the invasions grow. Negative consequences include injury and dietary or toxic impacts on people and animals, increase of vectors of zoonotic diseases, displacement of native plants through competition for limiting resources and allelopathy, and detrimental effects on the provision of ecosystem services by using excessive amounts of water and eliminating pasture grasses. This chapter proposes two approaches to assess the multidimensional impacts of *Prosopis juliflora* in a One Health context. First, an assessment of the impact of this invasive tree species on ecosystem service multifunctionality is presented, considering indicators of ecosystem services relating to all four dimensions of One Health. Second, the positive and negative aspects of *P. julilfora* are contrasted across the four dimensions to calculate a One Health score for different landscapes with varying degrees of invasion. Both approaches indicate that pristine or restored environments deliver positive benefits, but that heavily invaded sites are characterized by substantial changes in human, animal, plant and environmental health, a situation that is predominantly negative and unsustainable. Holistic assessments such as the approaches presented here will hopefully raise awareness of the gravity of the problem, thereby helping to secure the funding needed to contain and hopefully reverse the problems associated with these invasions.

9.1 Introduction

The impacts of invasive alien species are often considered in a sectoral context, as has been done for trees in the genus *Prosopis* by separately assessing their impacts on biodiversity (Chapter 5, this volume), livestock production (Chapter 7, this volume), non-timber products (Shackleton *et al.*, 2015a) and water resources (Chapter 6, this volume). To collate views on the impacts of invasive alien species on environmental or socio-economic aspects, the EICAT (Blackburn *et al.*, 2014) and SEICAT (Bacher *et al.*, 2018) schemes were developed. However, to be able

*Corresponding author: u.schaffner@cabi.org

© CABI International 2025. *The Ecology and Management of Invasive Prosopis Trees in Eastern Africa* (eds U. Schaffner *et al.*)
DOI: 10.1079/9781800623644.0009

to understand the overall effects of invasive alien species on socio-ecological systems more comprehensively, a more holistic approach may be needed.

One way to do this is to use the 'One Health' concept. While originally developed with a focus on public human health and the health of animals influencing the health of humans, the concept has more recently been widened, recognizing that 'the health of humans, domestic and wild animals, plants, and the wider environment (including ecosystems) are closely linked and interdependent' (One Health Commission, 2021). Considering biological invasions within the One Health (OH) framework is appropriate because the study and management of biological invasions and issues like emerging infectious diseases address similar biological phenomena in a rapidly changing world (Ogden *et al.*, 2019). Also, because invasive alien species often have multiple impacts across sectors (e.g. environment, agriculture, human and animal health, water), it has been argued that a biosecurity policy is needed that builds on the connections between human, animal, plant and environmental health to prevent and mitigate their impacts more effectively (Hulme, 2020, 2021).

This chapter considers the evidence for the impacts of the invasive tree *P. juliflora* on each of the four dimensions of OH, i.e. human, animal, plant, and environmental health, drawing on research findings from eastern Africa and including where appropriate the results of research carried out in other parts of Africa. We then use the example of the invasion by *P. juliflora* in eastern Africa to propose approaches guided by the OH concept for combining this evidence to arrive at a single score that could be applied to different management scenarios or landscapes along a continuum from uninvaded to fully invaded.

9.2 Underpinning Dimensions of the One Health Concept

9.2.1 Impacts on human health

Invasive *Prosopis* trees can affect human health both directly and indirectly. Residents in invaded areas in eastern Africa have repeatedly reported physical injuries caused by *Prosopis* thorns (e.g. Mwangi and Swallow, 2008; Maundu *et al.*, 2009; see also Chapter 8, this volume). There are records of hospitalization because of such injuries, as well as cases of amputations due to infections as a result of *Prosopis* thorns (Laxén, 2007), but quantitative data are scant.

Allergies are one of the most underestimated health costs in different parts of the world (Papadopoulos *et al.*, 2012). The pollen of *Prosopis* trees is allergenic, causing rhinitis, asthma and other forms of pollinosis among the sensitized population (Novey *et al.*, 1977; Hussain *et al.*, 2020). The pollen of *Prosopis* trees has at least 13 different allergenic fractions (Killian and McMichael, 2004). Clinical studies from the native and the invaded range revealed that between 18% and 50% of the population is sensitized against *Prosopis* pollen (Batista-Roche and Huerta-Ocampo, 2022).

In the native range, and to some extent also in the invaded range, local people use extracts from *Prosopis* trees for wound healing and to treat other health issues (Prabha *et al.*, 2014). Petri-dish experiments indicated some antibiotic properties of plant extracts (Henciya *et al.*, 2017). However, other metabolites of *Prosopis* have toxic effects. For example, the piperidine alkaloids found in *Prosopis* species are neurotoxins causing neuromuscular alterations and gliosis when ingested by humans or animals (Silva *et al.*, 2007).

Prosopis trees can also cause several important indirect effects on human health. Both field and laboratory studies have revealed that the presence of these trees can increase population densities of disease-vectoring insect species. In an experimental field study in Mali, removal of flowering branches from *Prosopis* trees reduced malaria-transmitting mosquito densities by 69.4% (Muller *et al.*, 2017). In Kenya, 40–60% of local residents perceive that the increase in malaria incidences is associated with the expansion of *Prosopis* thickets (Mwangi and Swallow, 2008). Also, Sissoko *et al.* (2019) showed that the mosquito species *Aedes aegypti*, which is a vector of yellow fever, dengue and zika and depends on natural sugar resources, is attracted by *Prosopis* flowers.

Outbreaks of leishmaniasis in eastern Africa have also been linked to *Prosopis* invasions

(Ngumbi *et al.*, 2010). For example, in the Merti Region in Kenya, leishmaniasis increased significantly at the time when *Prosopis* trees invaded the surrounding wetlands. The potential link between the presence of these trees and increased incidence of leishmaniasis is further supported by experimental field and laboratory studies. These revealed that *Prosopis* flowers are highly attractive to the sand fly species *Phlebototmus papatasi*, and plant DNA and sugar analyses in adult sandfly bodies showed that *Prosopis* trees are one of the four host plants of sandflies sampled in Kenya (Hassaballa *et al.*, 2021). Invasion by *Prosopis* trees in regularly flooded areas with alluvial soil may also change the breeding substrate of sand flies in a way that promotes population build-up, but this needs to be explored further.

Invasion by *Prosopis* trees in eastern Africa also has significant impacts on other aspects of human wellbeing, some of which are directly linked to human health. Dense thickets of *Prosopis* can increase the risk of ethnic conflicts. For example, ethnic conflicts in Afar region, Ethiopia, have increased after invasion by *Prosopis* trees, partly driven by reductions in dry season grazing or drought reserve areas that accompany invasion (Rettberg, 2010; Ilukor *et al.*, 2016). In Ng'ambo, Kenya, 50% of the people experienced some form of human conflict after invasion by *Prosopis* trees (Mwangi and Swallow, 2008).

Interestingly, invasion by *Prosopis* trees can also lead to less obvious changes in human wellbeing. Tabe-Ojong, Jr (2022) showed that people faced with invasion by *Prosopis* trees are less likely to believe that they have control over outcomes in their lives. Mindset-based approaches, such as the extent to which individuals believe that they control the events in their lives, as opposed to such events being determined by external factors, are increasingly recognized to be related to poverty and welfare outcomes (Tabe-Ojong, Jr, 2022).

Another indirect effect of *Prosopis* invasions on human health is that they can lead to a significant loss of native plant species with medicinal properties that have a long history of use in traditional medicine in eastern Africa (Linders *et al.*, 2019).

9.2.2 Impacts on animal health

Like humans, animals can suffer direct as well as indirect impacts on their health. Livestock and wild animals sustain flesh wounds from *Prosopis* thorns, which may render them lame (Maundu *et al.*, 2009). The invasive trees are also blamed for causing tooth decay in livestock as a result of feeding on *Prosopis* seed pods. By way of illustration, a toothless goat was taken to court during the course of a lawsuit, in which the Il Chamus community sued the Kenyan government for the introduction of *Prosopis* in Baringo County (Wabusya, 2013; see also Box 2.1, Chapter 2, this volume). Laboratory studies revealed that *Prosopis* browse caused the teeth and bones taken from slaughtered goats to crack or break into small pieces when soaked in aqueous suspensions of the meal from *Prosopis* pods and leaves, and that the effects were worse for the pods than the leaves (Maundu *et al.*, 2009).

The potential for *Prosopis* trees to provide fodder for livestock was one of the reasons for their introduction to eastern Africa. Their pods are high in protein and sugars and can provide fodder for livestock. A literature review revealed that pods could complement costlier feed ingredients such as grain by up to 10–50% of the diet, depending on the animal species (Sawal *et al.*, 2004). However, consumption of the pods has been reported to result in facial contortions, impacts to the rumen, and constipation among livestock (Mwangi and Swallow, 2008). A survey of farmers revealed that 79% of the respondents indicated that *Prosopis* trees had negative effects on mature goats, resulting in a loss of teeth and body condition, with negative impacts on the returns to the farmers (Mutavi, 2020). Also, livestock are prone to a disease called denervation atrophy when the proportion of *Prosopis* pods exceeds 20% in their diet (Tabosa *et al.*, 2006).

The most significant impact of invasion by *Prosopis* trees on livestock health is arguably their negative impact on the availability of herbaceous fodder. Linders *et al.* (2021) estimated that invasion by *Prosopis* trees in eastern Africa reduces the supply of herbaceous biomass to levels far below the minimum needed to sustain livestock, rendering the system unstable (Chapter 7, this volume).

Dense thickets of *Prosopis* trees also appear to increase wildlife–livestock conflicts. Residents in Afar, Ethiopia, reported increasing encounters with carnivores, such as hyenas, lions, leopards and jackals, presenting an increased threat to livestock and in some cases also children (Tessema, 2012; Mehari, 2015). This threat appears to be exacerbated when the increasingly difficult search for suitable pastures causes the household head to spend longer periods away from the family (Rogers *et al.*, 2017).

9.2.3 Impacts on plant health

Invasion by *Prosopis* trees interferes directly and indirectly with the growth of native plant species and crop plants. Numerous studies provide evidence that these invasions can displace native plant species (Chapter 5, this volume).

The cover of native woody species, perennial grasses and herbaceous plants declines with increasing *Prosopis* cover (Shackleton *et al.*, 2015b). Above 60% *Prosopis* cover, species richness and above-ground biomass of the herbaceous layer drop to almost zero (Linders *et al.*, 2019). The mechanisms behind these declines remain to be assessed under field conditions, but they are probably due to competition for light, water and nutrients, and possibly to allelopathic effects (van Wilgen *et al.*, 2022). Laboratory studies suggest that *Prosopis* trees have allelopathic effects on native plant species. For example, in the United Arab Emirates, aqueous extracts of fresh and old leaves of the invasive species *P. juliflora* revealed an inhibitory effect on the germination of native plant species, while the leaves and litter of the native species *Prosopis cineraria* did not (El-Keblawy and Abdelfatah, 2014).

However, field observations in different parts of eastern Africa indicate that annual plant species can be relatively common under the canopy of *Prosopis* trees, while perennial grasses and forbs are rarely found (Tessema *et al.*, 2017). This suggests that other factors, including competition for water, may also affect the relative proportions of *Prosopis* and native plant species. This is supported by a study by Schachtschneider and February (2013) who found that mortality of the native tree *Vachellia erioloba* was 50% greater in invaded plots than in plots where *Prosopis* trees had been removed along ephemeral river beds. They used stable isotopes of hydrogen and oxygen in water to argue that the increase in mortality and dieback was related to direct competition for groundwater (Chapter 6, this volume).

The effect of *P. juliflora* on soil organic carbon, soil erosion and other indicators of soil health is context-dependent. For example, soil organic carbon can be increased and soil erosion reduced when *P. juliflora* invades degraded grasslands; however, if pristine grassland is invaded, the supply of these services is not affected or actually reduced (Mbaabu *et al.*, 2020). As *P. juliflora* can grow on saline soils, it can be used to alleviate soil salinity and restore soil fertility (Bhojvaid and Timmer, 1998).

9.2.4 Impacts on environmental health

Prosopis trees have numerous negative impacts on indicators of environmental health. First, they cause drastic declines in plant species richness (Shackleton *et al.*, 2015b; Linders *et al.*, 2019; Mbaabu *et al.*, 2020) and also species richness of invertebrates (Steenkamp and Chown, 1996; Linders *et al.*, 2019), birds (Dean *et al.*, 2002) and large grazing mammals (Kebede and Coppock, 2015). *Prosopis* invasion in the arid savanna of South Africa resulted in a marked alteration of seed dispersal of native plants by birds, thereby initiating major changes in vegetation structure (Milton *et al.*, 2007).

One of the key impacts of *Prosopis* trees on the environment is through their effects on the hydrology of invaded areas (Chapter 6, this volume). Single trees of the evergreen *P. juliflora* can transpire up to 36 litres of water per day (Shiferaw *et al.*, 2021). In Afar, where in 2017 1.2 million ha of land was invaded by *P. julifora* at densities of up to 2000 trees per ha, Shiferaw *et al.* (2021) estimated that about half of the annual rainfall in the invaded range was abstracted by *Prosopis* trees. Water abstraction by the evergreen *P. juliflora* occurs throughout the year because deep-rooted trees can tap into groundwater at considerable depths. Dzikiti *et al.* (2013) determined that in South Africa, groundwater savings of up to 70 m^3 per ha per month could be achieved by clearing land invaded by *Prosopis* trees (see also Shiferaw *et al.*, 2023).

9.3 Using the One Health Concept to Assess the Overall Effects of *Prosopis* Invasions

9.3.1 The need for a multidimensional assessment

Many alien tree species that were introduced deliberately but have become invasive are capable of simultaneously generating benefits and causing harm (Richardson *et al.*, 2014), and *Prosopis* species generate substantial conflicts of interest in many areas. Both the benefits and the harmful impacts need to be considered when trying to find the best way for managing these conflict species (Zengeya *et al.*, 2017). Achieving agreement on an appropriate way forward can be very difficult because different interest groups often hold strongly opposing views (Woodford *et al.* 2016). Thus, any decision on a course of action will need to be based on a defensible and comprehensive assessment of all costs and benefits.

One relatively simple way to achieve this has been to draw up lists of positive and negative effects, or to count the number of publications reporting positive vs negative effects (Roy *et al.*, 2023). However, this approach is of questionable value because the relative contribution of each impact or benefit is not known, and simply listing publications can also lead to double counting.

A more realistic approach has been to use economic assessments to convert the various costs and benefits to a common currency (money) and to examine the outcomes in terms of financial benefit–cost ratios under different scenarios (e.g. De Wit *et al.*, 2001; Wise *et al.*, 2012). Using this approach, Wise *et al.* (2012) estimated that benefits generated by *Prosopis* trees in South Africa were greater than the costs (i.e. benefit–cost ratios were >1), but that this would become negative (i.e. benefit–cost ratios would be <1) within 5 years as the invasions continued to spread and the costs of negative impacts grew to a point where they exceeded the value of any benefits.

Finally, questionnaire surveys of people living in areas where their livelihoods are directly affected by *Prosopis* invasions have been used to assess opinions on the relative value of the invasive trees (e.g. Bekele *et al.*, 2022; Shiferaw *et al.*, 2022). These surveys provide a useful assessment of overall value because people have direct experience and understanding of the relative costs and benefits as they affect their wellbeing.

The OH approach has the potential to improve on these methods because it explicitly considers aspects that are often omitted from the methods outlined in sections 9.2.1–9.2.4 above. Considering the overall outcomes in terms of the health of people, animals, plants and the environment can lead to a more comprehensive assessment of the situation, provided that defensible scores can be assigned to the different elements. Below we outline possible approaches for using the OH concept to generate a multidimensional impact assessment of invasion by *Prosopis* trees.

9.3.2 Relating ecosystem service multifunctionality to One Health

The OH concept is built on the reasonable assumption that the health of humans, animals, plants and the wider environment is closely linked and interdependent. The OH concept can therefore guide assessments of the overall impact of invasions by jointly considering the status of the four health dimensions. Bardgett *et al.* (2021) (Box 9.1) made an attempt to assess the impact of *Prosopis* invasions by considering indicators relating to human, animal, plant and environmental health. Fuelwood, charcoal and microhabitat were included in the assessment as potential benefits related to the presence of *Prosopis* trees.

Ecosystem service multifunctionality can then be defined as the simultaneous supply of all ecosystem services of interest, relative to the demand by stakeholders or stakeholder groups. Irrespective of the stakeholder group assessed, the invasion of semi-arid and arid grasslands in eastern Africa by *Prosopis* trees led to losses of ecosystem service multifunctionality

[multifunctionality (MF) index; Box 9.1], with significant negative consequences for at least one indicator of all four health dimensions (e.g. disease regulation, fodder, groundwater, natural products and ecotourism) (Dzikiti *et al.*, 2013; Muller *et al.*, 2017; Linders *et al.*, 2019, 2020).

Until recently, the demand for ecosystem services by different stakeholder groups has been assumed or has been based on qualitative assessments (Darvill and Lindo, 2016). A next step in this process would be to actually ask the different stakeholder groups how they prioritize different ecosystem services (Bekele *et al.*, 2018; Manning *et al.*, 2018). Linders *et al.* (2021) combined ecological data on the impacts of invading *Prosopis* trees on 14 indicators of ecosystem service supply with data from a questionnaire-based survey of ecosystem service demands from eight different stakeholder groups. The two data sets were collected in the same invaded and degraded area in the lowlands of Baringo County, Kenya. As in Bardgett *et al.* (2021), the set of indicators used in their study included indicators relating to human (medicinal plants and protection against insect-vectored diseases), animal (fodder production of both herbaceous biomass and *Prosopis* pods), plant (soil organic carbon, woody biomass and crop production) and environmental (plant species richness and tourism) health. Linders *et al.* (2021) found that ecosystem service multifunctionality (defined here as the measure of overall ecosystem service supply relative to stakeholder demand) declined for most stakeholder groups as the level of invasion increased. However, charcoal producers and, marginally, developmental NGOs benefited from an increase in *Prosopis* cover.

The results from these surveys indicated that charcoal producers in particular emphasized their requirement for wood, the availability of which increases with *Prosopis* cover. Charcoal producers, who tend to own little land and few livestock, benefited from *Prosopis* invasion as producing charcoal needs little investment. This indicates that they have made opportunistic use of an emerging resource. Developmental NGOs gave relatively high priority to shade and honey production, whereas indicators for biodiversity,

Box 9.1. One Health and ecosystem service multifunctionality.

Based on the understanding that human, animal, plant and environmental health are all interlinked, the One Health concept aims to promote a multifunctional 'landscape' which supports all dimensions of health. To measure the codelivery of multiple ecosystem services, Manning *et al.* (2018) proposed an approach that integrates indicators of the supply of multiple ecosystem services into a single measure that covers all health dimensions.

Bardgett *et al.* (2021) assessed the effects of *P. juliflora* invasion on the supply of multiple ecosystem services using indicators related to human health (disease regulation, subsistence crop production as a proxy for food availability, fuelwood and charcoal, natural products as a proxy for medicinal plants, and microclimate), animal health (fodder and natural products), plant health (soil organic carbon as a proxy for soil health, and groundwater) and environment (ecotourism as a proxy for biodiversity). They found that the supply of most ecosystem services demanded by stakeholders is greatly reduced by this invasion, but that fuelwood and charcoal production are increased (see D2 in Box Fig. 9.1). The supply of other services, such as soil carbon storage, is context-dependent; if grassland previously degraded by overgrazing and other forms of unsustainable management becomes invaded, then soil conditions are improved. However, if healthy grassland is invaded, the supply of these services is reduced. Invasion by *P. juliflora* significantly reduces at least one ecosystem service related to each of the four health dimensions.

All stakeholders, apart from charcoal producers, lose ecosystem service benefits (D2 × SD in Box Fig. 9.1) if healthy grasslands are invaded by *P. juliflora* and therefore perceive it as degradation. Restoration options have differing impacts on stakeholder groups. Importantly, abandonment does not restore the supply of any ecosystem service and even incurs health-related ecosystem service losses, also because *P. juliflora* consumes substantial amounts of groundwater. Independent of stakeholder groups, the ecosystem multifunctionality index is always lowest for the scenario of uncontrolled invasion by *P. juliflora*. For a more detailed discussion of the calculation of the ecosystem multifunctionality index, see Manning *et al.* (2018) and Bardgett *et al.* (2021).

Continued

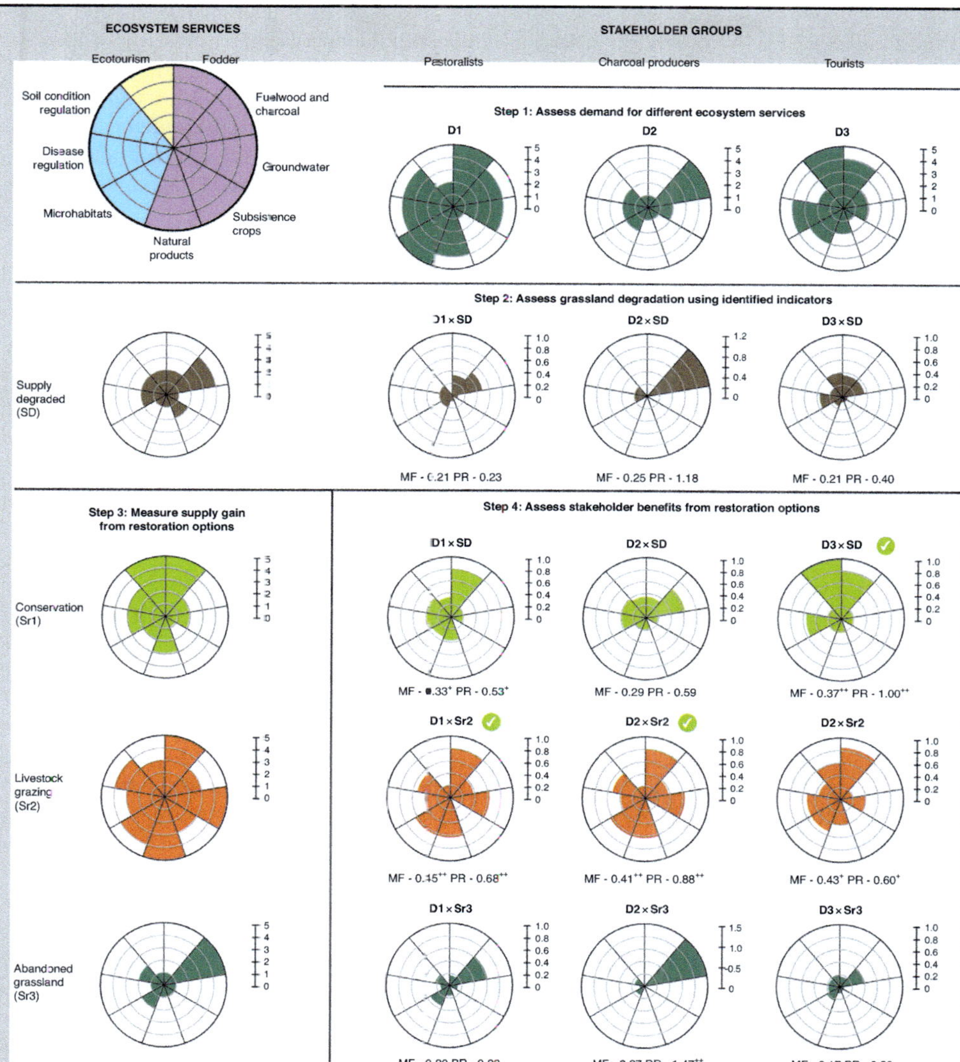

Box Fig. 9. 1. Effects of invasion by *Prosopis juliflora* on ecosystem service multifunctionality (MF) in degraded and restored grasslands in arid and semi-arid eastern Africa. ES, ecosystem services; SD, supply of ES; D, demand for ES; Sr, restoration options; MF, ecosystem multifunctionality (scaled between 0 and 1 to provide a comparable metric); +, ++, restoration options with the highest MF. Redrawn from Bardgett *et al*. (2021).

livestock production or cultural ecosystem services, including tourism, received low priority in their view. Although wood, shade and honey were all positively related to *Prosopis* cover in the heavily degraded study region, this does not mean that *Prosopis* trees are also beneficial to stakeholder groups who benefit from these resources over larger spatial scales. Native woody species that also provide these services (without the negative impacts caused by *Prosopis* trees) still dominate other areas in the region (e.g. various *Vachellia* species;

Mbaabu *et al.*, 2020). In the SEICAT scheme (Bacher *et al.*, 2018), the difference between MODERATE and MAJOR impacts is determined to a large extent by whether alternatives exist, as in this case.

9.3.3 Assigning health scores to landscapes

Another way in which the OH concept can help to assess the overall effects of *Prosopis* invasions is by assigning health scores to a continuum of landscapes from pristine through to degraded, degraded and lightly invaded, and heavily invaded by *Prosopis* trees. We regard a pristine, uninvaded landscape as a benchmark against which to assess health. For each landscape on the continuum, we draw on information from published studies to assign scores associated with positive and negative effects on human, animal, plant and environmental health (see sections 9.2.1–9.2.4 above). Scores are allocated as follows: Very Low effect: 0; Low effect: 0.25; Moderate effect: 0.5; High effect: 0.75; Very High effect: 1, and scores are positive for benefits and negative for harmful impacts. A health score is then calculated for each aspect for each landscape on the continuum by adding the score for positive and negative effects. An overall OH score is then calculated as the sum of scores for each aspect (the health of people, animals, plants and the environment) for each particular landscape.

Using this technique, health scores were estimated for each aspect in each landscape (Table 9.1; see Box 9.2 for a motivation for the allocation of scores). As expected, net OH scores are high in pristine, uninvaded ecosystems and decline substantially when ecosystems become degraded due to the removal of native trees, overgrazing and erosion (Nyssen *et al.*, 2015; Kirui *et al.*, 2021) (Fig. 9.1). The introduction of *Prosopis* trees alleviates the situation marginally, mainly by providing an additional source of fodder for livestock and possibly reducing erosion somewhat. However, in heavily invaded ecosystems almost all of the benefits associated with pristine or even degraded ecosystems are lost, and net OH scores decline markedly (Fig. 9.1). Moreover, our ability to improve the net OH score in a degraded landscape is significantly hampered in the presence of invasive *Prosopis* trees, as the sustainable management of densely invaded landscapes is very difficult and requires an integrated management approach, which includes biological, physical and chemical control (Sun *et al.*, 2022; Chapter 10, this volume).

9.4 Benefits of Managing *Prosopis* Invasions by Applying the One Health Concept

We illustrate the rationale of assessing the impacts and benefits of sustainable management of *P. juliflora* within OH using the Lorian Swamps in Eastern Isiolo, Kenya, as a case study.

The Lorian Swamps is an area of wetlands in the lower parts of the Ewaso Ng'iro water catchment. With a length of almost 200 kilometres and a width of up to 25 kilometres, the Lorian Swamps are probably the most important wetland in the semi-arid and arid zones of northern Kenya. They are home to a rich biodiversity and are at the core of traditional grazing management systems with wet-season and dry-season grazing areas on both sides and drought reserve areas further away. The Lorian Swamps are currently so heavily invaded by *P. juliflora* that the surface water flow is being redirected, other parts of the wetland are drying up, and large areas of the remaining surface water are hardly accessible any more due to impenetrable *Prosopis* thickets. Local people report many attacks by predators such as lions and hyenas that find cover in *Prosopis* thickets. Livestock that drink in the swamps also pick up seed pods, thereby contributing to the spread of *Prosopis* when migrating to the dry-season grazing areas. The villages bordering the swamps have also experienced recurrent leishmaniasis outbreaks. The density of leishmaniasis-transmitting sandflies has significantly increased since *Prosopis* invasion, which can be explained by a change in the availability and humidity of sandy soils, the breeding substrate of sandflies, and the attractiveness of *Prosopis* flowers for adult sandflies (see above). Thus, *Prosopis* invasion in the Lorian Swamps causes multiple impacts on all four dimensions of the OH concept.

Table 9.1. Components of an overall One Health score for four habitat types in East Africa, on a gradient from uninvaded to heavily invaded. See text and for a description of the scoring scheme and a detailed motivation for the allocation of scores to the component aspects considered by the One Health framework.

Aspect	Pristine environment	Degraded environment	Lightly invaded degraded environment	Heavily invaded environment
Human health	Positive: Very High (1) Negative: Very Low (0) Net score: 1	Positive: Low (0.25) Negative: Low (–0.25) Net score: 0	Positive: Moderate (0.5) Negative: Low (–0.25) Net score: 0.25	Positive: Very Low (0) Negative: Very High (–1) Net score: –1
Animal health	Positive: High (0.75) Negative: Very Low (0) Net score: 0.75	Positive: Low (0.25) Negative: Low (–0.25) Net score: 0	Positive: Moderate (0.5) Negative: Moderate (–0.5) Net score: 0	Positive: Low (0.25) Negative: Very High (–1) Net score: –0.75
Plant health	Positive: Very High (1) Negative: Very Low (0) Net score: 1	Positive: Moderate (0.5) Negative: Low (–0.25) Net score: 0.25	Positive: Moderate (0.5) Negative: Low (–0.25) Net score: 0.25	Positive: Very Low (0) Negative: Very High (–1) Net score: –1
Environmental health	Positive: Very High (1) Negative: Very Low (0) Net score: 1	Positive: Low (0.25) Negative: High (–0.75) Net score: –0.5	Positive: Moderate (0.5) Negative: High (–0.75) Net score: –0.25	Positive: Very Low (0) Negative: Very High (–1) Net score: –1
Overall One Health score	3.75	–0.25	0.25	–3.75

Box 9.2. Motivation for the allocation of One Health scores along a continuum of landscapes.

The motivation for assigning the scores in Table 9.1 to each dimension of OH is described briefly below. More details can be found in sections 9.2.1–9.2.4 in this chapter, and elsewhere in this volume (see Chapters 5–8 where the impacts of invasions on biodiversity, water resources, rangeland productivity and rural livelihoods are discussed in more detail).

Human health: In pristine environments, human health is assumed to be unaffected by degradation or invasion (Very High positive score), and negative effects arising from degradation and invasion are absent (Very Low negative score). In degraded ecosystems, people's access to resources such as medicinal plants, firewood and other harvested products is substantially reduced (Low positive score), and stress increases, impacting on health (Low negative score). By introducing *Prosopis* trees to degraded environments, some of the stresses associated with keeping livestock well fed are removed (Moderate positive score). The negative score remains Low, however, because although some of the stresses have been alleviated, the presence of *Prosopis* trees adds risks of injury and allergic reactions. When ecosystems become heavily invaded, the health benefits of ecosystem services become Very Low due to an increase in stress. Negative impacts are substantially increased in the form of injury and allergic reactions, reduced opportunities to harvest medicinal plants, water shortages, and conflicts with neighbours and wildlife for resources, leading to a Very High negative score.

Animal health: In pristine environments, animal health is assumed to be unaffected by degradation or invasion, but periodic droughts can lead to occasional shortages of fodder (High rather than Very High positive score). Negative effects arising from degradation and invasion are absent (Very Low negative score). Degraded environments become overgrazed and eroded, reducing the positive effects of native pastures substantially (Low positive score). Animals lose condition due to insufficient fodder (Low negative score). The initial introduction of *Prosopis* trees provides additional fodder and improves the health of animals somewhat (Moderate positive score), but the risk of injuries to livestock and wildlife from thorns increases the negative score to Moderate. In heavily invaded ecosystems the positive benefits for animals are Low. Some pods are available to feed on, but they are insufficient to meet all dietary needs (Low positive score). However, negative aspects far outweigh any benefits. Most grazing grasses are eliminated and thickets become impenetrable, preventing access to what grazing there may be, and to water; serious injuries increase, leading to mortality in livestock and wildlife; habitat changes lead to the exclusion of native mammals, birds and invertebrates (Very High negative score).

Plant health: Plant health is assumed to be Very High in pristine environments and negative effects are Very Low. In degraded environments, overgrazing and erosion reduce the positive factors that promote plant health in pristine environments (Moderate effects remain) and increase negative effects (Low). The introduction of *Prosopis* trees at low densities does not change the situation that exists under a degraded state. When invasions become dominant, all the positive benefits of factors promoting the growth and health of native plants are lost (Very Low positive effects). Native plants are eliminated due to competition, and reestablishment is inhibited through allelopathy. Attempts to control invasions using herbicides have additional detrimental effects on native plants, and overall negative effects are therefore Very High.

Environmental health: In pristine environments, the ability of ecosystems to provide a suite of services is assumed to be unaffected by degradation or invasion (Very High positive score) and negative effects are absent (Very Low negative score). In degraded ecosystems, the quantum of services is substantially reduced (Low positive effects) and the ecosystems lose biodiversity and stability, leading to High negative effects. The introduction of *Prosopis* trees at low densities improves the situation marginally (for example, by reducing erosion), leading to Moderate positive effects, but the negative effects of degraded ecosystems remain High. When invasions reach high densities, the ecosystems lose biodiversity and stability (e.g. decomposers, pollinators and other ecosystem regulators are eliminated) and hydrological processes are altered (e.g. *Prosopis* trees deplete groundwater, leading to death of native trees), so negative effects are Very High.

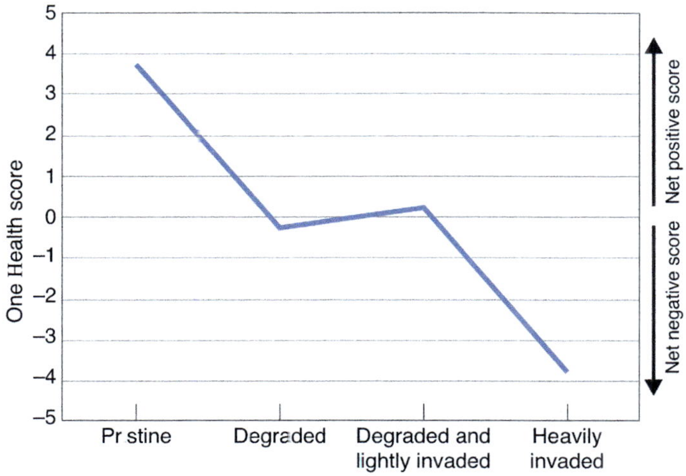

Fig. 9.1. Net One Health scores assigned to different landscapes on a continuum from pristine to degraded and heavily invaded by *Prosopis* trees.

When developing management strategies against *P. juliflora* in Isiolo County, we noticed that the impacts on biodiversity, including charismatic wildlife, did not necessarily raise major concerns among policy makers. More concern was expressed about the threats to a traditional grazing management scheme that extends over hundreds of kilometres and involves tens of thousands of people living in the surroundings of the swamps. Arguably the biggest concern, though, was raised about the potential indirect effects of *Prosopis* on human health through increased incidences of zoonotic diseases. The OH concept may thus also help in identifying those impacts that attract high levels of attention from policy makers and organizations supporting the wellbeing of rural people.

9.5 Implications for Management

Kenya's national strategy for the management and control of invasive *Prosopis* species (Ndufa *et al.*, 2024) notes that *P. juliflora* displaces native biodiversity, reduces habitat for wildlife, fodder for livestock and water availability, and negatively affects human health. This indicates that the problem of invasion by *Prosopis* trees is being considered from multiple perspectives, and that a comprehensive approach is needed to raise awareness of the full extent of the problem.

The approach proposed here is an attempt to develop such a multidimensional assessment. It indicates clearly that heavily invaded sites are characterized by substantial declines in human, animal, plant and environmental health, a situation that is unsustainable. The problem has reached a stage where the invasions are set to grow exponentially, which could result in the collapse of sustainable agriculture across vast tracts of land unless it is adequately addressed. Multidimensional assessments such as the case presented in this chapter will be necessary to raise awareness of the gravity of the problem, and to secure the funding that will be needed to contain and hopefully reverse the problem.

Brundu *et al.* (2020) proposed a set of global guidelines as a first step towards building global consensus on approaches that are needed when introducing and planting alien trees and for devising sustainable management strategies to maximize their benefits and minimize their negative impacts. The OH approach, demonstrated in this chapter for *Prosopis* in Africa, could be incorporated into these guidelines to provide objective indicators for triggering intervention or to manage the progress of various measures, and to develop multidimensional tools for assessing biosecurity risks of invasive alien species (Hulme, 2020).

Acknowledgements

This work was supported by the Swiss Programme for Research on Global Issues for Development (r4d), funded by the Swiss National Science Foundation and the Swiss Agency for Development and Cooperation, through the project 'Woody invasive alien species in East Africa: Assessing and mitigating their negative impact on ecosystem services and rural livelihood' (Grant Number: 400440_152085), and by the Swiss Agency for Development and Cooperation through the project 'Woody Weeds +' (Project number 7F.10651.01). BWvW thanks the National Research Foundation of South Africa for funding (Grant Numbers 150260 and 151653). DMR acknowledges support from the DSI-NRF Centre of Excellence in Invasion Biology, Mobility 2020 project no. CZ.02.2.69/0.0/0.0/18_053/0017850 (Ministry of Education, Youth and Sports of the Czech Republic) and long-term research development project RVO 67985939 (Czech Academy of Sciences).

References

Bacher, S., Blackburn, T.M., Essl, F., Genovesi, P., Heikkilä, J. *et al.* (2018) Socio-economic impact classification of alien taxa (SEICAT). *Methods in Ecology and Evolution* 9, 159–168.

Bardgett, R.D., Bullock, J.M., Lavorel, S., Manning, P., Schaffner, U. *et al.* (2021) Combatting global grassland degradation. *Nature Reviews Earth and Environment* 2, 720–735.

Batista-Roche, L.G. and Huerta-Ocampo, J.Á. (2022) Mezquite: Una planta alergénica de relevancia clínica en México. *TIP Revista Especializada En Ciencias Químico-Biológicas* 24(1).

Bekele, K., Haji, J., Legesse, B. and Schaffner, U. (2018) Economic value of dry land ecosystem services affected by *Prosopis juliflora*: Evidence from Ethiopia and Kenya. *Journal of Arid Environments* 158, 9–18.

Bekele, K., Linders, T.E.W., Eschen, R., Shiferaw, H., Haji, J. *et al.* (2022) How well do local stakeholders' perceptions of environmental impacts of an invasive alien plant species relate to ecological data? *Ecological Indicators* 137, 108748.

Bhojvaid, P.P. and Timmer, V.R. (1998) Soil dynamics in an age sequence of *Prosopis juliflora* planted for sodic soil restoration in India. *Forest Ecology and Management* 106, 181–193.

Blackburn, T.M., Essl, F., Evans, T., Hulme, P.E., Jeschke, J.M. *et al.* (2014) A unified classification of alien species based on the magnitude of their environmental impacts. *PLoS Biology* 12, e1001850.

Brundu, G., Pauchard, A., Pyšek, P., Pergl, J., Bindewald, A.M. *et al.* (2020) Global guidelines for the sustainable use of non-native trees to prevent tree invasions and mitigate their negative impacts. *NeoBiota* 61, 65–116.

Darvill, R. and Lindo, Z. (2016) The inclusion of stakeholders and cultural ecosystem services in land management trade-off decisions using an ecosystem services approach. *Landscape Ecology* 31, 533–545.

De Wit, M.P., Crookes, D.J. and van Wilgen, B.W. (2001) Conflicts of interest in environmental management: Estimating the costs and benefits of a tree invasion. *Biological Invasions* 3, 167–178.

Dean, W.R.J., Anderson, M.D., Milton, S.J. and Anderson, T.A. (2002) Avian assemblages in native acacia and alien *Prosopis* drainage line woodland in the Kalahari, South Africa. *Journal of Arid Environments* 51, 1–19.

Dzikiti, S., Schachtschneider, K., Naiken, V., Gush, M., Moses, G. *et al.* (2013) Water relations and the effects of clearing invasive *Prosopis* trees on groundwater in an arid environment in the Northern Cape, South Africa. *Journal of Arid Environments* 90, 103–113.

El-Keblawy, A. and Abdelfatah, M.A. (2014) Impacts of native and invasive exotic *Prosopis* congeners on soil properties and associated flora in the arid United Arab Emirates. *Journal of Arid Environments* 100, 1–8.

Hassaballa, I.B., Torto, B., Sole, C.L. and Tchouassi, D.P. (2021) Exploring the influence of different habitats and their volatile chemistry in modulating sand fly population structure in a leishmaniasis endemic foci, Kenya. *PLoS Neglected Tropical Diseases* 15(2), e0009062.

Henciya, S., Seturaman, P., James, A.R., Tsai, Y.H., Nikam, R. *et al.* (2017) Biopharmaceutical potentials of *Prosopis* spp. (Mimosaceae, Leguminosa). *Journal of Food and Drug Analysis* 25, 187–196.

Hulme, P.E. (2020) One Biosecurity: A unified concept to integrate human, animal, plant, and environmental health. *Emerging Topics in Life Sciences* 4, 539–549.

Hulme, P.E. (2021) Advancing one biosecurity to address the pandemic risks of biological invasions. *Bioscience* 71, 708–721.

Hussain, M.I., Shackleton, R.T., El-Keblawy, A., Del Mar Trigo Pérez, M. and González, L. (2020) Invasive mesquite (*Prosopis juliflora*), an allergy and health challenge. *Plants* 9, 141.

Ilukor, J., Rettberg, S., Treydte, A. and Birner, R. (2016) To eradicate or not to eradicate? Recommendations on *Prosopis juliflora* management in Afar, Ethiopia, from an interdisciplinary perspective. *Pastoralism: Research, Policy and Practice* 6, 1–8.

Kebede, A.T. and Coppock, D.L. (2015) Livestock-mediated dispersal of *Prosopis juliflora* imperils grasslands and the endangered Grevy's Zebra in northeastern Ethiopia. *Rangeland Ecology and Management* 68, 402–407.

Killian, S. and McMichael, J. (2004) The human allergens of mesquite (*Prosopis juliflora*). *Clinical and Molecular Allergy* 2, 1–5.

Kirui, O.K., Mirzabaev, A. and von Braun, J. (2021) Assessment of land degradation 'on the ground' and from 'above'. *SN Applied Sciences* 3, 1–13.

Laxén, J.P. (2007) *Is Prosopis a Curse or a Blessing? An Ecological Economic Analysis of an Invasive Alien Tree Species in Sudan*. University of Helsinki Tropical Forestry Reports. University of Helsinki, Helsinki, Finland.

Linders, T.E., Schaffner, U., Eschen, R., Abebe, A., Choge, S.K. *et al.* (2019) Direct and indirect effects of invasive species: Biodiversity loss is a major mechanism by which an invasive tree affects ecosystem functioning. *Journal of Ecology* 107, 2660–2672.

Linders, T.E., Bekele, K., Schaffner, U., Allan, E., Alamirew, T. *et al.* (2020) The impact of invasive species on social-ecological systems: Relating supply and use of selected provisioning ecosystem services supply to demand. *Ecosystem Services* 41, 101055.

Linders, T.E., Schaffner, U., Alamirew, T., Allan, E., Choge, S.K. *et al.* (2021) Stakeholder priorities determine the impact of an alien tree invasion on ecosystem multifunctionality. *People and Nature* 3, 658–672.

Manning, P., van der Plas, F., Soliveres, S., Allan, E., Maestre, F.T. *et al.* (2018) Redefining ecosystem multifunctionality. *Nature Ecology and Evolution* 2, 427–436.

Maundu, P., Kibet, S., Morimoto, Y., Imbumi, M. and Adeka, R. (2009) Impact of *Prosopis juliflora* on Kenya's semi-arid and arid ecosystems and local livelihoods. *Biodiversity* 10, 33–50.

Mbaabu, P.R., Olago, D., Gichaba, M., Eckert, S., Eschen, R. *et al.* (2020) Restoration of degraded grasslands, but not invasion by *Prosopis juliflora,* avoids trade-offs between climate change mitigation and other ecosystem services. *Scientific Reports* 10, 1–13.

Mehari, Z.H. (2015) The invasion of *Prosopis juliflora* and Afar pastoral livelihoods in the Middle Awash area of Ethiopia. *Ecological Processes* 4, 1–9.

Milton, S.J., Wilson, J.R.U., Richardson, D.M., Seymour, C.L., Dean, W.R.J. *et al.* (2007) Invasive alien plants infiltrate bird-mediated shrub nucleation processes in arid savanna. *Journal of Ecology* 95, 648–661.

Muller, G.C., Junnila, A., Traore, M.M., Traore, S.F., Doumbia, S. *et al.* (2017) The invasive shrub *Prosopis juliflora* enhances the malaria parasite transmission capacity of anopheles mosquitoes: A habitat manipulation experiment. *Malaria Journal* 16, 1–9.

Mutavi, S.K. (2020) Effects of *Prosopis juliflora* pod and leaf meal on physical characteristics of teeth and bones of goats in Kitui county, Kenya. *Livestock Research for Rural Development* 32, 163.

Mwangi, E. and Swallow, B. (2008) *Prosopis juliflora* invasion and rural livelihoods in the Lake Baringo area of Kenya. *Conservation and Society* 6, 140.

Ndufa, J.K., Choge, S. and Cheboiwo, J. (2024) *National Strategy for the Management and Control of the Invasive Prosopis Species*. Ministry of Environment and Forestry, Nairobi, Kenya.

Ngumbi, M.P., Kaburia, C.J., Anjili, O.C. and Haas, F. (2010) *Phlebotomus (Larroussius) orientalis* (Diptera: Psychodidae) as a probable secondary vector of visceral leishmaniasis in Kenya. *Journal of Vector Borne Disease* 47, 58–60.

Novey, H.S., Roth, M. and Wells, I.D. (1977) Mesquite pollen: An aeroallergen in asthma and allergic rhinitis. *Journal of Allergy and Clinical Immunology* 59, 359–363.

Nyssen, J., Poesen, J., Lanckriet, S., Jacob, M., Moeyersons, J. *et al.* (2015) Land degradation in the Ethiopian highlands. In: Paolo, B. (ed.) *Landscapes and Landforms of Ethiopia*. Springer, Dordrecht, The Netherlands, pp. 369–385.

Ogden, N.H., Wilson, J.R.U., Richardson, D.M., Hui, C., Davies, S.J. *et al.* (2019) Emerging infectious diseases and biological invasions: A call for a One Health collaboration in science and management. *Royal Society Open Science* 6(3), 181577.

One Health Commission (2021). Available at: www.onehealthcommission.org/en/news/one_health_news/?year=2021 (accessed 30 November 2024).

Papadopoulos, N.G., Agache, I., Bavbek, S., Bilo, B.M., Braido, F. *et al.* (2012) Research needs in allergy: An EAACI position paper, in collaboration with EFA. *Clinical and Translational Allergy* 2, 1–23.

Prabha, D.S., Dahms, H.U. and Malliga, P. (2014) Pharmacological potentials of phenolic compounds from *Prosopis* spp. – a review. *Journal of Coastal Life Medicine* 2, 918–924.

Rettberg, S. (2010) Contested narratives of pastoral vulnerability and risk in Ethiopia's Afar region. *Pastoralism: Research, Policy and Practice* 1, 48–273.

Richardson, D.M., Hui, C., Nuñez, M.A. and Pauchard, A. (2014) Tree invasions: Patterns, processes, challenges and opportunities. *Biological Invasions* 16, 473–481.

Rogers, P., Nunan, F. and Fentie, A.A. (2017) Reimagining invasions: The social and cultural impacts of *Prosopis* on pastoralists in Southern Afar, Ethiopia. *Pastoralism: Research, Policy and Practice* 7, 1–13.

Roy, H., Pauchard, A., Stoett, P., Renard Truong, T., Bacher, S. *et al.* (2023) Summary for policymakers of the thematic assessment report on invasive alien species and their control of the Intergovernmental Science-Policy Platform on Biodiversity and Ecosystem Services. 19 March 2024: IPBES Secretariat. Available at: https://doi.org/10.5281/zenodo.7430692 (accessed 30 November 2024).

Sawal, R.K., Ratan, R. and Yadav, S.B.S. (2004) Mesquite (*Prosopis juliflora*) pods as a feed resource for livestock: A review. *Asian-Australasian Journal of Animal Sciences* 17, 719–725.

Schachtschneider, K. and February, E.C. (2013) Impact of *Prosopis* invasion on a keystone tree species in the Kalahari desert. *Plant Ecology* 214, 597–605.

Shackleton, R.T., Le Maitre, D.C., van Wilgen, B.W. and Richardson, D.M. (2015a) Use of non-timber forest products from invasive alien *Prosopis* species (mesquite) and native trees in South Africa: Implications for management. *Forest Ecosystems* 2, 1–11.

Shackleton, R.T., Le Maitre, D.C., van Wilgen, B.W. and Richardson, D.M. (2015b) The impact of invasive alien *Prosopis* species (mesquite) on native plants in different environments in South Africa. *South African Journal of Botany* 97, 25–31.

Shiferaw, H., Alamirew, T., Dzikiti, S., Bewket, W., Zeleke, G. *et al.* (2021) Water use of *Prosopis juliflora* and its impacts on catchment water budget and rural livelihoods in Afar region, Ethiopia. *Scientific Reports* 11, 1–14.

Shiferaw, H., Alamirew, T., Dzikiti, S, *et al.* (2023) Water abstraction of invasive *Prosopis juliflora* and native Senegalia senegal trees: A comparative study in the Great Rift Valley area, Ethiopia. *Science of the Total Environment* 862, 160833.

Shiferaw, W., Demissew, S., Bekele, T. and Aynekulu, E. (2022) Community perceptions towards invasion of *Prosopis juliflora*, utilization, and its control options in Afar region, Northeast Ethiopia. *PLoS One* 17, e0261838.

Silva, A.M.M.D., Silva, A.R., Pinheiro, A.M., Freitas, S.R.V.B.D., Silva, V.D.A.D. *et al.* (2007) Alkaloids from *Prosopis juliflora* leaves induce glial activation, cytotoxicity and stimulate NO production. *Toxicon* 49, 601–614.

Sissoko, F., Junnila, A., Traore, M.M., Traore, S.F., Doumbia, S. *et al.* (2019) Frequent sugar feeding behavior by *Aedes aegypti* in Bamako, Mali makes them ideal candidates for control with attractive toxic sugar baits (ATSB). *PLoS One* 14, e0214170.

Steenkamp, H.E. and Chown, S.L. (1996) Influence of dense stands of an exotic tree, *Prosopis glandulosa* Benson, on a savanna dung beetle (Coleoptera: Scarabaeinae) assemblage in Southern Africa. *Biological Conservation* 78(3), 305–311.

Sun, Y., Müller-Schärer, H. and Schaffner, U. (2022) Fighting neobiota with neobiota: Consider it more often and do it more rigorously. *Biological Conservation* 268, 109506.

Tabe-Ojong, Jr, M.P. (2022) Ecological shocks and non-cognitive skills: Evidence from Kenya. *Ecological Economics* 194, 107330.

Tabosa, I.M., Riet-Correa, F., Barros, S.S., Summers, B.A., Simões, S.V.D. *et al.* (2006) Neurohistologic and ultrastructural lesions in cattle experimentally intoxicated with the plant *Prosopis juliflora*. *Veterinary Pathology* 43, 695–701.

Tessema, Y.A. (2012) Ecological and economic dimensions of the paradoxical invasive species *Prosopis juliflora* and policy challenges in Ethiopia. *Journal of Economics and Sustainable Development* 3, 62–70.

Tessema, Z.K., Ejigu, B. and Nigatu, L. (2017) Tree species determine soil seed bank composition and its similarity with understory vegetation in a semi-arid African savanna. *Ecological Processes* 6, 1–16.

van Wilgen, B.W., Zengeya, T.A. and Richardson, D.M. (2022) A review of the impacts of biological invasions in South Africa. *Biological Invasions* 24, 27–50.

Wabusya, E.N. (2013) Policy outcomes and community perceptions of a dryland invasive species: A case study of *Prosopis juliflora* in Baringo county, Kenya. Master's thesis, International Institute of Social Studies, The Hague, The Netherlands.

Wise, R.M., van Wilgen, B.W. and Le Maitre, D.C. (2012) Costs, benefits and management options for an invasive alien tree species: The case of mesquite in the Northern Cape. *Journal of Arid Environments* 84, 80–90.

Woodford, D.J., Richardson, D.M., MacIsaac, H.J., Mandrak, N.E., van Wilgen, B.W. *et al.* (2016) Confronting the wicked problem of managing biological invasions. *NeoBiota* 31, 63–86.

Zengeya, T., Ivey, P., Woodford, D.J., Weyl, O., Novoa, A. *et al.* (2017) Managing conflict-generating invasive species in South Africa: Challenges and trade-offs. *Bothalia* 47(2), a2160.

Part 3: Management

Parts 1 and 2 gave an overview of the introduction and spread of *Prosopis* trees to eastern Africa, of the areas at risk of being invaded, and of the impacts of these invasive species on environmental, economic and social aspects. This is a significant problem that is set to grow, and active and coordinated management will be required if the impacts are to be effectively reduced. Part 3 explores what it is possible to do, how it should be done, and how it can be coordinated.

In Chapter 10, Eschen *et al.* outline best practice management approaches that can be utilized to prevent introductions to new areas, detect new incursions and respond to them appropriately, and to contain invasions where they are well established. The chapter informs on the best way to select suitable management practices within a broader adaptive and context-aware management framework, and on how to monitor the effectiveness of these practices. The chapter concludes with the limitations of the 'management by utilization' approach and the essential role that biological control will have to play if the goal of containing these invasions is to have any reasonable chance of success.

Chapter 11 by Ehrensperger *et al.* builds on the findings of Chapter 4, and on the invasion patterns of *Prosopis juliflora*, to outline a spatially explicit management concept that considers the density of *Prosopis* tree invasions, the presence of pathways of spread, and of important assets needing protection from invasion. The chapter proposes a method to assist in the identification of appropriate interventions to achieve the goals of management, which will differ spatially depending on the suitability of the habitat for invasion, and whether the trees are absent, recent arrivals or well-established populations. Finally, it outlines technical options, such as the use of unmanned aerial vehicles (drones), in support of spatial management.

Chapter 12 by Kiteme *et al.* describes attempts to engage stakeholders across sectors and at different scales and to develop novel governance and institutional approaches to managing *Prosopis* tree invasions, based on a case study in Kenya. This study tested novel institutional arrangements at different scales by exploring the shortcomings of existing governance and institutional structures for managing natural resources in Kenya. The study established multi-stakeholder Local and County Implementation Groups, and these were engaged in deliberative decision processes to codesign management interventions at local and county scales and to promote their implementation. The initial efforts have led to social learning among the stakeholders involved and have helped to mainstream invasive species management into planning and budgetary processes at the county and the national scale.

10 A Review of Best Management Practices for the Control of Invasive *Prosopis* Trees

René Eschen[1]*, Beatrice Adoyo[2,3], Mickfanaka Mwihomeke[4,5], Charles J. Kilawe[6] and Brian W. van Wilgen[7]
[1]CABI, Delémont, Switzerland; [2]Department of Geography, Population, and Environmental Studies, University of Nairobi, Kenya; [3]Centre for Training and Integrated Research in ASAL Development, Nanyuki, Kenya; [4]Department of Land Resource Management and Agricultural Technology, University of Nairobi, Kenya; [5]CABI, Nairobi, Kenya; [6]Department of Ecosystems and Conservation, Sokoine University of Agriculture, Morogoro, Tanzania; [7]Centre for Invasion Biology, Department of Botany and Zoology, Stellenbosch University, South Africa

Abstract

The management of *Prosopis* trees should aim to prevent introduction, remove sparse infestations, prevent spread once the trees have become established in an area, or control locally and mitigate impacts in densely invaded areas. Several methods for management of *Prosopis* trees on the local or landscape scale exist, including mechanical, chemical and biological options, which can be combined into sustainable land management practices. Each of these practices has particular and often context-dependent advantages and disadvantages, which need to be considered when planning management interventions. This chapter presents an overview of the methods used to manage *Prosopis* trees and provides criteria for the selection of the most suitable combination of methods into context-specific sustainable land management practices. Historical management efforts have not been successful in lastingly extirpating *Prosopis* trees, and reestablishment has happened at many control sites in eastern Africa, but also in other parts of the world including South Africa and Australia. Some of the reasons for the lack of success include failure of private landowners to continue managing the species after government support ended, and conflicts between landowners and government. Moreover, none of the described control options will be successful in isolation and combinations of methods should be employed as part of sustainable *Prosopis* management strategies for invaded landscapes that are codeveloped with relevant stakeholders. The needs of people reliant on *Prosopis* trees for their livelihoods should be considered during the transition from invasion to more productive landscapes where tree density has been brought down to a tolerable level in the landscape. We strongly recommend including the use of classical biological control against invasive *Prosopis* trees as part of the management in eastern Africa.

*Corresponding author: R.Eschen@cabi.org

© CAB International 2025. *The Ecology and Management of Invasive Prosopis Trees in Eastern Africa* (eds U. Schaffner *et al.*)
DOI: 10.1079/9781800623644.0010

10.1 Managing *Prosopis* Trees

10.1.1 Selecting management objectives

Prosopis tree invasions have spatial patterns that require complementary management objectives (Chapter 11, this volume).

- Prevention is an adequate management objective in parts of a landscape where *Prosopis* trees have not yet established but are suitable for their establishment.
- Early detection and rapid response (EDRR) is appropriate for parts of a landscape where *Prosopis* trees occur only sparsely and can be locally extirpated.
- Control through containment and asset protection is needed in parts of the landscape where *Prosopis* trees occur in widespread, dense thickets.

In other words, the choice for these objectives depends on the presence, distribution and abundance of *Prosopis* trees in parts of a landscape and on the presence of important assets that may need protection from the invader (Fig. 10.1).

10.1.2 Setting SMART management objectives

Beyond the above spatial considerations, the choice of management objectives needs to be balanced against available means. It is important to set objectives that are specific,

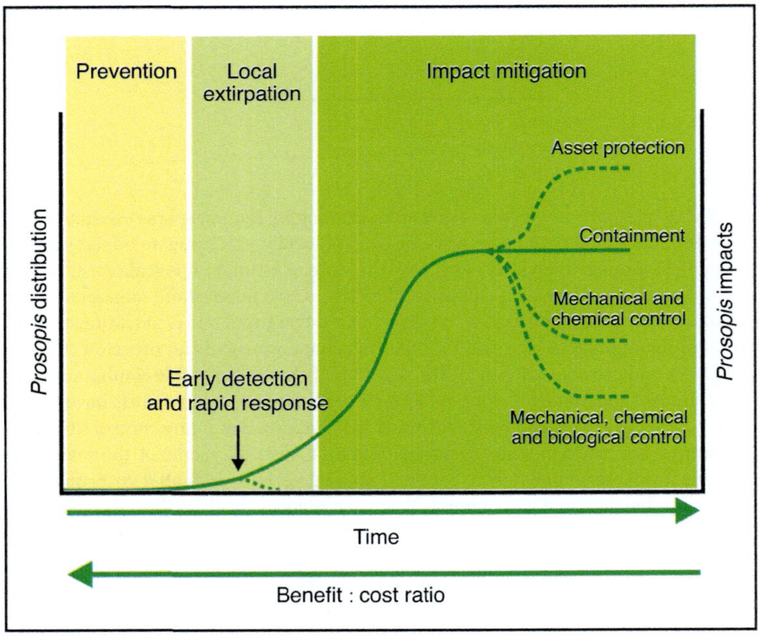

Fig. 10.1. Conceptual framework illustrating the stages of invasion by *Prosopis* trees at a landscape scale and appropriate management responses to each stage. The solid line indicates abundance of *Prosopis* and *Prosopis* impacts without management, whereas the effects of different management approaches are indicated by dashed lines. Abundance of *Prosopis* generally increases with time since the start of the invasion and the benefit–cost ratio decreases at the same time, indicating that prevention of establishment and early detection and rapid response (EDRR) are less costly than containment, asset protection and control of widely established species. Asset protection does not affect, containment stops the increase of, and control reduces the size of the infested area and overall impacts of *Prosopis* on the landscape scale, respectively.

measurable, achievable, realistic and time bounded (SMART), i.e. that can be achieved within the desired timeframe, given the resources that are available, and that can be specifically understood and measured. For example, a measurable objective would be to reduce the cover of an invasive species in a defined area to below 3% within 5 years, but an objective that simply states a desire to control invasive species would be unsuitable.

Whether management goals are realistic or not often depends on scale. For example, removal of sparse *Prosopis* tree invasions may require a substantial effort if the target area is large. Hence, if success (i.e. achieving local extirpation) is unlikely, it may be wise to focus on containing the spread of *Prosopis* trees where extirpation cannot be achieved, while designating a buffer zone where the aim is to implement EDRR to protect adjacent areas currently free of invasion. Similarly, if the management objective is to control *Prosopis* tree invasions in areas where complete removal is impossible, then alternative objectives should be considered, including preventing further spread from the invaded areas (containment) or protecting assets in areas where the *Prosopis* trees have become totally dominant and only localized removal is feasible.

10.1.3 Monitoring outcomes

Regular monitoring is as important as setting SMART objectives, because early corrective action could bring a project back on track if it becomes apparent that progress towards achieving the objective is falling behind or failing altogether. For that reason, monitoring must be part of any management practice to track the progress towards meeting clear and measurable goals, which could include local extirpation, containment of the invasion, reducing the invaded area and the density of invasions, or protecting assets. It is important to note that monitoring is not a management intervention per se (Sequeira and Griffin, 2013) and that it has no impact on the status of invasive species.

10.1.4 Using an adaptive management framework

Selection of objectives, setting of SMART goals and adequate monitoring need to take place within the wider scope of an adaptive management framework. Indeed, biological invasions are spatially and temporally dynamic processes occurring in complex ecosystems that are rarely fully understood (Holling and Walters, 1978; Wilhere, 2002). The abundance and spread of *Prosopis* species are not uniform across a landscape or region and can change over time (Nehrbass *et al.*, 2007; Shiferaw *et al.*, 2019b). Therefore, *Prosopis* tree management strategies and management objectives in a landscape need to adapt to these spatial and temporal patterns and dynamics (Chapter 11, this volume).

Under an adaptive management framework, the starting point is to agree on an objective and use best management practices, as understood at the time, to achieve that objective. It also recognizes that the objective may need to be adapted if it becomes clear that the chosen practice for control is not working. Adaptive management follows a structured process of scoping (to set a goal), implementation, monitoring and assessment. Depending on the outcomes, the chosen management practice can be continued if good progress is being made, or alternative practices can be employed if the current practice is not working. Alternative practices should also be employed if the current objective has been achieved. If, for example, EDRR has been implemented successfully, the area should be free of *Prosopis* trees and the objective for that area would revert to preventing reintroduction. Should none of the available management practices achieve the intended objective, then a new and more realistic objective would have to be identified and plans revised accordingly. This approach provides flexibility, which is essential when dealing with complex issues and incomplete understanding.

The adage is that to fail to plan is to plan to fail, but a failure to plan adaptively is a common feature of ecosystem management programmes. For example, one of the shortcomings of South Africa's Working for Water programme has been a lack of adequate and adaptive planning, and the practice of setting objectives in terms of inputs and outputs rather than outcomes (van Wilgen *et al.*, 2022). In the following sections of this chapter, we outline known effective methods for *Prosopis* management at different stages of invasion, criteria for selecting and combining methods into integrated control practices, as well as prospects for additional management methods that may be introduced and would particularly improve management of very large invasions.

10.2 Overview of Potential Management Approaches

Following the setting of management goals or a review of progress, the methods and tools needed to achieve the set objectives must be identified and implemented by stakeholders who need to have a good understanding of invasion processes and the ecology of *Prosopis* trees. The goals of management and the appropriate control methods to use depend on the stage of invasion (Fig. 10.1) and other relevant socio-economic contexts (Chapter 8, this volume) in any given area. Each of these methods is discussed in the sections below.

10.2.1 Measures to prevent spread

In eastern Africa, *Prosopis* trees have not yet established at many sites that would provide suitable habitat for their establishment (Eckert *et al.*, 2020). In such areas, the goal would be to prevent their introduction and establishment. Managing established trees is more difficult and costly than preventing their arrival in the first place, making prevention one of the most important and cost-efficient management objectives. Generally, spread can be influenced by reducing propagule pressure (managing seed production or survival) and management of vectors of spread. *Prosopis* trees can be spread by animals, water and human activities, and pathways of spread often follow linear elements in the landscape, such as water courses, livestock movement routes and roads (Shiferaw *et al.*, 2019b).

Livestock, including cattle, goats, camels and donkeys, feed on *Prosopis* pods and defecate the seeds as they move around (Alvarez *et al.*, 2017; Abbas *et al.*, 2018). It is therefore important to manage the movement of such animals from areas with *Prosopis* trees to uninvaded areas. Livestock should only be allowed to move out of invaded areas if viable seeds are no longer in their digestive tract, which can be achieved by holding animals in 'quarantine' areas on the edge of infested areas for 10 days. Such areas may be fenced or designated by communal grazing committees and enough water and fodder should be provided for the period the animals spend in the holding area. It would be essential to strengthen the capacities of indigenous or customary natural resource management institutions to entrench this practice. Strong indigenous institutions would require every household to manage their livestock according to the commonly set objectives such as quarantine for 10 days. Any *Prosopis* seedlings emerging in holding areas should be regularly removed.

The seeds of *Prosopis* trees may also be spread by wild animals and water, but it is difficult to manage these pathways, and it would be more effective and more feasible to control livestock and human vectors of spread.

Spread by humans (for example, when people plant seeds or seedlings and move pods for use as livestock feed) is not uncommon when people are unaware of the negative impacts of the species. Public perception about *Prosopis* trees and preferred management changes rapidly once the impacts of the species are known (Adoyo *et al.*, 2022; R. Eschen, personal observation). Public awareness should therefore be raised through campaigns that use printed and social media, radio and television (e.g. Weedbuster week in South Africa: https://botanicalsociety.org.za/weedbuster-week-alien-invasive-plants/, or the website Invasives South Africa: https://invasives.org.za/), as well as presentations during public meetings (including barazas, i.e. public community meetings in East Africa). In addition, legislation and regulations should be adopted that prohibit the sale and planting of invasive species of *Prosopis*.

If *Prosopis* pods are harvested and transported, crushing or milling of pods can contribute to reducing the spread of viable seeds, while still being useful as livestock fodder (Mahgoub *et al.*, 2005). Milling contributes to reducing the spread of viable seeds and the fodder can support livestock during droughts, recognizing that consumption of large quantities of *Prosopis* seeds is unhealthy (Chapter 7, this volume). Breaking the seed coat allows for the animal's digestive system to destroy the seeds that are milled, which is important as undamaged seeds will pass through the gut and germinate. It is also important to note that milling of seed does not mean that quarantine measures should be removed when moving livestock. Several host-specific and safe biological control agents destroy a significant proportion of the seeds in pods of *Prosopis* trees at different stages of seed development. Thus, biological control can effectively reduce the number of viable seeds produced (see below).

Reducing the number of viable seeds in the system is likely to reduce the average dispersal distance of seeds (Harper, 1977), and thus reduce the rate of spread of invasive plants (Impson *et al.*, 2024), although this has not been demonstrated for *Prosopis* species yet.

10.2.2 Measures to achieve local extirpation

Areas of suitable habitat where *Prosopis* trees do not yet occur need to be constantly monitored to detect any recent introductions. Monitoring and surveillance may involve the use of remotely sensed images and scouting by people living and working in areas that are prone to invasion. While remote sensing has been used as a tool for monitoring large areas and describing the evolution of the invasion (van den Berg *et al.*, 2013; Vidhya *et al.*, 2017; Mbaabu *et al.*, 2019), surveillance and detection of individual and small trees at a stage when EDRR is possible may be done by scouts, such as herders while grazing their livestock. Should such a presence be detected, it would be important to attempt to remove all specimens before they mature, flower, set seed and begin to spread, so the objective of management would be to achieve local extirpation.

Successful EDRR can be achieved through regular and structured surveillance of the target area and then using the mechanical and chemical methods outlined below (if biological control agents are used in the region, then these should be fostered so as to reduce spread into the area that is currently clear of *Prosopis* trees). EDRR has been employed during the Woody Weeds project (www.woodyweeds.org; Chapter 14, this volume) in Ruko Conservancy in Baringo, Kenya, an area jointly managed by two communities that is at the early stages of invasion from surrounding areas. Game wardens and other community members were trained on identification of *Prosopis* and the *Prosopis* trees in the area were mapped. Then, the trees were removed by community members using hand tools during a workday organized with project support by the Conservancy.

Simberloff (2009) identified four additional features that characterize successful attempts at extirpation.

- Allocating sufficient resources to achieve extirpation, including ongoing surveys and follow-up, if necessary.
- Giving managers the authority to enforce cooperation and to ensure that no landowners allow *Prosopis* trees or seedlings to persist.
- A good understanding of the biology and ecology of the target species.
- Energetic, optimistic and persistent project managers.

In advanced stages of invasion, extirpation at the landscape scale may not be possible any more. In such cases, methods to reduce the spread and density of the trees in the landscape (such as biological control) as well as asset protection and local extirpation (manual, chemical control and rehabilitation), and combinations thereof, are more appropriate. Below, we provide an overview of methods that are available for these situations as well as integrated management.

10.2.3 Mechanical control

Mechanical tools to remove individual *Prosopis* trees or small populations range from simple hand tools, such as machetes, axes and hoes, to powerful machines for removal from larger areas, such as digger-excavators and bulldozers (Abdulahi *et al.*, 2017).

In the early stages of establishment, *Prosopis* seedlings may be uprooted by hand or by using 'tree poppers' or machetes. Removing a *Prosopis* tree manually is tedious and time-consuming, especially in hot climates where they are invasive (Fig. 10.2). However, despite the high cost compared to the value of land that is not used for crop production, it is a common method where inexpensive labour is in ample supply (Wise *et al.*, 2012; Abdulahi *et al.*, 2017). One example is the South African Working for Water programme, established by the government in 1995, which had the dual aims of assisting landowners to bring invasive species under control and creating employment in impoverished rural areas (Wilgen and Wannenburgh, 2016). Between 1998 and 2020, the programme spent ZAR 587 million (expressed in 2020 values, around USD 33.5 million) clearing ~250,000 ha of *Prosopis* invasions, but this effort only reached ~14% of the estimated invaded area and invasions continued to spread (van Wilgen *et al.*, 2022).

It is important that the chosen removal method kills the trees, as the stumps easily coppice or resprout which results in the need to continue removing the regrowth and fails to control the trees (Box 10.1). Hence, mechanical control of *Prosopis* trees must remove the growing points, i.e. the root stock, to at least 30 cm below the soil surface. Uprooting often results in significant soil disturbance. This may stimulate the germination of *Prosopis* seeds in the soil and it is important that emerging seedlings are regularly removed for at least 2 years after felling the adult trees (Eschen *et al.*, 2021). Another 'mechanical' way of removing *Prosopis* trees is to remove the branches and burn the stump with a fire that is fueled with dry manure or small *Prosopis* branches, but the kill rate is comparatively low (~ 65%, U. Schaffner, personal communication), and success is determined by the intensity and duration of the burning.

10.2.4 Chemical control

Chemical control of *Prosopis* trees can be achieved using various active ingredients and application

Fig. 10.2. Removal of *Prosopis* trees, especially individual trees or small infestations, can be achieved by mechanical means through manual uprooting (A) or the use of a 'tree popper' (B) to pull small trees from the ground. Photos courtesy of René Eschen.

methods, as revealed in studies in different parts of the invaded range, including India (Shanwad *et al.*, 2015), East Africa (Eschen *et al.*, 2023), South Africa (Osmond *et al.*, 2003) and the Middle East (Qasem, 2007). Herbicides need to be absorbed to affect the growth or survival of the treated trees, and it is important to understand the physiology of the plant and to ensure that cut stump and chemical treatment is implemented during the correct season and the correct time of day.

Box 10.1. Unsuccessful management of *Prosopis* trees by utilization.

The net economic impact of invasive *Prosopis* trees across landscapes in the invaded range is negative (Bekele *et al.*, 2018b; Shiferaw *et al.*, 2019a), but many stakeholders (donors and governments included) perceive an economic value of utilizing these invasive trees. Utilization of invasive alien species is prominent in eastern Africa, where many people in invaded areas are looking for alternative ways to diversify their livelihoods as a strategy for coping with natural calamities such as climate change, and the sale of *Prosopis* charcoal provides cash income (Wise *et al.*, 2012; Shackleton *et al.*, 2014; Linders *et al.*, 2020). *Prosopis* trees were introduced into Baringo County, Kenya, in the 1980s through donor-funded projects and government initiatives, with the aim of reducing land degradation and providing fuel wood, timber and fodder (Mwangi and Swallow, 2005). A time series of *Prosopis* distribution revealed a continuous increase in *Prosopis* cover from 882 to 18,792 ha between 1988 and 2016 (Mbaabu *et al.*, 2019; Box Fig. 10.1).

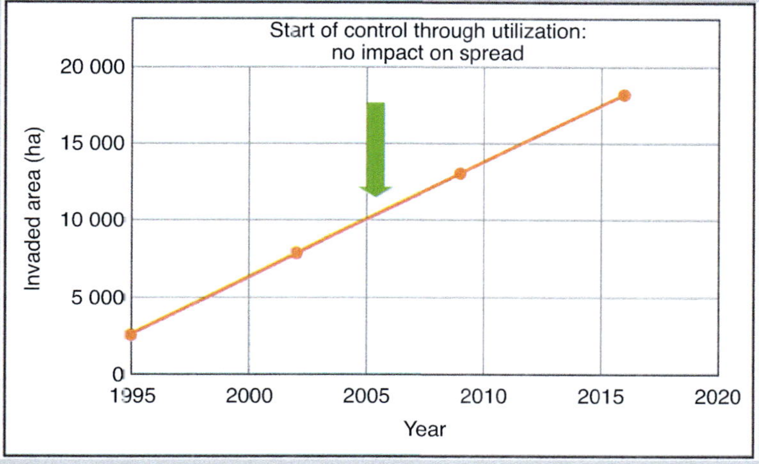

Box Fig. 10.1. The increase in the area invaded by *Prosopis* trees in Baringo County, Kenya, from 1988 to 2016, showing that the introduction of a policy of control through utilization had no effect on the invaded area, which continued to increase. Adapted from Mbaabu *et al.* (2019).

'Control through utilization' was promoted by the Kenyan government (Choge *et al.*, 2022) as a strategy aiming to combine the use of *Prosopis* trees as an income-generating and charcoal-producing activity with reducing the abundance and spread of the species. Unfortunately, the typical practice of charcoal burners is to harvest the branches and stems of *Prosopis* trees for charcoal and leave the stumps in the ground, resulting in coppicing and the formation of impenetrable thickets. As this practice does not kill the trees, the coppicing stems continue to produce seeds but can only be harvested for producing charcoal after several years (Box Fig. 10.2).

Thus, utilization does not control *Prosopis* trees, as it does not sustainably reduce the abundance and spread of the species, leading to reestablishment on the cleared land (Adoyo *et al.*, 2022). Organizations promoting utilization of *Prosopis* trees should therefore distinguish between utilization as an income-generating activity, which it is, and control of the invasive alien species, which it is not.

Continued

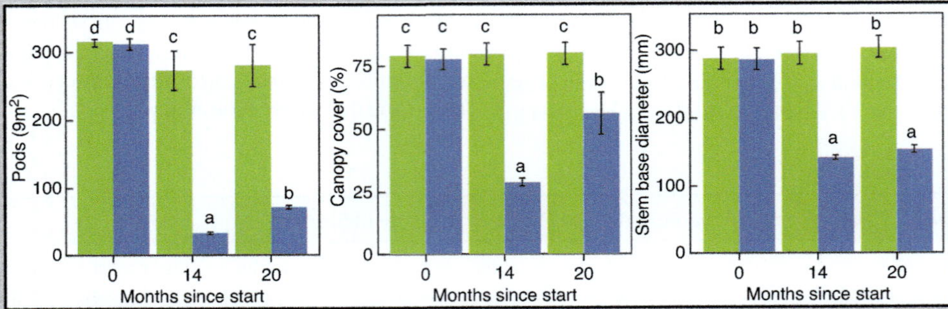

Box Fig. 10. 2. Evolution of *Prosopis juliflora* canopy cover, the number of pods per plot of 9 m^2 and stem base diameter following removal of above-ground biomass for charcoal production ('control through utilization') on ten farms in Baringo, Kenya. Green and blue bars indicate untreated control and control through utilization, respectively, and error bars indicate SE. From M. Mwihomeke, unpublished.

Herbicides may be applied to the lower 50–75 cm of the stem or stems of a tree ('basal bark') when the plants are actively growing, taking care to apply the herbicide to the entire circumference of each stem (Fig. 10.3). This method kills the trees while leaving the above-ground biomass standing and is a rapid way of treating small, isolated stands (Osmond *et al.*, 2003). Alternatively, trees can be killed and removed by cutting the stem close to the soil using a chainsaw and applying herbicide to the cut surface immediately ('cut stump'; Fig. 10.3). This method requires at least three people working together to cut and remove the trees but leaves no trees standing. The thicker cut branches can be used to make charcoal and smaller branches can be used as firewood or for fencing. The stumps remain in the soil using this method, but the land becomes accessible for livestock and grasses can be sown on the cleared land.

While the use of herbicides is considered environmentally detrimental by some, especially when applied as a foliar spray, chemical treatments that can be applied in targeted, environmentally safe ways by trained technicians have been applied successfully for management of invasive *Prosopis* trees in Australia and South Africa (Osmond *et al.*, 2003). It is important to recognize that the environmental damage from the trees due to loss of groundwater and herbaceous plants important for grazing by far exceeds any environmental damage caused by chemicals that target the plant and break down quickly in the soil. Products with the same active ingredients should be registered for use against *Prosopis* trees in eastern African countries as they have been proven to be effective and environmentally safe elsewhere.

In a field study in Ethiopia, Kenya and Tanzania, a test of cut stump and basal bark application of herbicides with triclopyr as active ingredient found that both methods were effective in the three countries, despite differences in the products and formulations used (Eschen *et al.*, 2023). The effectiveness of basal bark applications in Kenya was lower than in the other two countries, which was probably the result of the greater stem diameter of the trees in the Kenyan sites, some of which were old trees that had been coppiced after being cut previously, suggesting that basal bark treatment of thicker trees may require a higher dosage of the active ingredient. Cut stump treatment was highly effective in all the three countries, killing all trees. It uses less herbicide than basal bark treatments but requires more personnel, so could be costlier.

Herbicides can provide effective control of *Prosopis* tree invasions but there are few registered herbicides for *Prosopis* trees in eastern African countries. Triclopyr is registered for use against trees in Tanzania and registration of herbicides against *Prosopis* trees in Kenya is under way.

Fig. 10.3. Cutting stems (A) to immediately apply herbicide to the stump (B) and herbicide application to the entire bark at the base of trees (C) are effective, safe methods to kill *Prosopis* trees. Photos courtesy of B. Megersa.

10.2.5 Biological control

One of the major reasons why some alien plants become invasive when introduced to new environments is that they are freed from the coevolved pests and pathogens that keep them in check in their native range (the so-called 'enemy release hypothesis'; Keane and Crawley, 2002). Biological control seeks to cancel out this advantage by introducing some of the target plant's natural enemies, and in so doing remove the plant's competitive advantage and therefore its tendency to invade. Only natural enemies (biological control agents) that are host-specific (i.e. proven, after rigorous structured research, to attack only the targeted species, or a restricted group of closely related species), safe (i.e. they have no known or predicted non-target effects) and preferably very damaging to the target plant are used for biological control (Heimpel and Mills, 2017).

Using biological control to manage plant invasions, either alone or in combination with other methods, is an extremely attractive option, because it is relatively cheap, safe and, above all, sustainable. It is globally accepted practice to conduct extensive studies in a quarantine facility to ensure that any prospective agent will not damage non-target plants. A biological control agent is typically only released once it has been proved to be sufficiently host-specific. As a result, biological control has an excellent record of safety, with hundreds of different agents released globally without affecting native species or ecosystems (van Wilgen *et al.*, 2013), and procedures to predict potential non-target effects are expected to improve even further in future (Hinz *et al.*, 2019).

Biological control involves an initial investment to locate and assess the safety of candidate biological control agents (Moran *et al.*, 2013) but once introduced, the agents are largely self-sustaining and control is achieved at very little cost in perpetuity. Several studies in Australia and South Africa have demonstrated highly favourable returns on investment, often in excess of 100:1 (Page and Lacey, 2006; van Wilgen and De Lange, 2011).

Several biological control agents have been released against invasive *Prosopis* species in Australia and South Africa (van Klinken, 2014; Kleinjan *et al.*, 2021) (Table 10.1). In South Africa, biological control was originally restricted to the use of seed-feeding insects rather than more destructive agents, so that the putative benefits of the trees could be retained (Zimmermann, 1991). This proved to be ineffective and the biological control programme was expanded to include flower-feeding insects in 2001. Wise *et al.* (2012) conducted an economic evaluation of *Prosopis* trees in South Africa and concluded that the cost of impacts would soon exceed the value of benefits as the invasions grew. This provided justification for considering potential agents that directly impact vegetative growth, which began in 2014 and led to the release of a leaf-tying gelechiid *Agnippe* species (previously known as *Evippe* species) in February 2021 (Kleinjan *et al.*, 2021). *Agnippe* species had earlier been deployed against *Prosopis* trees in Australia, where it had proved to be effective (van Klinken, 2014). The moths spread across a large area and reached high and sustained defoliation rates of ca. 90% despite interannual variations in rainfall, resulting in asynchronous flushing, very low seed production and tree death (van Klinken, 2012). Its effectiveness has yet to be assessed in South Africa.

The prospects for finding additional agents in future are good, because *Prosopis* trees support a rich assemblage of largely unexplored and potentially damaging phytophages in their native ranges (Kleinjan *et al.*, 2021). Experience in South Africa has demonstrated that, even with substantial spending on mechanical and chemical treatments, *Prosopis* invasions have continued to grow. It is now largely accepted that biological control offers the only real hope of achieving widespread control of these aggressive invaders at a landscape scale (Kleinjan *et al.*, 2021). Biological control will only reduce the number and density of the trees, however, and will not eliminate the species from the landscape and biomass will remain available for use. Should biological control prove as effective in Africa as it has been in Australia, the prospects for integrated control (where the use of mechanical and chemical treatments become feasible because of biological control) would be much improved.

10.2.6 Rehabilitation of cleared sites

To prevent reestablishment of *Prosopis* trees or invasion by other weeds, it would be important for cleared sites to be rehabilitated to a suitable alternative vegetation cover. While sites that were lightly invaded could be left to self-recover (passive restoration), heavily invaded sites would require further interventions to establish such a cover (active rehabilitation; see Holmes *et al.*, 2020). Rehabilitation efforts would be helped by the development of lists of indigenous tree and herbaceous species for each region from where *Prosopis* trees are to be removed. These species can then be planted and nurtured as appropriate. This is especially important because characteristic, perennial grassland species may not become established following the removal of

Table 10.1. Biological control agents that have been released against invasive *Prosopis* species in Australia and South Africa. Data from Winston *et al.* (2014), van Klinken *et al.* (2009) and Zachariades (2021).

Biological control agent	Feeding guild	Notes	Recorded impact
Algarobius bottimeri Kingsolver (Coleoptera: Chrysomelidae: Bruchinae)	Seed feeder	Released in South Africa in 1990 and Australia in 1997 but failed to establish.	None
Algarobius prosopis (LeConte) (Coleoptera: Chrysomelidae: Bruchinae)	Seed feeder	Released in South Africa in 1987; established and redistributed to Namibia. Also released in Australia in 1996. Established in several north-eastern African countries and the Arabian Peninsula, including Egypt, Kenya and Tanzania, from unknown source.	Substantial damage to seeds on the ground and in the soil (Roberts, 2006; van Klinken and White, 2014).
Coelocephalapion gandolfoi Kissinger (Coleoptera: Brentidae: Apioninae)	Seed feeder	A permit has been granted for its release in South Africa, but it has not yet been released.	None
Evippe sp. #1 (now known as *Agnippe* sp. #1) (Lepidoptera: Gelechiidae)	Leaf miner, tier and feeder	Released in Australia in 1998; established. Released in South Africa in 2021, establishment yet to be confirmed.	Very effective at reducing the vigour and seed output of adult plants. Can cause death of stressed trees.
Neltumius arizonensis (Schaeffer) (Coleoptera: Chrysomelidae: Bruchinae)	Seed feeder	Released in South Africa in 1993; established.	1–10% seed destruction; *A. prosopis* is more effective, partly because it can access more life stages.
Prosopidopsylla flava Burckhardt	Sap sucker	Released in Australia in 1998; tenuously established.	None

Prosopis trees (Nerlekar *et al.*, 2022). The desired ecosystem could include continuous crop production in agricultural areas, functional rangelands where free-ranging livestock occur (Chapter 7, this volume) or natural ecosystems in protected areas (Chapter 5 this volume). If the cleared land is used for crop production, regular weeding can be used to remove any new seedlings. Similarly, increased vegetation cover in restored grasslands would help to prevent the reestablishment of *Prosopis* trees on the cleared land. In addition, crop or forage production, or restoring grasslands can help to support local livelihoods once the trees are gone.

Growing crops without a fallow period, and ensuring that the land is weeded regularly, can deplete the seed bank of *Prosopis* trees within a few years (R. Eschen, unpublished data; O. Kayaa, unpublished data). In addition crop production can be a lucrative activity, which can lead to significant net financial benefits within a few years (Fig. 10.4) and is preferred by stakeholders over any benefits provided by *Prosopis* trees (Bekele *et al.*, 2018a; Eschen *et al.*, 2024).

The restoration of ecosystem functioning and structure is rarely considered after control of invasive species due to lack of funding to support active restoration activities. As a result,

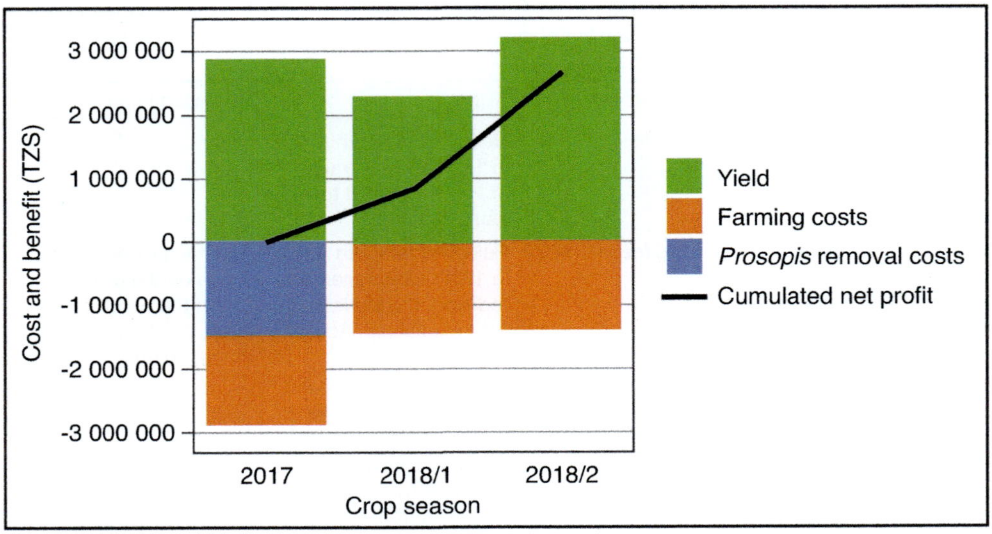

Fig. 10.4. Evolution of costs and benefits on a single farm in Kahe, Tanzania, where an SLM practice was implemented on 1 ha in three subsequent seasons. The practice consisted of manually uprooting *Prosopis* trees and subsequently continuously growing crops (maize and beans) for sale in the nearby town of Moshi. Data provided by M. Mijay.

many sites where invasive species have been removed are left to passive regeneration (van Wilgen *et al.*, 2022). Although the vegetation that establishes after control of invasive alien species may be dominated by other alien species (Clewley *et al.*, 2012), natural regeneration can in many other cases restore degraded ecosystems (Lohbeck *et al.*, 2021). The degree to which unassisted natural regeneration of grasslands happens following removal of *Prosopis* trees may depend on the length of time that the trees have dominated and whether the seed bank still contains native species (Holmes *et al.*, 2020). For example, Linders *et al.* (2019) found that the effect of *Prosopis* trees on herbaceous biomass was more pronounced in Afar, Ethiopia, than in Baringo, Kenya, which appears to be the result of the long history of land degradation in Baringo. Conversely, diverse vegetation developed at sites in Kenya and Tanzania following control and grassland restoration practices, even in plots where other grasses were sown (Eschen *et al.*, 2023), indicating that the seed bank contained viable seeds of native plants, despite *Prosopis* trees having been present for between 5 and 10 years at both locations (Fig. 10.5).

A prerequisite for successful vegetation recovery is protection from grazing for 1–2 years until the vegetation has become established, followed by ensuring reduced grazing pressure (Wairore *et al.*, 2015; Mureithi *et al.*, 2016). A comprehensive overview of ecosystem restoration following removal of invasive plants is provided in recent work from South Africa (Holmes *et al.*, 2020).

10.3 Utilization as a Control Method

The high cost of controlling *Prosopis* trees may be a barrier to control on large areas, especially for households that have low income or capital (Eschen *et al.* 2024). Making charcoal from the removed biomass, including large roots, provides cash income that can cover some of the cost and act as motivation to clear *Prosopis*-invaded land, even if net financial benefit may only be realized after two or more crop harvests (e.g. O. Kaaya, unpublished data, Fig. 10.4). Utilization of *Prosopis* trees as a resource has been promoted in many countries, including in

Fig. 10.5. Productive, diverse grassland (A) can establish naturally from the seedbank following removal of *Prosopis* trees (seen in B, prior to removal), provided the area is protected from grazing, such as here in Ethiopia and Tanzania. Photos courtesy of René Eschen.

eastern Africa, as a control measure (Chapter 2, this volume). However, utilization as the sole management approach has not been shown to be effective as the abundance of the species is not reduced and the rate of spread is unchanged (Box 10.1).

In areas where complete removal of *Prosopis* trees is unfeasible, utilization may continue to provide a source of income, as in such heavily invaded areas many households have diversified their income sources by generating income from charcoal sales which supports their traditional livestock and/or crop-dominated incomes (Linders *et al.*, 2020). Where utilization is part of management, it should ensure permanent removal of *Prosopis* trees, and some households may lose an income source. In the long run, it is important to replace removed trees by planting indigenous or non-invasive tree species that will provide similar benefits once they have grown to a suitable size. In the short term, support for households and other stakeholders that rely on *Prosopis* trees for a substantial part of their livelihoods should be provided during the implementation of management plans. Such solidarity is important to ensure broad community support for and sustainability of management of *Prosopis* invasions on a large scale.

10.4 Context-Dependent Sustainable Land Management Practices

Implementation of any of the methods mentioned above in isolation is unlikely to be sufficient to achieve management goals, and a combination of methods needs to be integrated into sustainable land management (SLM) practices that manage the species and prevent (re-)invasion (van Klinken *et al.*, 2009). Implementation of SLM practices to manage *Prosopis* tree invasions can contribute to the long-term production and ecological potential, as well as the provision of ecosystem services by the landscape through rational use of land and its resources in a way that meets the demands of multiple stakeholder groups (Alemu, 2016).

Several SLM practices may be suitable for managing *Prosopis* trees at different stages of the invasion. Such practices may be adopted or adapted to local conditions, by ensuring that the preferences and knowledge of stakeholders are considered in the selection process. Practices may also be defined based on the land tenure system, local knowledge and experience, but it is advisable to test their effectiveness and acceptance prior to recommending these as part of a landscape-level management plan. Thus, selection of the most suitable SLM practices should involve a diverse group of stakeholders who represent different (local and subnational) interests and who bring different experiences and knowledge (Pärli, 2023). Together, for example through a process of multicriteria decision making, these stakeholders should agree on the most appropriate SLM practices to achieve stated management goals to ensure financial feasibility and sociocultural acceptance (Schwilch *et al.*, 2012).

In the case of managing *Prosopis* trees, selection of the most appropriate tool for mechanical removal depends on the size of the trees, resource availability and the size of the area from which the trees are to be removed. Mechanical control using simple tools is the most common practice used by small-scale farmers to control *Prosopis* trees in Africa (Shackleton *et al.*, 2014). However, while mechanical control is effective for removing invading *Prosopis* trees, it is also the costliest method (Shackleton *et al.*, 2014). For example, manual removal of *Prosopis* trees from 100 m^2 plots took ca. 3–10 times longer than treating the trees on the same area using basal bark or cut stump application of herbicides (Eschen *et al.*, 2023). Thus, these chemical methods allow clearing of *Prosopis* trees from a larger area in the same amount of time. However, the latter two methods leave the dead stems or stumps in the area, limiting the options for subsequent use of the land as tilling or ploughing is not possible. Moreover, if the cut stump method is selected, application of herbicides and operation of chainsaws require training to ensure safety whereas using simple tools for mechanical removal can be done by anyone. Finally, herbicides for use on *Prosopis* trees are not yet registered and available in all countries.

Classical biological control is a cost-efficient, long-term method for large, landscape or country-wide invasions, since hardly any costs are incurred once the agent is released and established. The initial investment in research, propagation and release is also small compared to mechanical and chemical methods, but decisions about the introduction of agents are taken by national governments and not by individual landowners and communities. At present, no biological agents against *Prosopis* trees have been released in eastern Africa (although the seed-feeding bruchid beetle *Algarobius prosopis* is present), yet biological control needs to be applied, in combination with other solutions, to achieve large-scale control of invasive *Prosopis* trees.

One SLM practice for mechanical control of *Prosopis* was developed and tested in arable land in Kahe, Tanzania, for 2 years (2018–2020) as part of the Woody Weeds project. The practice was identified by participants representing various local stakeholder groups as well as project scientists. The practice involved cutting the *Prosopis* tree using a machete or/and axe and digging out the rootstock using a hand hoe. The removed biomass was used to make charcoal or firewood, and annual crops were cultivated on the cleared land. Farm management included irrigation and ongoing removal of emerging *Prosopis* seedlings. The density of *Prosopis* trees was significantly lower after four cropping seasons (2 years; Fig. 10.6). Farmers involved in the trials reported that the implementation of the SLM was costly, especially during the first cropping cycle (Fig. 10.4). Data from one farmer who implemented the practice independently on a nearby field that was heavily invaded by *Prosopis* confirmed the high initial cost of implementing the SLM practice but showed that net profit can be made in the following years (Fig. 10.4). However, when people continue to implement the SLM practice over time, their perceptions of the approach improve (R. Eschen, unpublished data).

10.5 Prospects for Future Control

Efforts to control *Prosopis* invasions in eastern Africa have not been successful to date, as revealed by the continuous spread of the species in areas where it has been introduced (Mbaabu *et al.*, 2019; Shiferaw *et al.*, 2019b) (Box 10.1) and the small number of instances where *Prosopis* tree removal has resulted in the lasting disappearance of the trees in Baringo, Kenya (Adoyo *et al.*, 2022) (Box 10.1), Australia (van Klinken *et al.*, 2009) and South Africa. A recent analysis of spatiotemporal trends of *Prosopis* in Baringo, Kenya, revealed that management efforts are confined to small areas where land

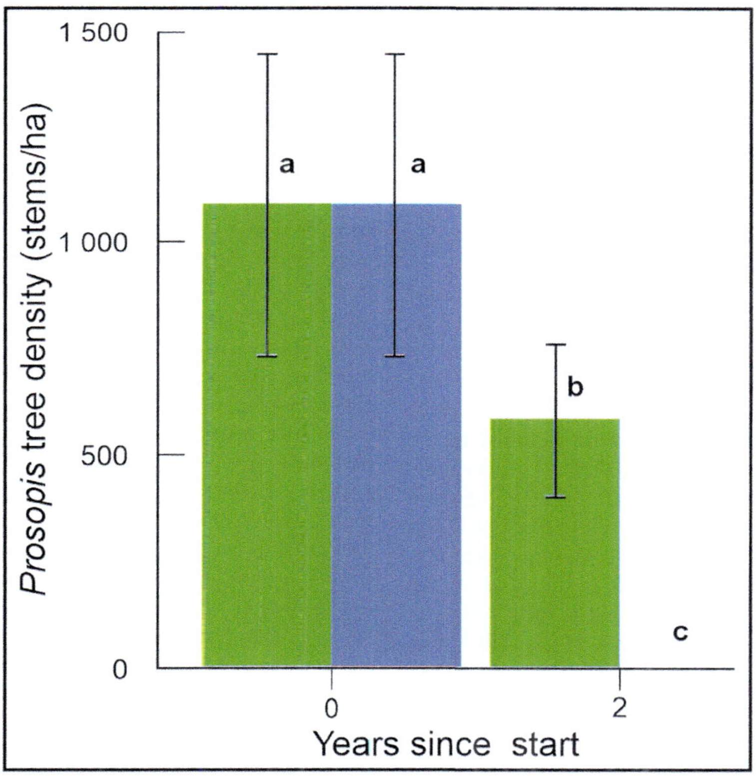

Fig. 10.6. Change in the density of *Prosopis* trees after the application of mechanical control (SLM) at the start of experiment and after four cropping seasons (2 years). Green and blue bars indicate (1) the effect of cutting aboveground *Prosopis* biomass followed by crop cultivation (standard local practice) and (2) uprooting of *Prosopis* followed by continuous cultivation (the SLM), respectively. Error bars indicate one SE. The letters a, b and c indicate significant differences between *Prosopis* tree densities.". Redrawn from Malila (2021).

users expect to derive economic benefits (Adoyo *et al.*, 2022). Unmanaged areas, meanwhile, act as sources of seed for reinvasion of the managed areas and a coordinated *Prosopis* management plan at a landscape or regional level should be developed and implemented (Epanchin-Niell *et al.*, 2010). However, control efforts in the Australian Pilbara region, which were initiated with financial support from the government and in consultation with private and government stakeholders, reduced the density of *Prosopis* trees, but eradication was not achieved due to early withdrawal of funds, long-lived seed banks and conflicts between the government and landowners (van Klinken *et al.*, 2009).

These examples show that local eradication of *Prosopis* is rarely possible and highlight the need for combinations of control options on the landscape scale that are part of a management strategy, as any single management method is unlikely to be effective in controlling *Prosopis*. Combinations of control options may include effective biological control agents, chemical treatment and mechanical removal of isolated stands or on valuable land, as well as management of livestock to minimize seed dispersal (van Klinken *et al.*, 2009).

In South Africa, the total area invaded by *Prosopis* trees increased from 127,821 ha to 1,473,953 ha between 1974 and 2007, a mean annual increase of 7.4% (van den Berg *et al.*, 2013). Formal attempts to bring these invasions under control were initiated in the 1980s and involved two distinct approaches – biological control and, later in 1995, government support to farmers in the form of teams that conducted mechanical clearing operations. Several authors identified weaknesses in the approach adopted, including a failure to plan, set goals and monitor progress (van Wilgen and Wannenburgh, 2016; van Wilgen *et al.*, 2022). They recommended that a national strategy should be based on conservation triage, focusing on clearly defined priority sites, improving planning and monitoring, and increasing operational efficiency.

In addition to these two initiatives Shackleton *et al.*, (2017) proposed a national strategy to optimize the management of *Prosopis* species in South Africa. The strategy identified different control objectives (prevention of spread to unoccupied areas, local eradication, containment and asset protection), which were based on the degree of invasion in each of the 234 local municipalities in the country. The implementation of the strategy would have to be coordinated at a national level but it has yet to be formally adopted.

Work to implement an integrated, coordinated approach to managing invasive *Prosopis* trees in South Africa has been done in terms of the development of the national strategy, formation of a working group representative of the multiplicity of stakeholders, supporting landowners with farm-scale planning for management of *Prosopis* trees, development of a mass-rearing centre that will employ people to rear the *Agnippe* biological control agent, identifying opportunities to utilize biomass from *Prosopis* in construction, carbon fixing (biochar) and biogas – heat and power generation, and application for funding to support the projects above.

A survey of farmers in South Africa found some evidence that government-supported programmes to control invasive *Prosopis* trees were not as effective as control undertaken and funded by the farmers themselves (Barnard, 2021). However, farming in the areas where *Prosopis* trees are invasive is typically only marginally profitable, and farmers can seldom afford the expensive control interventions needed to contain or reverse the spread of *Prosopis* tree invasions. Where farmers do spend their own money on control, they are more likely to be more careful and effective because it is a personal investment. On the other hand, government-funded interventions can often be less effective for a range of reasons (e.g. McConnachie *et al.*, 2012; Kraaij *et al.*, 2017). Because of these factors, Barnard (2021) postulated that government funding should be used to carry out the initial clearing, and that this should then be followed up by the farmers themselves. Furthermore, this dual funding model was only likely to be financially sustainable if biological control could be introduced so that the effort required for ongoing control would be reduced to affordable levels.

Given the already widespread occurrence of *Prosopis* trees in parts of eastern Africa (Mbaabu *et al.*, 2019; Shiferaw *et al.*, 2019b) and potential for spread to a much larger area (Eckert *et al.*, 2020), it is important to focus on containing the species to its current distribution. Further spread should be minimized and efforts made to remove individual trees and small populations from areas with sparse infestations (EDRR), as long as this is still possible using mechanical or targeted chemical means, combined with preventing reintroduction to those areas. In areas where the invasion is well established, it will be impossible to remove the species entirely. Local eradication should be undertaken where the cleared land can be used for agriculture, grazing or other land uses that will prevent reinvasion, but this would only be possible on relatively small areas because of the prohibitive costs or environmental impacts of using herbicides on large areas.

Even if landscape-level or subnational management plans could be successfully implemented, *Prosopis* trees will remain present in large parts of eastern Africa, and their impacts may only be reduced by introducing effective classical biological control agents. Host-specific and safe agents have been identified through rigorous scientific research and have been safely introduced in Australia and South Africa where they contribute to reducing the spread and impacts of *Prosopis* in areas where it would be impossible to control the species by other means (Zimmermann, 1991; Zachariades *et al.*, 2011).

While *Prosopis* control on small scales may be part of ongoing management practices and may not require additional or only small amounts of resources, effective management should sustainably be integrated into landscape-level management plans. Such sustainable large-scale management requires coordination among stakeholders and significant resources. Integration of *Prosopis* tree management in national invasive species management strategies, such as the recent Ethiopian, Kenyan and Tanzanian strategies (Chapter 2, this volume), and mainstreaming in subnational and local land use planning enables prioritization of research and management approaches, such as biological control, and budget allocation to support research and implementation of management activities.

10.6 Implications and Management Recommendations

Defining the objectives of management and developing a monitoring plan for tracking progress are vital for successful management of invasive alien species. The nature and extent of the invasion determine the most suitable and effective method for managing the spread and abundance of *Prosopis* trees. However, integrating several management methods into sustainable practices, for example a combination of removal, restoration and rehabilitation, is needed for achieving management goals ranging from prevention of spread into uninvaded areas to control and asset protection in densely invaded areas. Moreover, by adopting a coordinated approach that targets a landscape rather than a plot-level scale, management is more likely to be successful. This is because it can deal more effectively with the complex nature of *Prosopis* invasions that spread rapidly across boundaries, and the diversity of affected stakeholders and required actors. The appropriateness and selection of best practices should thus be decided by groups of stakeholders through joint decision-making processes.

Further, the relatively low cost of and benefits derived from early invasion management are an incentive to manage invasions sooner rather than later. Leveraging the perceived benefits of management efforts through cost-effective management practices is therefore a promising pathway for promoting participatory control of *Prosopis* tree invasions. Some efficient and cost-effective tools that are currently unavailable (e.g. herbicides or biological control agents) should be prioritized for registration or approval in eastern African countries, as they allow for safe control of *Prosopis* invasions on larger areas and with fewer resources than what is possible using manual removal.

Acknowledgements

This work was supported by the Swiss Programme for Research on Global Issues for Development (r4d), funded by the Swiss National Science Foundation and the Swiss Agency for Development and Cooperation, through the project 'Woody invasive alien species in East Africa: Assessing and mitigating their negative impact on ecosystem services and rural livelihood' (Grant Number: 400440_152085). RE and MM were supported by CABI with core financial support from its member countries. CABI is an international intergovernmental organization, and we gratefully acknowledge the core financial support from our member countries (and lead agencies) including the United Kingdom (Foreign, Commonwealth and Development Office), China (Chinese Ministry of Agriculture and Rural Affairs), Australia (Australian Centre for International Agricultural Research), Canada (Agriculture and Agri-Food Canada), Netherlands (Directorate-General for International Cooperation) and Switzerland (Swiss Agency for Development and Cooperation). See https://www.cabi.org/about-cabi/who-we-work-with/key-donors/ for full details. This work also received funding from the Darwin Initiative (Project DARCC013). BWvW thanks the National Research Foundation of South Africa for funding (Grant Numbers 150260 and 151653).

References

Abbas, A.M., Mancilla-Leytón, J.M. and Castillo, J.M. (2018) Can camels disperse seeds of the invasive tree *Prosopis juliflora*. *Weed Research* 58, 221–228.

Abdulahi, M.M., Ute, J. and Regasa, T. (2017) *Prosopis juliflora* L: Distribution, impacts and available control methods in Ethiopia. *Tropical and Subtropical Agroecosystems* 20, 75–89.

Adoyo, B., Schaffner, U., Mukhovi, S., Kiteme, B., Mbaabu, P.R. *et al.* (2022) Spatiotemporal trajectories of invasive tree species reveal the importance of collective action for successful invasion management. *Journal of Land Use Science* 17, 487–504.

Alemu, M.M. (2016) Sustainable land management. *Journal of Environmental Protection* 07, 502–506.

Alvarez, M., Leparmarai, P., Heller, G. and Becker, M. (2017) Recovery and germination of *Prosopis juliflora* (Sw.) DC seeds after ingestion by goats and cattle. *Arid Land Research and Management* 31, 71–80.

Barnard, J.J. (2021) Prosopis invasion in the Northern Cape: Remote sensing analysis of management action effectiveness. MSc thesis, University of Stellenbosch, Stellenbosch.

Bekele, K., Haji, J., Legesse, B. and Schaffner, U. (2018a) Economic impacts of *Prosopis* spp. invasions on dryland ecosystem services in Ethiopia and Kenya: Evidence from choice experimental data. *Journal of Arid Environments* 158, 9–18.

Bekele, K., Haji, J., Legesse, B., Shiferaw, H. and Schaffner, U. (2018b) Impacts of woody invasive alien plant species on rural livelihood: Generalized propensity score evidence from *Prosopis* spp. invasion in afar region in Ethiopia. *Pastoralism* 8, 1–17.

Choge, S., Mbaabu, P.R. and Muturi, G.M. (2022) Management and control of the invasive *Prosopis juliflora* tree species in Africa with a focus on Kenya. In: Puppo, M.C. and Felker, P. (eds) *Prosopis as a Heat Tolerant Nitrogen Fixing Desert Food Legume: Prospects for Economic Development in Arid Lands*. Academic Press, New York, USA, pp. 67–81.

Clewley, G.D., Eschen, R., Shaw, R.H. and Wright, D.J. (2012) The effectiveness of classical biological control of invasive plants. *Journal of Applied Ecology* 49, 1287–1295.

Eckert, S., Hamad, A., Kilawe, C.J., Linders, T.E., Ng, W.T. *et al.* (2020) Niche change analysis as a tool to inform management of two invasive species in Eastern Africa. *Ecosphere* 11(2), e02987.

Epanchin-Niell, R.S., Hufford, M.B., Aslan, C.E., Sexton, J.P., Port, J.D. *et al.* (2010) Controlling invasive species in complex social landscapes. *Frontiers in Ecology and the Environment* 8, 210–216.

Eschen, R., Bekele, K., Mbaabu, P.R., Kilawe, C.J. and Eckert, S. (2021) *Prosopis juliflora* management and grassland restoration in Baringo county, Kenya: Opportunities for soil carbon sequestration and local livelihoods. *Journal of Applied Ecology* 58, 1302–1313.

Eschen, R., Bekele, K., Jumanne, Y., Kibet, S., Makale, F. *et al.* (2023) Experimental *Prosopis* management practices and grassland restoration in three Eastern African countries. *CABI Agriculture and Bioscience* 4, 21.

Eschen, R., Kaaya, O.E., Kilawe, C.J. *et al.* (2024) Adoption of a sustainable land management practice for invasive *Prosopis juliflora* in East Africa. *CABI Agriculture and Bioscience* 5, 113.

Harper, J.L. (1977) *Population Biology of Plants*. Academic Press, London, UK.

Heimpel, G.E. and Mills, N.J. (2017) *Biological Control*. Cambridge University Press, Cambridge, UK.

Hinz, H.L., Winston, R.L. and Schwarzländer, M. (2019) How safe is weed biological control? A global review of direct nontarget attack. *Quarterly Review of Biology* 94, 1–27.

Holling, C.S. and Walters, C. (1978) *Adaptive Environmental Assessment and Management*. John Wiley and Sons, Chichester, UK.

Holmes, P.M., Esler, K.J., van Wilgen, B.W. and Richardson, D.M. (2020) Ecological restoration of ecosystems degraded by invasive alien plants in South African Fynbos: Is spontaneous succession a viable strategy?. *Transactions of the Royal Society of South Africa* 75, 111–139.

Impson, F.A.C., Hoffmann, J.H., Impson, O.R., Kleinjan, C.A. and Moran, V.C. (2024) Densities of a perennial invasive tree, *Acacia cyclops*, decline in the 20 years since inception of biological control with two seed-reducing agents, a flower-galling midge and a seed-feeding weevil. *Biological Control* 189, 105442.

Keane, R.M. and Crawley, M.J. (2002) Exotic plant invasions and the enemy release hypothesis. *Trends in Ecology & Evolution* 17, 164–170.

Kleinjan, C.A., Hoffmann, J.H., Heystek, F., Ivey, P. and Kistensamy, Y. (2021) Developments and prospects for biological control of *Prosopis* (Leguminosae) in South Africa. *African Entomology* 29, 859–874.

Kraaij, T., Baard, J.A., Rikhotso, D.R., Cole, N.S. and van Wilgen, B.W. (2017) Assessing the effectiveness of invasive alien plant management in a large fynbos protected area. *Bothalia* 47, a2105.

Linders, T.E., Bekele, K., Schaffner, U., Allan, E., Alamirew, T. *et al.* (2020) The impact of invasive species on social-ecological systems: Relating supply and use of selected provisioning ecosystem services. *Ecosystem Services* 41, 101055.

Linders, T.E.W., Schaffner, U., Eschen, R., Abebe, A., Choge, S.K. *et al.* (2019) Direct and indirect effects of invasive species: Biodiversity loss is a major mechanism by which an invasive tree affects ecosystem functioning. *Journal of Ecology* 107, 2660–2672.

Lohbeck, M., Rother, D.C. and Jakovac, C.C. (2021) Editorial: Enhancing natural regeneration to restore landscapes. *Frontiers in Forests and Global Change* 4, 735457.

Mahgoub, O., Kadim, I.T. and Forsberg, N.E. (2005) Evaluation of meskit (*Prosopis juliflora*) pods as a feed for goats. *Animal Feed Science and Technology* 121, 319–327.

Malila, B.P. (2021) Socio-economic and ecological implications of sustainable land management practices for controlling invasive alien species in Moshi rural and Muheza districts, Tanzania. Master's thesis, Sokoine University of Agriculture, Morogoro, Tanzania.

Mbaabu, P.R., Ng, W.T., Schaffner, U., Gichaba, M., Olago, D. *et al.* (2019) Spatial evolution of *Prosopis* invasion and its effects on LULC and livelihoods in Baringo, Kenya. *Remote Sensing* 11, 1217.

McConnachie, M.M., Cowling, R.M., van Wilgen, B.W. and McConnachie, D.A. (2012) Evaluating the cost-effectiveness of invasive alien plant clearing: A case study from South Africa. *Biological Conservation* 155, 128–135.

Moran, V.C., Hoffmann, J.H. and Zimmermann, H.G. (2013) 100 years of biological control of invasive alien plants in South Africa: History, practice and achievements. *South African Journal of Science* 109(9/10), Art. #a0022.

Mureithi, S.M., Verdoodt, A., Njoka, J.T., Gachene, C.K., Warinwa, F. *et al.* (2016) Impact of community conservation management on herbaceous layer and soil nutrients in a Kenyan semi-arid savannah. *Land Degradation & Development* 27, 1820–1830.

Mwangi, E. and Swallow, B. (2005) *Invasion of Prosopis juliflora and Local Livelihoods: Case Study from the Lake Baringo Area of Kenya*. ICRAF Working Paper 3. World Agroforestry Centre, Nairobi, Kenya.

Nehrbass, N., Winkler, E., Müllerová, J., Pergl, J., Pyšek, P. *et al.* (2007) A simulation model of plant invasion: long-distance dispersal determines the pattern of spread. *Biological Invasions* 9, 383–395.

Nerlekar, A.N., Mehta, N., Pokar, R., Bhagwat, M., Misher, C. *et al.* (2022) Removal or utilization? Testing alternative approaches to the management of an invasive woody legume in an arid Indian grassland. *Restoration Ecology* 30, e13477.

Osmond, R., March, N., Cobon, R., Campbell, S. and van Klinken, R.D. (2003) The mesquite control toolbox. In: *Best Practice Manual: Mesquite. Control and Management Options for Mesquite (Prosopis Spp.) in Australia*. State of Queensland (Department of Natural Resources and Mines), Brisbane, Australia, pp. 37–60.

Page, A.R. and Lacey, K.L. (2006) Economic impact assessment of Australian weed biological control. *CRC for Australian Weed Management*. Canberra, Australia.

Pärli, R. (2023) How input, process, and institutional factors influence the effects of transdisciplinary research projects. *Environmental Science & Policy* 140, 80–92.

Qasem, J.R. (2007) Chemical control of *Prosopis farcta* (banks and sol.) Macbride in the Jordan valley. *Crop Protection* 26, 572–575.

Roberts, A.P. (2006) Biological control of alien invasive mesquite species (*Prosopis*) in South Africa: The role of introduced seed-feeding bruchids. PhD thesis, University of Cape Town, Cape Town, South Africa.

Schwilch, G., Bachmann, F. and Graaff, J. (2012) Decision support for selecting SLM technologies with stakeholders. *Applied Geography* 34, 86–98.

Sequeira, R. and Griffin, R. (2013) The biosecurity continuum and trade: Pre-border operations. In: McKirdy, S. and Gordh, V. (eds) *The Handbook of Plant Biosecurity: Principles and Practices for the Identification, Containment and Control of Organisms That Threaten Agriculture and the Environment Globally*. Springer, Dordrecht, The Netherlands, pp. 119–148.

Shackleton, R.T., Le Maitre, D.C., Pasiecznik, N.M. and Richardson, D.M. (2014) *Prosopis*: A global assessment of the biogeography, benefits, impacts and management of one of the world's worst woody invasive plant taxa. *AoB PLANTS* 6, 1–8.

Shackleton, R.T., Le Maitre, D.C., van Wilgen, B.W. and Richardson, D.M. (2017) Towards a national strategy to optimise the management of a widespread invasive tree (*Prosopis* species; mesquite) in South Africa. *Ecosystem Services* 27, 242–252.

Shanwad, U.K., Chittapur, B.M., Honnalli, S.N., Shankergoud, I. and Gebremedhin, T. (2015) Management of *Prosopis juliflora* through chemicals: A case study in India. *Journal of Biology, Agriculture and Health Care* 5, 30–38.

Shiferaw, H., Bewket, W., Alamirew, T., Zeleke, G., Teketay, D. *et al.* (2019a) Implications of land use/land cover dynamics and *Prosopis* invasion on ecosystem service values in Afar region, Ethiopia. *Science of the Total Environment* 675, 354–366.

Shiferaw, H., Bewket, W. and Eckert, S. (2019b) Performances of machine learning algorithms for mapping fractional cover of an invasive plant species in a dryland ecosystem. *Ecology and Evolution* 9, 2562–2574.

Simberloff, D. (2009) We can eliminate invasions or live with them. Successful management projects. In: Langor, D.W. and Sweeney, J. (eds) *Ecological Impacts of Non-Native Invertebrates and Fungi on Terrestrial Ecosystems*. Springer, Berlin, Germany, pp. 149–157.

van den Berg, E.C., Kotze, I. and Beukes, H. (2013) Detection, quantification and monitoring of *Prosopis* in the Northern Cape province of South Africa using remote sensing and GIS. *South African Journal of Geomatics* 2, 68–81.

van Klinken, R.D. (2014) Mesquite (*Prosopis* species) biological control in Australia. In: *Proceedings of the XIV International Symposium on Biological Control of Weeds, Kruger National Park, South Africa*, University of Cape Town, pp. 100–101.

van Klinken, R.D. and White, A.J. (2014) The role of pre- and post-dispersal seed predation in determining total seed loss. *Basic and Applied Ecology* 15(7), 581–589.

van Klinken, R.D., Campbell, S.D., Miniappan, R., Reddy, G.V.P. and Raman, A. (2009) *Prosopis Species*. Cambridge University Press, Cambridge, UK, pp. 353–379.

van Klinken, R.D. (2012) *Prosopis* spp. – mesquite. In: Cullen, J., Julien, M. and McFadyen, R. (eds) *Biological Control of Weeds in Australia. CSIRO*. Melbourne, Australia, pp. 477–485.

van Wilgen, B.W. and De Lange, W.J.D. (2011) The costs and benefits of biological control of invasive alien plants in South Africa. *African Entomology* 19, 504–514.

van Wilgen, B.W., Moran, V.C. and Hoffmann, J.H. (2013) Some perspectives on the risks and benefits of biological control of invasive alien plants in the management of natural ecosystems. *Environmental Management* 52, 531–540.

van Wilgen, B.W., Wannenburgh, A. and Wilson, J.R.U. (2022) A review of two decades of government support for managing alien plant invasions in South Africa. *Biological Conservation* 274, 109741.

van Wilgen, B.W. and Wannenburgh, A. (2016) Co-facilitating invasive species control, water conservation and poverty relief: Achievements and challenges in South Africa's working for water programme. *Current Opinion in Environmental Sustainability* 19, 7–17.

Vidhya, R., Vijayasekaran, D. and Ramakrishnan, S.S. (2017) Mapping invasive plant *Prosopis juliflora* in arid land using high resolution remote sensing data and biophysical parameters. *Indian Journal of Geo-Marine Sciences* 46, 1135–1144.

Wairore, J.N.U., Mureithi, S.M., Wasonga, O.V. and Nyberg, G. (2015) Characterization of enclosure management regimes and factors influencing their choice among agropastoralists in North-Western Kenya. *Pastoralism: Research, Policy and Practice* 5, 1–10.

Wilgen, B.W. and Wannenburgh, A. (2016) Co-facilitating invasive species control, water conservation and poverty relief: Achievements and challenges in South Africa's working for water programme. *Current Opinion in Environmental Sustainability* 19, 7–17.

Wilhere, G.F. (2002) Adaptive management in habitat conservation plans. *Conservation Biology* 16, 20–29.

Winston, R.L., Schwarzländer, M., Hinz, H.L., Day, M.D., Cock, M.J. *et al.* (2014) *Biological Control of Weeds: A World Catalogue of Agents and Their Target Weeds*. USDA Forest Service, Morgantown, USA.

Wise, R.M., van Wilgen, B.W. and Le Maitre, D.C. (2012) Costs, benefits and management options for an invasive alien tree species: The case of mesquite in the Northern Cape, South Africa. *Journal of Arid Environments* 84, 80–90.

Zachariades, C. (2021) A catalogue of natural enemies of invasive alien plants in South Africa: Classical biological control agents considered, released and established, exotic natural enemies present in the field, and bioherbicides. *African Entomology* 29, 1077–1142.

Zachariades, C., Hoffmann, J.H. and Roberts, A.P. (2011) Biological control of mesquite (*Prosopis species*) (Fabaceae) in South Africa. *African Entomology* 19, 402–415.

Zimmermann, H.G. (1991) Biological control of mesquite, *Prosopis* spp. (Fabaceae), in South Africa. *Agriculture, Ecosystems & Environment* 37, 175–186.

11 Spatially Explicit Management of *Prosopis* Tree Invasions in Eastern Africa

Albrecht Ehrensperger[1]*, Beatrice Adoyo[2,3], Sandra Eckert[1], René Eschen[4], Urs Schaffner[4] and Ross Shackleton[5,6]

[1]Centre for Development and Environment (CDE), University of Bern, Switzerland; [2]Centre for Training and Research in ASAL Development, Nanyuki, Kenya; [3]Department of Geography, Population, and Environmental Studies, University of Nairobi, Kenya; [4]CABI, Delémont, Switzerland; [5]Swiss Federal Institute for Forest Snow and Landscape Research WSL, Birmensdorf, Switzerland; [6]Centre for Invasion Biology, Department of Botany and Zoology, Stellenbosch University, South Africa

Abstract

Management plans that clearly identify spatial priorities are crucial for effectively managing invasive plant species, such as *Prosopis juliflora*, especially when resources are limited and optimal policy implementation needs to be ensured. However, developing such plans requires the consideration of multiple spatial factors. This chapter outlines key factors that influence management options and proposes a decision support tool to aid the selection of appropriate management objectives for specific spatial management units while accounting for these factors. This tool consists of four decision steps that incorporate occurrence and abundance of *Prosopis* tree cover, suitable range for *Prosopis* invasion, pathways of spread and the presence of significant assets that need protection. By following these steps, 11 specific management objectives, divided into four broad management categories (awareness creation, prevention, early detection and rapid response, and control), can be identified. The tool is intended for participatory processes involving various stakeholder groups in affected areas, as knowledge about the local context is crucial for its effective utilization. Additionally, its ability to support spatially explicit management strategies depends on the availability of reliable and up-to-date spatial data on the distribution and suitable habitat range of invasive *Prosopis* trees, on potential pathways of spread, and on the location of assets needing protection. Promising solutions exist to overcome two challenges affecting the future utilization of the decision support tool. The first is unmanned aerial vehicles (UAVs) that might be an alternative or complement to satellite-based earth observation to generate information on the distribution of *Prosopis* invasions. While not suitable for very large areas, UAVs could fill a gap for the mapping of areas up to 100 km^2. The second involves spatial modelling in a geographic information system that might be an option to provide quick overviews of the expanse of various management objectives at subnational to national levels.

*Corresponding author: albrecht.ehrensperger@unibe.ch

© CAB International 2025. *The Ecology and Management of Invasive Prosopis Trees in Eastern Africa* (eds U. Schaffner *et al.*)

DOI: 10.1079/9781800623644.0011

11.1 Introduction

The rapid spread of invasive alien *Prosopis* trees and increasing awareness of their detrimental impacts (Chapters 5–9, this volume) have contributed to putting *Prosopis* invasions on the national political agendas of numerous countries where these trees are invasive (van Klinken and Campbell, 2001; Australian Weeds Committee, 2012; Shackleton *et al.*, 2017).

Considering the scale of the problem, policy and management responses are needed in a timelier manner. For example, in eastern Africa, regions such as the Afar lowlands in Ethiopia and the Lorian Swamps in Kenya have lost immense stretches of valuable grazing land and wetlands to the invasive trees, and many local communities in these areas are overwhelmed by the magnitude of the invasion (Shiferaw *et al.*, 2019b; Chapter 4, this volume). Accordingly, national governments in the region have acknowledged the necessity for strategies at higher decision-making levels: in Ethiopia a National *Prosopis* Strategy (NPS) was initiated and implemented in 2017 (Ministry of Livestock and Fisheries, 2017); in Kenya a NPS was also recently adopted (2023) by the Ministry of Environment and Forestry (2022) to guide the management of *Prosopis* invasions in the country; in Tanzania, a National Invasive Alien Species Strategy and Action Plan was developed in 2019, and *Prosopis* was identified as one of the top priority targets (Anonymous, 2019).

The Kenyan NPS was developed from 2020 to 2022 with support from the Kenya Forestry Research Institute, the Kenya Forest Service, the Kenya Wildlife Service and a group of international scientists (Ministry of Environment and Forestry, 2022). Goal 3.1 of the NPS highlights the need to 'minimise impacts of established invasions and enhance the capacity and commitment of the country to completely manage *Prosopis juliflora* invasions'. Importantly, the Kenyan NPS specifies that these objectives should be achieved based on spatially explicit management plans (i.e. plans that include maps providing overviews of the current and potential coverage of *Prosopis* trees, as well as guidance on the most efficient spatial arrangement of management objectives). Thus, the NPS echoes earlier recommendations from researchers to optimize management actions across different scales with the help of spatial planning (Grice *et al.*, 2011; Wiesmann *et al.*, 2011; Forsyth *et al.*, 2012; Vaz *et al.*, 2020).

For spatially explicit management plans to be realistic, efficient and feasible, they need to consider the highly complex spatiotemporal dynamics of biological invasion processes especially of woody alien species such as *Prosopis* (Pyšek and Hulme, 2005; van den Berg *et al.*, 2013; Hall *et al.*, 2018). In addition, management plans need to consider various factors that might restrict management options such as the size (extent) and accessibility of an invaded area (van Wilgen *et al.*, 2011), administrative boundaries cutting across invaded areas, etc. Further, the need to prioritize actions in relation to the value attributed by various groups of stakeholders to areas that are either invaded or threatened by the invasion should be accounted for (Forsyth *et al.*, 2012). Finally, broader specificities of the social-environmental context in which the invasion takes place must be considered for the design and implementation of spatially explicit invasive species management plans (Larson *et al.*, 2011; Yletyinen *et al.*, 2021). These various factors are presented in more detail in the second part of this chapter.

The Kenyan NPS is a detailed overarching strategy document presented at the national scale that systematically outlines strategic objectives, broader initiatives and expected outcomes (Ministry of Environment and Forestry, 2022). Even though the NPS has a high level of detail, it does not provide guidance on how to integrate complex spatiotemporal dynamics and social-environmental factors into the effective management of *Prosopis* invasions. Hence, the institutions in charge of implementation (national and subnational authorities, with their respective partners) need to decide on appropriate methods and approaches, depending on invasion dynamics, management feasibility, asset valuation and contextual characteristics. Owing to the complexity of biological invasion processes, decision support tools are needed to guide these implementing institutions in codesigning effective management strategies. Such tools must be easily understandable and provide clear guidelines for spatial priority setting and the selection of management objectives.

Based on the above, the overall goal of this chapter is to provide practical guidance for the development of spatially explicit management plans at various scales to support institutions in charge of implementing invasive *Prosopis* tree management in eastern Africa or similar strategies for invasive plant management globally. This overall goal is addressed through two specific objectives: (1) to provide a comprehensive overview of spatially relevant factors that might influence or codetermine the selection of management objectives and should therefore be reflected in codesign processes, and (2) to propose a decision support tool for the spatial prioritization of management objectives in the contexts of codesign processes with stakeholders at subnational to local scales. The methodological framework draws on the management objectives described in other chapters of this volume (particularly Chapters 10 and 12), i.e. prevention, early detection and rapid response (EDRR), and control (either containment or asset protection).

11.2 Factors Influencing the Spatially Explicit Management of *Prosopis* Invasions

Below, we list factors that can influence and are relevant for the spatially explicit management of *Prosopis* tree invasions. The key factors considered important include spread rates and invasion dynamics (e.g. habitat suitability, pathways and land use), spatial variables influencing the ability to cope with *Prosopis* tree invasions (e.g. invasion size and available capital/means) and context-specific factors (e.g. land tenure types, infrastructure and governance systems).

11.2.1 The spread of *Prosopis* trees

There are several factors influencing the spread and spatial dynamics of *Prosopis* invasions, such as habitat suitability (Chapter 4, this volume). *Prosopis* trees only spread to and establish in habitats that are suitable for their growth (in eastern Africa, these are arid and semi-arid lowland areas particularly along water courses and in flood plains; Mbaabu *et al.*, 2019). Unsuitable habitats can, *de facto*, be considered low-priority areas for managing *Prosopis* invasions. However, there are areas where the suitability of habitat for *Prosopis* trees is uncertain, i.e. their establishment is less probable but cannot be totally excluded. Additionally, such areas that are currently not or only partly suitable might become suitable over time due to climate change (Eckert *et al.*, 2020; Sintayehu *et al.*, 2020). Also, species distribution models based on native range data may not necessarily provide an accurate projection of habitat suitability for populations in the invaded range (e.g. Broennimann *et al.*, 2007). Therefore, regular updates of suitability mapping and regular monitoring of potential invasion areas are necessary.

Another factor determining the dynamics and patterns of invasion processes is the presence of various pathways of spread. In the case of *P. juliflora*, these are very often ephemeral and perennial rivers, livestock and wildlife migration routes, and roads (Chapter 4, this volume). River water as well as heavy rainfall transport *Prosopis* pods and their seeds downstream or flood previously uninvaded areas, where they may germinate. Livestock and some wildlife ingest *Prosopis* pods and release their seeds, sometimes in distant places, in their faeces (Shiferaw *et al.*, 2004). Finally, cars and trucks might become vectors of spread along roads, either deliberately by transporting pods for various uses or unintentionally when seeds are inadvertently transported. In addition, the dung from livestock transported along roads can also be an issue.

Human factors can also influence the speed and patterns of spread, such as the deliberate transport of *Prosopis* seeds or the planting of trees in new areas. Further, invasions of *Prosopis* trees are likely to progress faster on land that is seen as unimportant. This is because local communities have fewer incentives to manage *Prosopis* trees on such land than on land they consider to be of high value. Unsolved conflicts related to land use and land tenure might have a similar effect as the conflicting parties tend to leave the land idle and do not invest in management during such conflicts, thus opening the way to invasive species (Adoyo *et al.*, 2022).

11.2.2 Ability to cope with *Prosopis* invasions

It is important to include in spatially explicit management plans variables that might foster or hinder the ability of stakeholders to respond to *Prosopis* invasions. One such variable is the size of the invaded area and the density of *Prosopis* trees on it. Beyond a certain extent or number of trees – a so-called threshold value (Panetta, 2023) – it becomes virtually impossible to remove these trees with the means and methods available to local communities due to a high probability of reinvasion. In such situations the only option available to local stakeholders is to focus on preventing further spread on the fringes of the densely invaded area and along pathways of spread (i.e. via containment, in combination with early detection and rapid response – EDRR) (Panetta and Cacho, 2014).

Another important spatial aspect that might influence *Prosopis* tree management is the accessibility of the invaded area. Remote areas or inaccessible terrain significantly increase the effort required to implement management practices and are therefore more likely to be avoided in the implementation of management plans, particularly when resources are limited. Local stakeholders in charge of clearing such areas would most certainly make a cost–benefit assessment and prioritize areas that are easier to access if available resources (finances and/or labour) do not allow them to cover the entire territory.

11.2.3 Presence of important assets

The urgency to implement management practices might differ from one area to another, even when their habitat suitability, pathways of spread, size of invasion and accessibility are similar. For example, owing to the potential loss to be incurred in case of invasion, stakeholders are likely to perceive areas with important assets and areas providing important ecosystem services as needing management interventions more urgently than others.

However, not all stakeholder groups value and perceive assets – and the land that harbours them – in the same way. For example, dry season grazing areas are crucial assets for pastoralists, the loss of which severely affects their livelihoods and increases the risk of losing livestock during dry spells. Farmers are more concerned about the loss of high-quality agricultural land and access to water for irrigation. Environmental organizations and the tourism industry might be inclined to prioritize *Prosopis* removal from protected areas, while traders rely on transport infrastructure that is kept clear of *Prosopis* trees.

Hence, like other sustainability concerns, *Prosopis* management is inherently normative or, in other words, decisions about how to manage *Prosopis* trees are not just based on scientific facts but also on the concerned stakeholders' values and beliefs. Consequently, a spatial prioritization of management objectives is, by definition, a process requiring codesign and consensus building about priorities, required action and related responsibilities among concerned stakeholders (O'Bryan *et al.*, 2023).

11.2.4 Context-specific drivers

There are many other factors influencing the management of *Prosopis* trees. As they are context-dependent and vary widely from one region to another, we have not included them in our generic decision support tool. These factors depend on access to technology and on dynamic social, economic, political and ecological conditions, such as legal frameworks, government incentives, market integration and natural hazards, which codetermine actions and strategies of local actors (Wiesmann *et al.*, 2011). Below, we focus on those that were identified as being of particular importance in the context of *P. julilfora* management through focus group discussions and interviews with concerned land users in Baringo County, Kenya (Adoyo *et al.*, 2022).

Land tenure arrangements and land rights are important factors to be considered when translating spatially explicit management objectives into concrete practices, particularly in terms of stakeholder mobilization. In areas dominated by private land ownership, individual land users play a central role as actors for the implementation of management practices. On communal land, decisions on how to prioritize management actions and share the management effort must be taken at the community level. Collective governance aspects are also important in communal settings (Ostrom, 1990). Additionally, on communal land, individual land users might fail to perceive or understand the importance of their own management contribution and the potential loss they might incur in case of invasion, which brings in specific management considerations related to the 'tragedy of the commons' (Hardin, 1998). In turn, lack of awareness can be a serious challenge when mobilizing community members. One problem that can arise on privately owned land is accumulation, by certain individuals in the community, of land plots that are scattered in various locations. In Baringo, Kenya, Adoyo *et al.* (2022) noted that, in such situations, land users tend to neglect the management of *Prosopis* invasions on land that is further away from their homesteads.

Market dynamics can have a direct impact on available management options and priorities. For example, in times of declining agricultural commodity prices, commercial farmers might choose to temporarily abandon some of their land because of reduced profits when selling their farm produce. Such unattended land parcels can subsequently become invaded by *Proposis* trees. Such was the case in Baringo following the decrease of commodity prices in 2013: several parcels of land in several irrigation schemes south of Lake Baringo were temporarily abandoned, leading to a partial invasion by *Prosopis* trees (Fig. 11.1) (Adoyo *et al.*, 2022). Land abandonment and subsequent invasion by *Prosopis* might also be an issue when the economic structure within a community changes, for example when an increasing share of the population moves to the secondary or tertiary economic sectors. Stakeholders in charge of designing a spatially explicit strategy for managing *Prosopis* invasions must be aware of such social and economic trends and anticipate their implications for stakeholder mobilization and other aspects related to the implementation of the strategy.

11.3 A Decision Support Tool for the Management of *Prosopis* Invasions

11.3.1 Principles

Strategies for managing *Prosopis* invasions must consider factors that influence the spatial and temporal dynamics of invasion processes and factors that determine the ability of stakeholders to respond to these processes, and must guide the identification of priorities in relation to the protection or rehabilitation of important assets. Considering all these simultaneously is a complex task since different groups of stakeholders are likely to identify diverging priorities and value the factors described above in different ways. Hence, tackling such a complex task requires the use of a deliberative multicriteria evaluation, i.e. a structured decision-making process that combines multicriteria decision analysis with deliberations and consensus building (Proctor and Drechsler, 2006; Liu *et al.*, 2011; Forsyth *et al.*, 2012; O'Bryan *et al.*, 2023). In turn, such processes require adequate guidelines and decision support tools to be goal oriented and efficient.

Using the invasion management curve described in Victorian Government (2010), we developed a decision support tool adapted from Grice *et al.* (2011) for identifying, prioritizing and mapping objectives for managing *Prosopis* invasions. We capitalized on experience gained during several workshops in Baringo, Isiolo, Marsabit and Tana River counties in Kenya (Chapter 13, this volume) to gradually adapt the decision support tool. For example, we included the identification of pathways of *Prosopis* spread into the selection process, as well as the localization of important assets to help distinguish areas requiring containment from those rather requiring asset protection. The number and definition of the resulting management objectives also underwent several adaptation steps.

Our decision support tool (Fig. 11.2) is based on a logical sequence of four evaluation steps, with each step having a 'yes' or 'no' outcome determining the next step in the sequence. The tool proposes 11 management objectives that can be grouped into four management types: (1) awareness raising, (2) prevention, (3) EDRR, and (4) control.

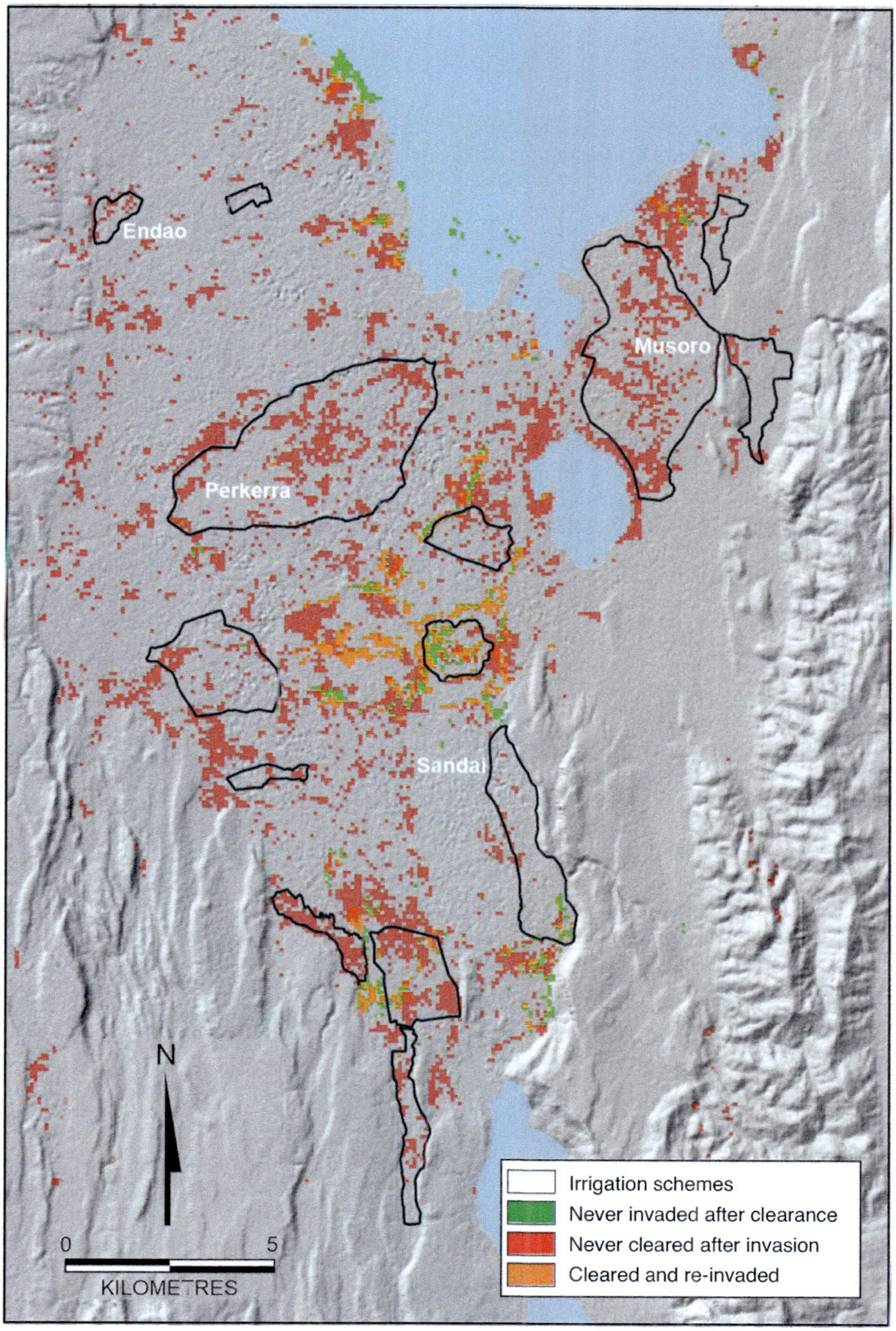

Fig. 11.1. Invasion by *Prosopis juliflora* from 1996 to 2016 in and around irrigation schemes south of Lake Baringo. Note that many areas were cleared of *P. juliflora* and then abandoned and re-invaded when a drop in commodity prices rendered maintenance of these areas unaffordable. *Prosopis* invasion trajectories from Adoyo *et al*. (2022) based on data from Mbaabu *et al*. (2019). Map by authors.

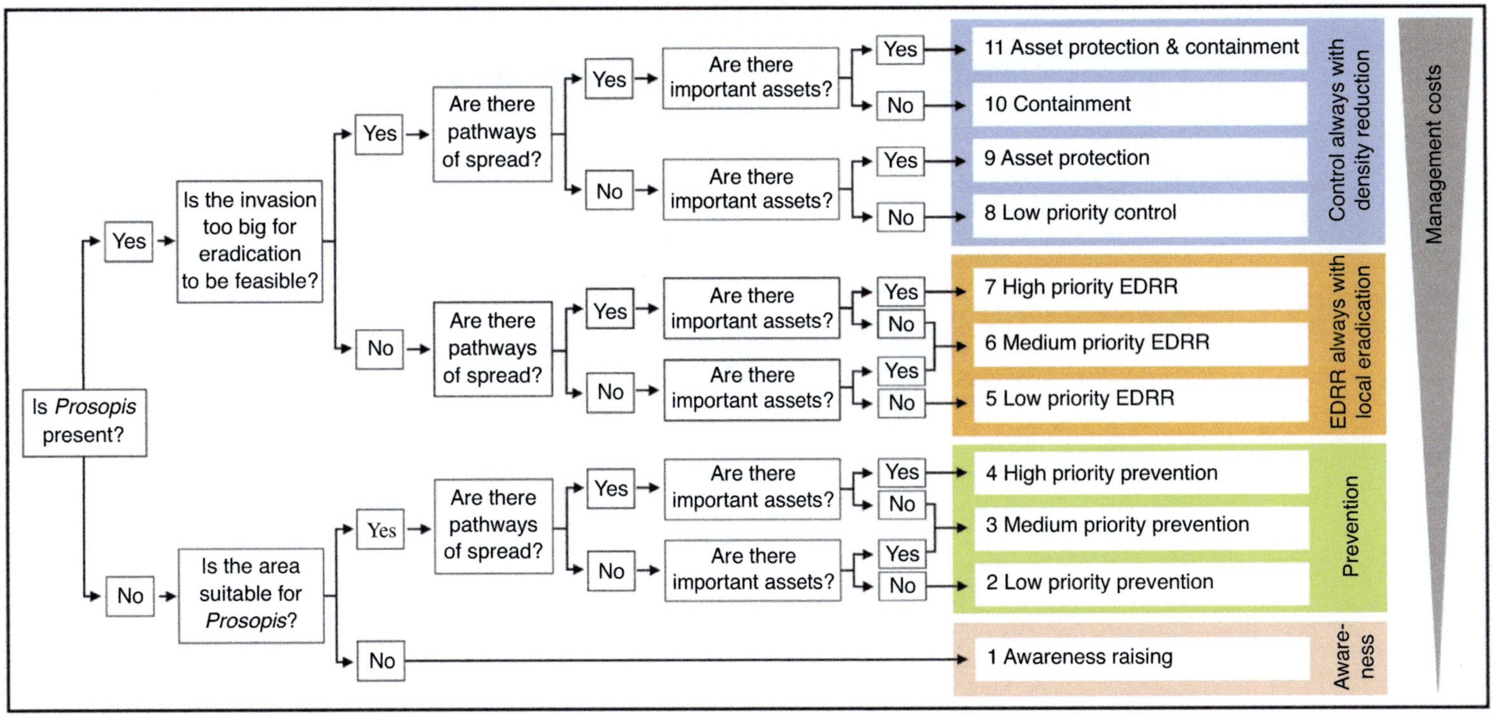

Fig. 11.2. Decision support tool for the selection and mapping of objectives and priorities related to the management of *Prosopis juliflora* invasions. Objectives 1–11, belonging to four management types, are arranged roughly in order of increasing management effort or costs per area. EDRR, early detection and rapid response. Developed by the authors based on the decision tree of Grice *et al*. (2011).

11.3.2 Selection of management types and objectives

Below, we outline the logical link between the four evaluation steps of the decision support tool and corresponding 11 management objectives belonging to four management types or options (see also Chapter 12, this volume). It is important to note that the decision support tool and the proposed management objectives are idealized models that need careful interpretation when translated into a management plan and/or concrete management activities. Often, a single invaded area includes a variety of situations and the selection of only one management objective might not be sufficient, and a more nuanced approach might be needed.

Awareness (objective 1): if *Prosopis* trees are unlikely to ever establish in an area, preventive measures can be reduced to creating awareness among communities (objective 1; Fig. 11.2) since people from the area might have activities in other places where *Prosopis* has established. For that reason, they could become vectors of spread when travelling to or from these areas.

Prevention (objectives 2–4) is the main management option in areas where *Prosopis* has not yet established but might do so, owing to the habitat suitability of the area. The absence or presence of pathways of spread and, where pathways are present, the existence or not of important assets are the factors that determine whether prevention should be tackled with low, medium or high priority. Prevention is a low-cost management option as it does not require the heavy equipment, chemicals or labour force required for eradication or containment. Yet it is often neglected because in the early stages of invasion, stakeholders often do not realize the need for preventive measures as they might not be aware of the dangers and potential impacts of an invasion (Leung *et al.*, 2002; Cuthbert *et al.*, 2022).

The EDRR (objectives 5–7) is the main management option in areas where *Prosopis* trees have established but the size of the invasion is small enough for local eradication to be feasible. The main purpose of EDRR is to bring the area back to the 'prevention' level. The priority with which EDRR must be implemented depends on the presence or absence of pathways of spread and of important assets. EDRR is the most cost-effective management option in areas where *Prosopis* trees have only recently established. However, it must be tackled swiftly before the spread or growing size of individual trees makes removal difficult.

Control (objectives 8–11) is the only available management option when the extent of an invasion is too large for local eradication to be feasible. Areas where pathways of spread are absent and no important assets exist would be of low priority (objective 8) if available resources do not allow management across the entire invaded area. The presence of important assets changes the focus of the control measures towards asset protection (objective 9), while the presence of pathways of spread calls for a focus on containment (objective 10). When both pathways of spread and assets are found in a 'control' area, asset protection needs to be combined with containment (objective 11).

11.3.3 Input data requirements

The proposed decision support tool alone is insufficient to adequately support the codesign of a spatially explicit management strategy. In addition to the tool, spatial data layers are required, which show the distribution and density of the invader and the expanse of its suitable range, as well as other spatial data showing water courses, water bodies, roads, livestock migration routes or any other pathway for spreading *Prosopis* trees. Finally, in-depth knowledge of the context, the factors that might influence invasion dynamics, priorities for action (i.e. the location of important assets) and variables influencing the ability of stakeholders to act are all crucial elements to guide the development of a management strategy. This is why the decision support tool should always be used in the frame of a participatory process involving diverse stakeholder groups who have the required context knowledge.

Whereas some spatial layers are open access, substantial research input may be required to generate the spatial layers of *Prosopis* invasions and suitable habitat ranges. In eastern Africa, Ng *et al.* (2017), Mbaabu *et al.* (2019) and Shiferaw *et al.* (2019a) calculated the distribution of *P. juliflora* invasions in Ethiopia and Kenya using satellite images and remote sensing algorithms (Chapter 4, this volume). Owing to the speed of invasion and associated changes in land cover, such models need regular updates to reflect reality, thus requiring a substantial investment, particularly for large areas.

Box 11.1. Using the decision support tool in codesign processes.

In November 2021, stakeholders in Baringo County, Kenya, went through a 3-day exercise aiming to identify management objectives for different parts of the county (Box Fig. 11.1). The codesign process was structured in three steps.

Box Fig. 11. 1. County Implementation Groups members identify the locations of important assets (A), discuss most appropriate management practices in selected intervention areas (B) and present their mapping results to other group members in order to harmonize intervention strategies over the entire county (C). Photos courtesy of Albrecht Ehrensperger.

Step 1: After an introduction on biological invasions and their management, participants got familiar with overview maps printed out in A0 format. Five groups were formed according to the participants' knowledge about the different parts of the county. They used small stickers and pens to map assets (as points, lines or surfaces) and pathways of *Prosopis* spread directly onto the map. Information about each element added to the map was recorded on paper. Results were presented to the other groups whose suggestions were included into a final version.

Step 2: Participants placed a transparent paper on the A0 map and drew areas that were assigned to general management objectives onto that paper. They used the decision support tool (Fig. 11.2) as a guide and recorded important information about each management unit in a table. The maps of the five groups were placed side by side on the ground so as to represent the entire county, and each group presented their map. Differences along the border between the groups were discussed and resolved.

Step 3: Participants used the mapping done in step two to identify hotspots needing particular attention in terms of management, for example because of the presence of important assets needing protection or rehabilitation or because of the presence of pathways of spread requiring regular surveillance. These areas were enlarged onto A3 maps and participants used them to map suggestions for more precise management practices (Chapter 12, this volume) in line with the management objectives identified in the previous step.

After the workshop, the mapping results were digitized and printed. The prints were used as inputs for a subsequent community validation process, during which stakeholders commented on and proposed improvements of the maps. These inputs formed the basis for a second workshop aimed at adapting and consolidating the spatial management strategy for the county. At the end of the second workshop, the revised Spatial Management Plan (Box Fig. 11.2) was approved by all workshop participants and shared with the government of Baringo County.

Continued

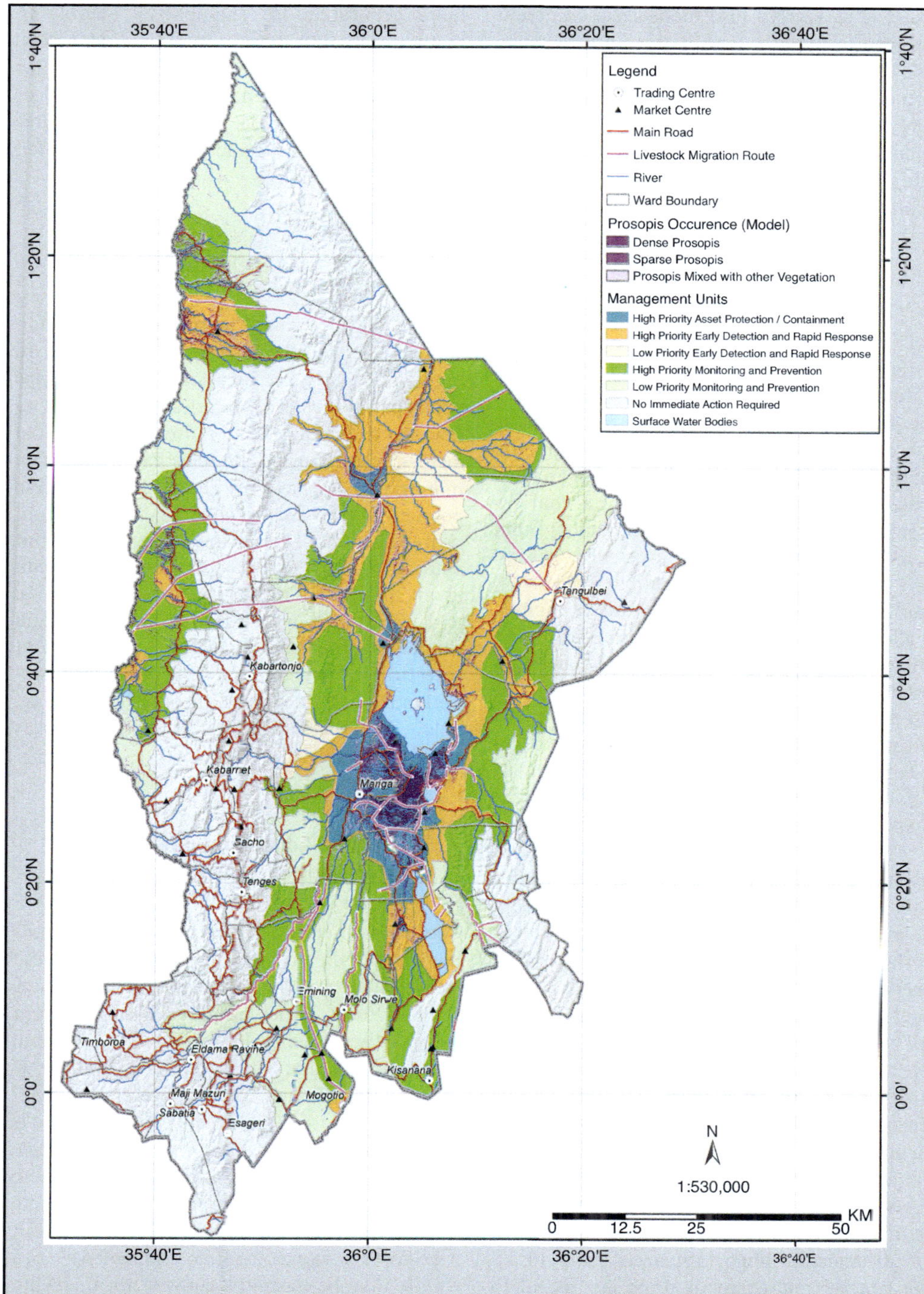

Box Fig. 11. 2. Map of Baringo County, Kenya, indicating the location of areas requiring different approaches to the management of invading *Prosopis* trees.

Eckert *et al.* (2020) also modelled the current and future spatial extent of the suitable range for *P. juliflora* in eastern Africa using several agro-climatic indicators and climate change scenarios (Chapter 4, this volume). This regional model is an excellent basis to support the codesign strategies for managing *Prosopis* invasions throughout the region.

11.3.4 Dealing with scale issues

In real-life situations, the answers to the four evaluation steps of the decision support tool are not always a clear-cut 'yes' or 'no'. For example, stakeholders might be uncertain about the habitat suitability of the assessed region (particularly when accounting for potential changes in suitable habitat due to climate change) and therefore a combination of awareness raising with preventive measures might be necessary. Similarly, a heavily invaded area might include smaller areas that are easily accessible and others that are more remote, and the assets in this area might be clustered at one end of it but absent from others. Thus, such an area might require a combination of different control objectives (objectives 8–11). The valuation of important assets could vary between different stakeholder groups, adding further complexity.

The scale at which a management strategy is designed has a direct effect on these ambiguities where the larger the area under consideration, the more heterogeneous the spatial patterns of the elements considered (e.g. pathways or assets) will be and therefore the more challenging it will be to find consensus, and the likelier it will be to have to combine several management objectives. The above-mentioned stakeholder workshops in Kenya showed that a multiscale approach helped in tackling this challenge. Participants started with mapping general management units for large areas, without dwelling too much on accuracy and detail, then went on to refining this mapping for smaller but strategically important areas (Box 11.1). The general management units are useful to develop plans at national or subnational scales, to get a sense of the areas of each management objective and of the costs involved, and help to identify management hotspots. Zooming in on the latter can then help with mapping the units more accurately, for example by identifying management objectives for individual assets, achieving higher granularity inside areas that were assigned homogenous management objectives at a higher scale, and introducing first ideas in terms of specific practices associated with these objectives.

11.4 Outlook

Above, we described a structured decision support tool for the codesign of strategies to manage *Prosopis* tree invasions. This tool helps participants involved in codesign processes to manage the complexity of overlapping spatial factors that influence the selection of adequate management objectives and practices. We pointed to two major challenges involved in using this tool: (1) securing the required input data for the codesign process, particularly data about the current distribution of *Prosopis* trees; and (2) dealing with the issues of scale and simplifying the identification of management objectives for large areas for the purpose of subnational or national planning or priority setting. Below, we outline ideas on how these challenges could be addressed in future.

11.4.1 Accessing data on the distribution of *Prosopis* invasions

The codesign process outlined above depends heavily on the availability of reliable spatial information on the current distribution of *Prosopis* trees, which requires data collection using either satellite-based sensors (Mbaabu, unpublished data; Ng *et al.*, 2017; Shiferaw *et al.*, 2019c; Eckert, 2021; Page *et al.*, 2022) or ground-based data collection methods (Osunkoya *et al.*, 2020). Depending on the scale of the study, one may choose either of these methods or a combination of both. Remote sensing can cover vast areas in a short time but it requires advanced technical knowledge, and results may be inaccurate owing to modelling challenges, particularly when *Prosopis* trees are young, the invasion is just starting or the trees occur mixed with other vegetation (see Chapter 4, this volume, and references cited above in this paragraph). Additionally, owing to rapid changes in the distribution of invasions, regular updates are needed, for example every 5 years. This requires specialized expertise and a dedicated workflow within a national mapping authority or similar body.

On the other hand, ground-based data collection provides highly accurate information but is extremely time-consuming and therefore only practical in small areas. Due to its resource intensity and time requirement, it might not be possible to cover large areas or to repeat ground-based approaches at regular intervals (e.g. every 5 years) to monitor changes in the distribution of *Prosopis* trees. Citizen science, for example a mobile application with which citizens can report the occurrence of *Prosopis* using their mobile phones, might be an option. But such an approach depends greatly on widespread awareness among a population that is sufficiently convinced about the necessity to manage *Prosopis* for them to invest the time and effort required for citizen surveillance.

The use of unmanned aerial vehicles (UAVs) or drones for the mapping of *Prosopis* distribution could be an interesting complementary way of obtaining data in future planning processes. It could become a middle ground approach between the use of satellite images and ground-based approaches and might help to facilitate the generation of baselines or to support regular monitoring at intermediate scales, for example in river catchments or (community) conservancies, as well as at ward or district level. In good weather conditions, multirotor UAVs can cover an area ranging from 4000 to 6000 ha per day, while fixed-wings UAVs can cover larger areas. An added advantage is that UAVs allow for the collection of data in areas that may be inaccessible or difficult to reach with ground-based methods, such as steep terrain or dense vegetation (Page *et al.*, 2022).

However, even though UAVs can cover larger areas in a shorter time than ground-based methods, the 'conventional' use of UAVs relying on the recording of multispectral high-resolution images generates massive amounts of data and requires lengthy and complex post-processing steps. Consequently, this approach tends to be too costly and time-consuming for actors involved in the implementation of *Prosopis* tree management who require near real-time information at low cost. Hence, novel, replicable, standardized and field-ready approaches are needed that help to overcome these challenges and adequately support the management of *Prosopis* invasions. Real-time detection techniques using a drone equipped with an onboard nanocomputer (e.g. Raspberry Pi or Jetson) would allow heavy image data downloading to be bypassed and limit data flow to the real-time transmission of coordinates. This approach requires a model that has been trained – using a very large number of images of *Prosopis* and other trees - to recognize and simultaneously map *Prosopis*

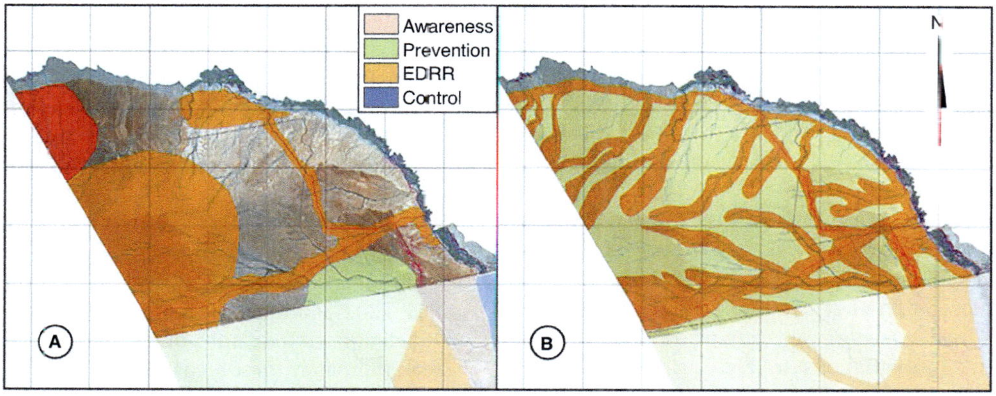

Fig. 11.3. Mapping of management objectives in the northern part of Tana River County, Kenya. Map A is the result of a mapping exercise during a first workshop with the County Implementation Group. Map B is the result of a mapping exercise during a community validation workshop. Orange, EDRR; green, prevention; blue, control; red, protected area.

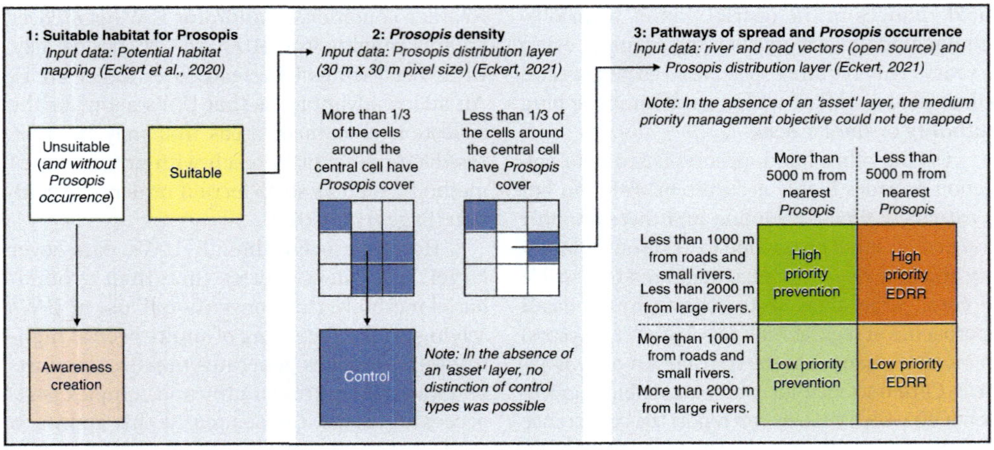

Fig. 11.4. Simplifying the decision support tree and translating it into spatial rules for the automated mapping of management objectives.

trees at different growth stages. Simultaneous localization and mapping algorithms and the Internet of Things have been used successfully in forest ecology and weed management (Panetta and Cacho, 2014; Osunkoya *et al.*, 2020; Page *et al.*, 2022; Yan *et al.*, 2023). But more research is needed to fully understand their capabilities and limitations for the management of *Prosopis* invasions.

11.4.2 Scale challenge

The codesign workshops conducted with local stakeholders in Kenya revealed that participants are quite confident when mapping small areas with a high degree of detail and proposing concrete management practices. However, the mapping of larger areas (for example, an entire county or a large share of it) would require a higher level of abstraction and generalization, and therefore opens the range of possible spatial interpretations. Consequently, mapping results at such scales can yield quite contrasting results depending on who does the mapping, what prior information they have and how the exercise is moderated (see an example of such diverging interpretations in Fig. 11.3). Eventually, such different interpretations can be resolved through discussion and consensus building, but the several iterations required to get there are time-consuming and the best option is not always easy to determine.

One way of overcoming this difficulty could be to use spatial modelling in a geographical information system to generate a first rule-based version of the map on management objectives as an input for the codesign process. This would allow the mapping of large areas quickly so that participants could be asked to critically review the mapping result, before using it to identify areas where more detailed mapping should be done as part of a codesign process. Hereafter, we propose a preliminary version of such a modelled map of *Prosopis* management objectives (see Fig. 11.5) based on the decision support tool presented above. The assumptions and thresholds used for the different variables included in the model need to be critically reviewed and adjusted to the specific needs and characteristics of the area for which the model is generated.

Figure 11.4 shows how the decision support tool can be translated into spatial rules for the mapping of management units, based on three analytical steps.

1. Areas that are unsuitable for *Prosopis* trees according to the habitat layer of Eckert *et al.* (2020) are assigned a value of zero and eliminated from subsequent calculation steps. Ultimately, these areas were assigned to the 'awareness creation' management type.
2. The density of *Prosopis* trees was calculated for the remaining area using the pixel-based *Prosopis* distribution layer (Eckert, 2021). For this example, we used a neighbourhood

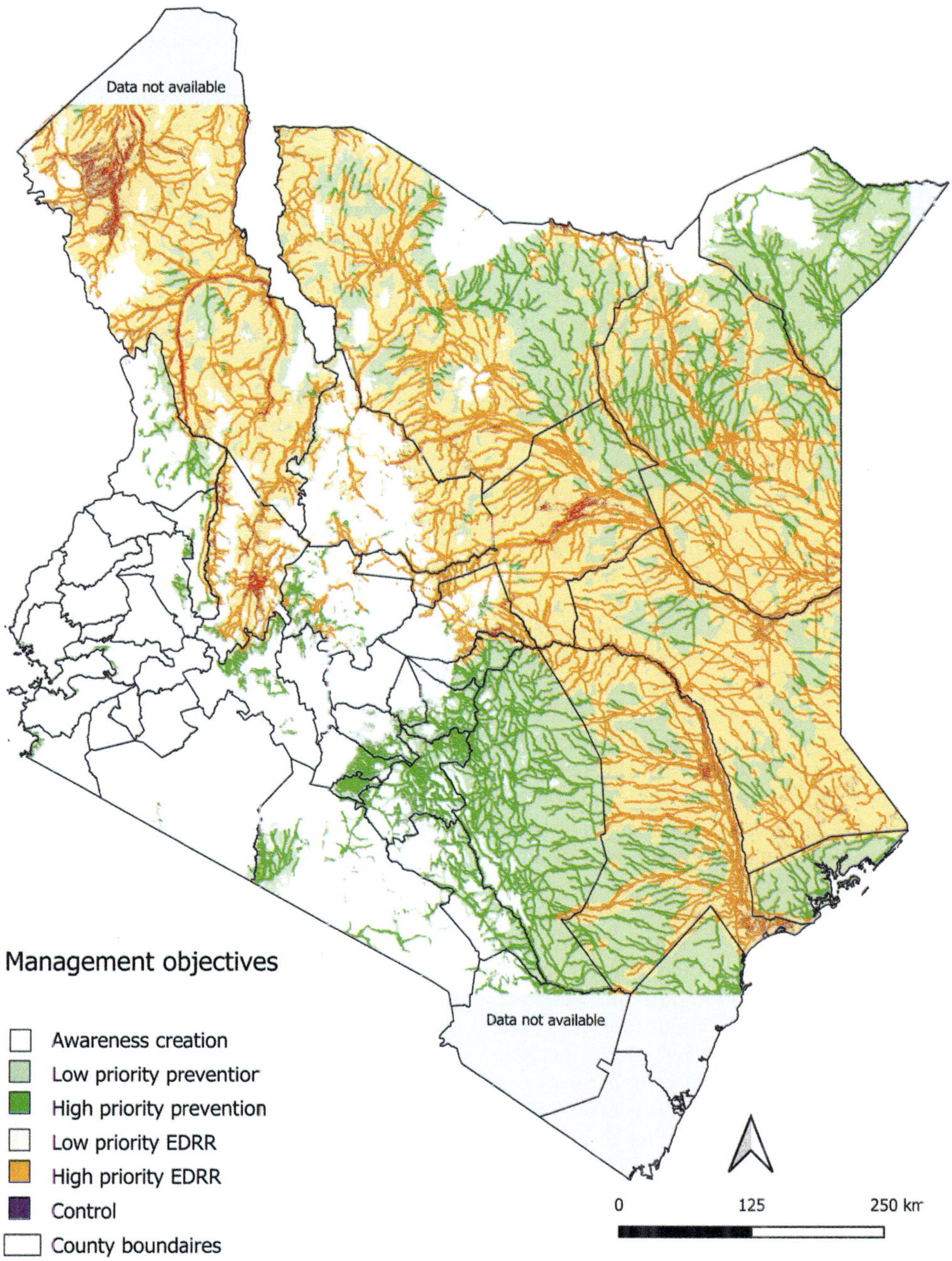

Fig. 11.5. A map of Kenya showing the spatial distribution of areas where different management objectives should be pursued.

analysis with a nine-cell moving window to count how many cells around the central cell are classified as '*Prosopis* tree cover'. If at least one-third of the cells in the neighbourhood of the central cell have *Prosopis* tree cover, the density in that cell was assumed to be high and the management option 'control' was assigned to it. The other cells with fewer than three cells in their neighbourhood having *Prosopis* cover were assumed to have a low *Prosopis* density and were kept for further analysis in the next step. No further distinction in terms of various control types was made, owing to the lack of a spatial layer showing the location of assets. It should further be noted that the threshold of one-third used for the moving window analysis is arbitrary and should be adjusted depending on characteristics of the context. It should also be noted that this analysis does not actually return a density value for the central cell of the moving window, but rather a proxy for a landscape-level density, whereas the 'landscape' is a square of 90 × 90 metres.

3. In the remaining areas, distance from the nearest *Prosopis* tree invasion can be used to distinguish between 'prevention' and 'EDRR' (i.e. local eradication). In this example, we opted for a threshold of 5 km using a proximity analysis on the *Prosopis* distribution model. Additionally, we used distance from roads and rivers to distinguish between high priority and low priority. We generated buffers of 1000 to 2000m around these features depending on their size and merged the results. Areas within the buffer were considered 'high priority' while those outside the buffer were considered 'low priority' because they are further away from pathways of spread. The 'medium-priority EDRR' management objective was omitted owing to the lack of a spatial layer showing the location of assets. As for the second step, the thresholds used in this third step need to be adapted to the characteristics of the assessed area.

The results from this analysis, conducted at national scale for Kenya, are shown in Fig. 11.5. Extracts from this map for individual counties could be a valuable basis for discussion during codesign events with stakeholders for the management of *Prosopis* invasions.

11.5 Conclusion

We have presented a decision support tool to guide the policy-mandated spatial prioritization of *Prosopis* management in Kenya. This framework draws on key factors including the occurrence of *Prosopis* trees, pathways of spread, access for management and locations of important assets. We also discussed the importance of codesign and comapping, the issues of scale and data availability and proposed pathways for overcoming the challenges associated with establishing spatially explicit management plans.

We hope this information will be of use to institutions and stakeholders in charge of managing *Prosopis* invasions in other counties of Kenya, in other countries in the eastern African region, or even in other regions of the world where *Prosopis* tree invasions are an issue. In Kenya, the National *Prosopis* Strategy provides the required policy framework under which county authorities can lay the foundation for an effective management of *Prosopis* invasions as part of their integrated county development plans. Accordingly, they might mobilize the required government funds to conduct the type of codesign processes that we describe in this chapter. With this, they will secure the buy-in from local stakeholders that is necessary to translate the management plan into concrete action.

Acknowledgements

This work was supported by the Swiss Programme for Research on Global Issues for Development (r4d), funded by the Swiss National Science Foundation and the Swiss Agency for Development and Cooperation, through the project 'Woody invasive alien species in East Africa: Assessing and mitigating their negative impact on ecosystem services and rural livelihood' (Grant Number:

400440_152085), and by the Swiss Agency for Development and Cooperation through the project 'Woody Weeds +' (Project number 7F.10651.01). Thanks also go to the Centre for Training and Integrated Research in ASAL Development (CETRAD) and the Kenya Forestry Research Institute (KEFRI) which facilitated the workshops in the context of which the codesign methodology for the elaboration of spatial *Prosopis* management plans was tested and refined. Finally, we wish to thank all the partners of the Woody Weeds research project for their support and participation.

References

Adoyo, B., Schaffner, U., Mukhovi, S., Kiteme, B., Mbaabu, P.R. *et al.* (2022) Pathways towards the sustainable management of woody invasive species: Understanding what drives land users' decisions to adopt and use land management practices. *Land* 11, 550.

Anonymous (2019) *National Invasive Species Strategy and Action Plan (2019–2029)*. Office of the Vice President, Dodoma, Tanzania.

Australian Weeds Committee (2012) Mesquite (*Prosopis* spp.) strategic plan 2012–17. Australian Government Department of Agriculture, Fisheries and Forestry, Canberra.

Broennimann, O., Treier, U.A., Müller-Schärer, H., Thuiller, W., Peterson, A.T. *et al.* (2007) Evidence of climatic niche shift during biological invasion. *Ecology Letters* 10, 701–709.

Cuthbert, R.N., Diagne, C., Hudgins, E.J., Turbelin, A., Ahmed, D.A. *et al.* (2022) Biological invasion costs reveal insufficient proactive management worldwide. *Science of the Total Environment* 819, 153404.

Eckert, S. (2021) *Map of Current Prosopis Invasion Status in Northern Kenya*. Centre for Development and Environment (CDE) and Wyss Academy for Nature, University of Bern, Switzerland.

Eckert, S., Hamad, A., Kilawe, C.J., Linders, T.E., Ng, W.T. *et al.* (2020) Niche change analysis as a tool to inform management of two invasive species in Eastern Africa. *Ecosphere* 11, e02987.

Forsyth, G.G., Le Maitre, D.C., O'Farrell, P.J. and van Wilgen, B.W. (2012) The prioritisation of invasive alien plant control projects using a multi-criteria decision model informed by stakeholder input and spatial data. *Journal of Environmental Management* 103, 51–57.

Grice, A.C., Clarkson, J.R. and Calvert, M. (2011) Geographic differentiation of management objectives for invasive species: A case study of *Hymenachne amplexicaulis* in Australia. *Environmental Science and Policy* 14, 986–997.

Hall, K.M., Albers, H.J., Alkaee Taleghan, M. and Dietterich, T.G. (2018) Optimal spatial-dynamic management of stochastic species invasions. *Environmental and Resource Economics* 70, 403–427.

Hardin, G. (1998) Extensions of 'the tragedy of the commons'. *Science* 280(5364), 682–683.

Larson, D.L., Phillips-Mao, L., Quiram, G., Sharpe, L., Stark, R. *et al.* (2011) A framework for sustainable invasive species management: Environmental, social, and economic objectives. *Journal of Environmental Management* 92, 14–22.

Leung, B., Lodge, D.M., Finnoff, D., Shogren, J.F., Lewis, M.A. *et al.* (2002) An ounce of prevention or a pound of cure: Bioeconomic risk analysis of invasive species. *Proceedings of the Royal Society B* 269(1508), 2407–2413.

Liu, S., Sheppard, A., Kriticos, D. and Cook, D. (2011) Incorporating uncertainty and social values in managing invasive alien species: A deliberative multi-criteria evaluation approach. *Biological Invasions* 13, 2323–2337.

Mbaabu, P.R., Ng, W.T., Schaffner, U., Gichaba, M., Olago, D. *et al.* (2019) Spatial evolution of prosopis invasion and its effects on LULC and livelihoods in Baringo, Kenya. *Remote Sensing* 11, 1217.

Ministry of Environment and Forestry (2022) *National Invasive Species Strategy and Action Plan (2019–2029)*. Ministry of Environment and Forestry, Dodoma, Tanzania.

Ministry of Livestock and Fisheries (2017) National Strategy on *Prosopis juliflora* Management (2017). Available at: http://woodyweeds.org/wp-content/uploads/2017/03/Prosopis-Strategy-Ethiopia.pdf (accessed 21 June 2023).

Ng, W.T., Rima, P., Einzmann, K., Immitzer, M., Atzberger, C. *et al.* (2017) Assessing the potential of Sentinel-2 and Pléiades data for the detection of *Prosopis* and *Vachellia* spp. in Kenya. *Remote Sensing* 9, 74.

O'Bryan, C.J., Rhodes, J.R., Osunkoya, O.O., Lundie-Jenkins, G., Mudiyanselage, N.A. *et al.* (2023) Setting conservation priorities in multi-actor systems. *BioScience* 73, 522–532.

Ostrom, E. (1990) *Governing the Commons: The Evolution of Institutions for Collective Action*. Cambridge University Press, Cambridge, UK.

Osunkoya, O.O., Lock, C. and Buru, J.C. (2020) Spatial extent of invasiveness and invasion stage categorisation of established weeds of Queensland, Australia. *Australian Journal of Botany* 68, 557–573.

Page, M.T., Perotto-Baldivieso, H.L., Ortega-S, J.A., Tanner, E.P., Angerer, J.P. *et al.* (2022) Evaluating mesquite distribution using unpiloted aerial vehicles and satellite imagery. *Rangeland Ecology and Management* 83, 91–101.

Panetta, F.D. (2023) A proposed method for evaluating management feasibility when determining weed control priorities after major fires and floods. *Research in Ecology* 5, 28–39.

Panetta, F.D. and Cacho, O.J. (2014) Designing weed containment strategies: An approach based on feasibilities of eradication and containment. *Diversity and Distributions* 20, 555–566.

Proctor, W. and Drechsler, M. (2006) Deliberative multicriteria evaluation. *Environment and Planning C: Government and Policy* 24, 169–190.

Pyšek, P. and Hulme, P.E. (2005) Spatio-temporal dynamics of plant invasions: Linking pattern to process. *Ecoscience* 12, 302–315.

Shackleton, R.T., Le Maitre, D.C., van Wilgen, B.W. and Richardson, D.M. (2017) Towards a national strategy to optimise the management of a widespread invasive tree (*Prosopis* species; mesquite) in South Africa. *Ecosystem Services* 27, 242–252.

Shiferaw, H., Teketay, D., Nemomissa, S. and Assefa, F. (2004) Some biological characteristics that foster the invasion of *Prosopis juliflora* (Sw.) DC. at Middle Awash Rift Valley Area, north-eastern Ethiopia. *Journal of Arid Environments* 58, 135–154.

Shiferaw, H., Bewket, W. and Eckert, S. (2019a) Performances of machine learning algorithms for mapping fractional cover of an invasive plant species in a dryland ecosystem. *Ecology and Evolution* 9, 2562–2574.

Shiferaw, H., Bewket, W., Alamirew, T., Zeleke, G., Teketay, D. *et al.* (2019b) Implications of land use/land cover dynamics and prosopis invasion on ecosystem service values in Afar region, Ethiopia. *Science of the Total Environment* 675, 354–366.

Shiferaw, H., Schaffner, U., Bewket, W., Alamirew, T., Zeleke, G. *et al.* (2019c) Modelling the current fractional cover of an invasive alien plant and drivers of its invasion in a dryland ecosystem. *Scientific Reports* 9, 1576.

Sintayehu, D.W., Egeru, A., Ng, W.T. and Cherenet, E. (2020) Regional dynamics in distribution of *Prosopis juliflora* under predicted climate change in Africa. *Tropical Ecology* 61, 437–445.

van den Berg, E.C., Kotze, I. and Beukes, H. (2013) Detection, quantification and monitoring of *Prosopis* in the Northern Cape province of South Africa using remote sensing and GIS. *South African Journal of Geomatics* 2, 68–81.

van Klinken, R. and Campbell, F. (2001) The biology of Australian weeds. *Prosopis Species. Plant Protection Quarterly* 16, 2–20.

van Wilgen, B.W., Dyer, C., Hoffmann, J.H., Ivey, P., Le Maitre, D.C, *et al.* (2011) National-scale strategic approaches for managing introduced plants: Insights from Australian acacias in South Africa. *Diversity and Distributions* 17, 1060–1075.

Vaz, A.S., Ribeiro, J., Honrado, J.P. and Vicente, J.R. (2020) Stakeholders' perceptions towards non-native acacias and implications for their management in Portugal. *Forestry: An International Journal of Forest Research* 93, 557–566.

Victorian Government. (2010) *Invasive Plants and Animals: Policy Framework*. Victoria Department of Primary Industries, Melbourne, Australia.

Wiesmann, U., Ott, C., Ifejika Speranza, S., Kiteme, B.P., Müller-Böker, U. *et al.* (2011) A human actor model as a conceptual orientation in interdisciplinary research for sustainable development. In: Wiesmann, U. and Hurni, H. (eds) *Research for Sustainable Development: Foundations, Experiences, and Perspectives*. University of Bern, Bern, Switzerland, pp. 231–256.

Yan, F., Guan, T., Ullah, M.R., Gao, L. and Fan, Y. (2023) A precise forest spatial structure investigation using the SLAM+AR technology. *Frontiers in Ecology and Evolution* 11, 1152955.

Yletyinen, J., Perry, G.L., Burge, O.R., Mason, N.W. and Stahlmann-Brown, P. (2021) Invasion landscapes as social-ecological systems: Role of social factors in invasive plant species control. *People and Nature* 3, 795–810.

12 Institutional Settings, Stakeholder Engagement and Decision-Making Processes in Invasive Species Management

Boniface Kiteme[1*], Beatrice Adoyo[1], Albrecht Ehrensperger[2] and Urs Schaffner[3]

[1]Centre for Training and Integrated Research in ASAL Development, Nanyuki, Kenya; [2]Centre for Development and Environment, University of Bern, Switzerland; [3]CABI, Delémont, Switzerland

Abstract

Effective management of invasive species, which is part of natural resource management, requires a coordinated approach by all stakeholders in the invaded landscape. This chapter describes attempts to engage stakeholders across sectors and at different scales and to develop novel governance and institutional approaches to address the challenges caused by the invasive alien tree *Prosopis juliflora* in eastern Africa. The chapter focuses on the work carried out in Kenya, as it is the only country in which research projects were conducted to test novel institutional approaches at different scales in selected counties. The chapter first explores the shortcomings of existing governance and institutional structures in Kenya regarding natural resource management and then describes the creation of Local Implementation Groups and County Implementation Groups. These multi-stakeholder institutions were engaged in deliberative decision processes to codesign *Prosopis juliflora* management at local and county scales and to promote its implementation. The initial efforts have led to social learning among the stakeholders involved and have helped to mainstream invasive species management into planning and budgetary processes at county and national scales but it is still too early to assess the impacts of the novel institutional arrangements and of the stakeholders' commitment to the management of *Prosopis juliflora* in the long term.

12.1 Introduction

Ecological systems exhibit complex hierarchical arrangements, with diverse processes occurring at various spatial and temporal scales. Human systems, including governance and regulatory frameworks, reflect such complexity. However, human institutions often fail to align with the biogeophysical processes underlying natural resource provision. Consequently, decisions made within specific jurisdictional boundaries can trigger biogeophysical repercussions at other scales (Cumming *et al.*, 2006).

Lockwood *et al.* (2010) argued that conventional governance systems and institutional structures in charge of natural resources management (NRM) are ill-equipped to handle this complexity, as they are segmented into

*Corresponding author: b.kiteme@africaonline.co.ke

© CAB International 2025. *The Ecology and Management of Invasive Prosopis Trees in Eastern Africa* (eds U. Schaffner *et al.*)

individual sectors and scales. This segmentation hinders the understanding of complexity and fails to reconcile the needs and demands of stakeholders operating in various sectors and at various scales. For instance, rural communities depend on natural resources for their livelihoods, yet their needs may conflict with broader sustainability objectives, such as climate change mitigation, ecosystem restoration or biodiversity conservation (Berkes, 2007). Solutions to reconcile the needs of stakeholders across sectors and scales, and to resolve mismatches between ecological processes and institutional arrangements require fostering knowledge sharing, mutual learning and collaborative decision making among stakeholders.

Multi-stakeholder platforms are increasingly promoted as an NRM governance arrangement that helps to integrate top-down processes with more bottom-up democratic structures to foster an inclusive approach to, and sustained engagement in, NRM (Ros-Tonen *et al.*, 2018). For example, the Voluntary Guidelines on the Responsible Governance of Tenure of Land, Fisheries and Forests (FAO, 2022) encourage states to establish multi-stakeholder platforms and frameworks at local, national and regional levels or to use existing platforms and frameworks to collaborate on the implementation of these guidelines. Multi-stakeholder platforms can offer an enabling environment for stakeholders acting at local, subnational, national and regional levels to jointly reflect on and negotiate land use objectives, conduct decision processes and monitor progress towards desired outcomes. While creating and maintaining such platforms is challenging, the number of well-documented cases is increasing (Ratner *et al.*, 2022; ElDidi *et al.*, 2024).

This chapter describes efforts to address complex decision-making processes relating to NRM across sectors and scales in Kenya's dryland regions. The specific context is the management of the invasive tree *Prosopis juliflora*, introduced in eastern Africa in the 1950s and 1960s to combat land degradation and provide firewood and livestock fodder (Chapter 2, this volume). However, it now negatively impacts biodiversity, agriculture, pastoralism, water availability and other ecosystem services (Chapters 5–7 and 9, this volume). The objective of the work, conducted as part of the Woody Weeds (2015–2021) and Woody Weeds + projects (2021–2024), was to develop a novel governance and planning approach for the effective management of *P. juliflora* (see Section 14.3, Chapter 14, this volume, for a full description of the Woody Weeds project). Using Baringo, Isiolo and Tana River counties as pilot study areas, it outlines a pathway for implementing the National *Prosopis* Strategy, approved by the Kenyan Cabinet in 2024. We conclude by discussing the applicability of this approach for managing *Prosopis* species elsewhere and for invasive alien plant management in general.

12.2 Approach

12.2.1 Project areas

In 2010, Kenya enacted a new constitution that devolved significant planning, decision-making and administrative authority to county governments. We took advantage of the opportunities offered by this devolution process to test a novel approach to NRM and governance in three counties – Baringo, Isiolo and Tana River – heavily invaded by *P. juliflora* (Chapter 4, this volume). Our efforts were facilitated by the findings of the Woody Weeds research project, which were shared with relevant stakeholders through scientific publications, policy briefs and extension materials, prompting the government of Kenya to revisit its previous strategy of 'management by utilization' and to develop a new National *Prosopis* Strategy (NPS).

Vegetation in the study region includes grasslands, savannas, shrublands and riparian ecosystems, with limited cultivated crops. Baringo County's ecosystems primarily consist of grasslands, savannas and some forests. Isiolo and Tana River counties feature large areas of grassland, significant shrubland and extensive riparian ecosystems along the Ewaso Ng'iro and Tana rivers, including the Lorian Swamps in Isiolo and the Tana River delta in Tana River. The region experiences a bimodal rainfall pattern, with long rains from March to May and short rains from October to December. Annual rainfall ranges from 1000 to 1500 mm in the highlands to approximately 600 mm in the lowlands of Baringo, 300–350 mm in central Isiolo, 150–250 mm in eastern Isiolo and 370 mm in the Tana River delta. The dominant livelihoods in the study region are pastoralism and agropastoralism.

12.2.2 Assessment of existing governance structures

Kenya largely follows a conventional governance approach that relies on division of work between sectors and governance scales. Thus, it was important for us to understand the existing governance and institutional structures in Kenya regarding NRM, to explore the shortcomings of these structures and to identify suitable entry points for novel governance approaches (see Section 12.3 for an assessment of the existing governance structures).

12.2.3 Stakeholder engagement

The stakeholder engagement process, which was tested in the Woody Weeds + project, is at the core of our novel approach, as it supports institutions at the national, subnational and local scales to operationalize the NPS. The process was used to engage stakeholders in developing spatially explicit County *Prosopis* Management Plans (Chapter 11, this volume), integrating them with existing planning and budgetary processes such as the County Integrated Development Plans (CIDPs) and implementing the management plans through local projects.

A broad range of stakeholders from different sectors and operating at different scales are affected by, or can affect, the invasion by *P. juliflora* (see Chapter 10, this volume, for a detailed map of these different stakeholder groups). They include land and resource users (e.g. pastoralists, farmers and charcoal producers), state actors (e.g. representatives from the sublocation administration to national government departments, staff from state research, environmental and educational institutions) and non-state actors (e.g. civil society organizations, NGOs and international organizations). We aimed to engage these stakeholder groups in codeveloping and implementing a broadly supported *P. juliflora* management approach that can be well aligned with existing NRM governance structures in the three target counties.

12.3 Policy and Institutional Arrangements for *Prosopis* Management in Kenya

Many laws and policies shape the current institutional arrangements for NRM and NRM governance in Kenya (Table 12.1). These have been shaped through a historical process of over 100 years of policy and legal reforms, a process that intensified after the 2010 Constitution of Kenya. Typically promulgated reactively, these policies and laws often fail to cover the full spectrum of actions needed to address a problem holistically.

In 2010, the Constitution of Kenya created a two-level governance structure: the national government and county governments. Both levels have specific obligations towards NRM as defined in the Constitution's fourth schedule. The national government is obligated to protect the environment and natural resources. On the other hand, the county governments are obligated to (a) implement national policies on natural resources and environmental conservation, and (b) coordinate and assist resource users and local authorities to develop the capacity for the effective performance of their respective functions and powers, and for participation in resource governance at local level (Muigua, 2018; Government of Kenya, 2021; Kameri-Mbote *et al.*, 2023).

This two-level structure has shaped the current institutional landscape and the laws and policy frameworks covering land and land administration, environment, water resources, forest resources, biodiversity resources, mineral resources, aquatic resources and agriculture and livestock resources. Each law is backed by policies and regulations, notably the Sessional Paper No. 3 of 2009 on National Land Policy, Sessional Paper No. 01 of 2017 on National Land Use Policy, Forest Policy 2014, Sessional Paper No. 01 of 2021 on National Water Policy and Sessional Paper No. 01 of 2020 on Wildlife Policy. These policies and laws are under the responsibility of

Table 12.1. Elements of a government's governance framework, guiding its actions and ensuring the rule of law in society. Understanding the distinctions between these components is crucial for both citizens and policy makers to navigate the complex landscape of governance effectively.

Term	Description
Government policy	Refers to a broad set of principles, objectives and guidelines that a government adopts to address specific issues or achieve overarching goals. The primary purpose of government policy is to articulate the government's vision and intentions, serving as a framework for decision making and action.
Government strategy	A detailed plan of action designed to achieve specific objectives outlined in government policies. The primary purpose of government strategy is to outline the steps and resources required to realize the objectives set out in policies.
Laws	Formal, binding rules established by a legislative body (i.e. parliament) that govern the behaviour of individuals, organizations and government itself. The primary purpose of laws is to provide a structured legal framework that defines acceptable conduct and consequences for violations.
Regulations	Rules and directives issued by government agencies to implement and enforce laws. They provide detailed guidance on how laws should be applied in practice. The primary purpose of regulations is to ensure the consistent and effective application of laws, translating legal provisions into practical requirements for individuals and businesses.
Institutions	Social structures of rules and norms that shape and constrain social behaviour, such as organizations, laws, customs, etc.
Institutional arrangements	Policies, systems and processes that organizations use to legislate, plan and manage their activities efficiently and to effectively coordinate with others to fulfil their mandate.

the respective government ministries that have established semi-autonomous implementation structures down to the county level, such as the National Environment Management Authority, the National Land Commission and the Kenya Wildlife Service (Table 12.2).

Others like the Water Resources Authority and the Kenya Forest Service have structures – Basin and Sub-basin offices (Water Resources) – at the level of catchments or ecoregions that cut across county boundaries and have links with organizations at grassroots level such as the Water Resources Users Associations (WRUAs), Community Forest Associations (CFAs) and Charcoal Production Associations that are recognized by the law. Likewise, the county governments departments have structures at sub-county and ward level. However, the capacities of these grassroot level institutions are usually not geared towards the sustainable management of *P. juliflora* or other invasive species.

The informal and indigenous institutions also play an important role in resource governance at community level, the continued erosion of their authority notwithstanding. These include, for example, the Nchuri Ncheke of the Meru community or the Jarsa Dedha of the Boran communities (World Bank, 2005; Kiteme and Wiesmann, 2008; Government of Kenya, 2016; Mburugu and Macharia, 2016). The mandate of Nchuri Ncheke is to act as the judiciary enforcer of the rules and regulations for conserving the environment, to resolve conflicts and maintain peace. Likewise, Jarsa Dedha enforces laws and provisions which are governed by the Gada council, the supreme Boran governance structure (World Bank, 2005; Mburugu and Macharia, 2016). Although the authority of these informal institutions has been grossly eroded by frequent revision of legal frameworks, conflicts between modern law and customary practices, and changing norms and cultures, they remain important actors in enhancing participation, inclusivity, empowerment and accountability in sustainable management and governance of natural resources in the areas where they still exist.

Despite their apparent coherence, the two-level governance structure and institutional arrangements discussed above have weaknesses

Table 12.2. Laws and policies that shape the current institutional arrangements for environmental and natural resources management and governance in Kenya.

Law	Responsible agencies	Overall purpose/ intention	Relevance for invasive species management
Land Act 2012	Ministry of Lands, Housing and Urban Development	Sustainable administration and management of land and land-based resources.	Has a direct bearing on county spatial planning, land use planning, land resources potential mapping and assessment, and determination of tenure arrangements.
Land Registration Act 2012	Ministry of Lands, Housing and Urban Development	Registration of interests in all public and private land, and registration and recording of community interests.	Invasive species affect land and land resources including people and their activities operating under all land tenure arrangements.
Community Land Act 2016	National Land Commission and other relevant departments and agencies	Recognition, protection and registration of community land rights.	Invasive species are most prevalent in rangelands most of which is governed by structures and regulations established under the Community Land Act.
Physical and Land Use Planning Act 2019	Department of Physical Planning	Planning, use, regulation and development of land and for connected purposes.	(Spatially explicit) invasive species management plans are expected to align with the County Spatial Plans that are prepared under the provisions of this Act.
Land Law (Amendment) Act 2016	National Land Commission	Amends the Land Act, Land Registration Act and National Land Commission Act, to provide for sustainable management of land resources.	Preparation and implementation of the County Spatial Plans that integrate county invasive species management plans stated above.
National Land Commission Act 2012	National Land Commission	Land management and administration, and oversight land use planning in the country, at national and county levels.	Preparation and implementation of the County Spatial Plans that integrate county invasive species management plans stated above.
Environmental Management and Coordination Act 1999	National Environment Management Authority (NEMA)	Coordination of environmental management; guides implementation of the National Biodiversity Strategy and Action Plan (NBSAP).	Management of invasive species may require development and installation of facilities which require Environmental Impact Assessment (EIA); this law provides for that.
Water Act 2016	Relevant departments and agencies under the Ministry of Water, Sanitation and Irrigation, e.g. Water Resources Authority or Water Sector Trust Fund	Development and management of water resources; provides also for the formation of Water Resources Users Associations (WRUAs) to foster community participation in water management and governance.	Invasive species threaten availability and accessibility of water resources. The Water Resources Authority is responsible for managing water resources; the Water Sector Trust Fund supports initiatives for water and sanitation services and can be a potential financier of invasive species management. WRUAs can enhance legitimacy in the design and implementation of promising invasive species management practices.

Continued

Table 12.2. Continued

Law	Responsible agencies	Overall purpose/ intention	Relevance for invasive species management
Forest Conservation and Management Act 2016	Ministry of Environment, Climate Change and Forestry; Kenya Forest Service	Conservation and management of public, community and private forests and areas of forest land that require special protection; community participation through Community Forest Associations (CFAs).	Leveraging on the debate of 30% national forest cover and the provision of the National Forest Fund to mainstream and mobilize resources to support SLM practices for the management of invasive species. CFAs can enhance legitimacy in the design and implementation of promising SLMs for invasive species management.
Wildlife Management, Conservation Act 2013, and associated Regulations	Kenya Wildlife Service	Protection, conservation and management of wildlife resources on public, community and private land, and Kenya territorial waters.	Invasive species threaten already endangered and fragile ecosystems, habitats and animal species. This law provides for the protection of these areas and gives the Kenya Wildlife Service the crucial role of ensuring this happens.
Mining Act 2016	Ministry of Mining, Blue Economy and Maritime Affairs, State Department for Mining	Management of mineral resources.	Mining disturbs land and promotes invasive species.
Fisheries Management and Development Act 2016	Ministry of Agriculture, Livestock and Fisheries, State Department for Fisheries	Management of aquatic resources.	Invasive species threaten water bodies/wetlands that support fisheries development.
Agriculture, Fisheries and Food Authority Act 2013	Agriculture and Food Authority	Management and administration of matters of agriculture and the preservation, utilization and development of agricultural land and related matters.	Invasive species threaten agricultural land especially within irrigation schemes in invaded areas. The Authority should be one of the key actors in the management of invasive weeds.
Crops Act 2013	Relevant agencies and departments in national and county government responsible for agriculture, livestock and fisheries development	Accelerate the growth and development of agricultural crops and enhance agricultural productivity.	Invasive species impact crop production and crop land productivity in agricultural lands that are already invaded. Concerned state agencies and departments are therefore important actors in the effort to manage invasive species.
National Environment Policy 2013	Ministry of Environment, Climate Change and Forestry, including relevant agencies and departments in water, environment and natural resources at county and national level	Has a broad range of measures and actions, responding to key environmental issues and challenges, and provides the framework for an integrated approach to planning and sustainable management of natural resources in the country.	Invasive species threaten freshwater and wetland ecosystems. The SLM practices for management of invasive species can benefit from the framework for integrated approaches provided by the policy during their implementation.

that hinder institutional collaboration, multisectoral integration and coordination, which are crucial for sustainable NRM practices. Each department's planning and implementation approaches for NRM and governance remain siloed, focusing on their respective sector goals and objectives, even when their guiding policy frameworks advocate for multisectoralism. Few, if any, legislative instruments and policies are backed by well-articulated implementation strategies that are needed to guide genuine and inclusive multiscale stakeholder participation and to promote the accountability of authorities.

Invasive species like *P. juliflora* affect multiple sectors, including agriculture, livestock, water, tourism, health and energy (Chapters 5–7, this volume). Yet, while the Kenya Forestry Research Institute oversees *Prosopis* management at the national level, no institution was in place prior to the Woody Weeds project to gather actors from the affected sectors and across scales around a coordinated implementation of *Prosopis* management as part of NRM. The lack of multi-stakeholder institutions partly explains why many NRM initiatives from national or county governments face implementation challenges and never fully realize their objectives. In contrast to that, cross-sector and cross-scale engagement processes promote vertical and horizontal integration among actors and stakeholders.

12.4 *Prosopis juliflora* Management in Kenya, Stakeholder Engagement and Institutional Realignment

The *P. juliflora* management initiative of the Woody Weeds and Woody Weeds+projects addressed the above outlined lack of institutional collaboration in NRM by developing new implementation structures: Local Implementation Groups (LIGs) were created to cater for bottom-up and County Implementation Groups (CIGs) for top-down implementation mechanisms. However, these two governance structures, which are briefly presented below, differ in their manner of formation, composition and the terms of reference under which they operate.

12.4.1 Local Implementation Groups

The LIGs were formed as platforms for connecting the Woody Weeds researchers and partners with local stakeholders to interact and foster mutual trust and joint learning, and to codesign solutions for effective management of *Prosopis*. This involved the identification, evaluation and selection of sustainable land management (SLM) practices for test implementation and learning about best practices and their impacts.

Unlike other similar grassroot institutions such as the WRUAs and CFAs (see Section 12.3), LIGs include members across sectors and scales, giving them their unique thematic, geographic and demographic inclusivity without which *P. juliflora* management would be difficult to implement. LIGs include stakeholders who live in the affected rural environment, external stakeholders such as policy makers and administrators covering different sectors, researchers and development professionals with different professional expertise. All LIG members should have experience in and knowledge about the specific rural environment, as it affects or may be affected by the invasion.

The Woody Weeds experience in Baringo County showed that, for ease of coordination, the optimal size of a LIG ranges from 15 to 20 members. They include 3–5 land resource users, e.g. farmers, pastoralists, charcoal producers; 2–3 representatives from different sectors of national and county governments (e.g. environment, agriculture, forest, water and development), including local administration; 2–3 local leaders (chiefs and opinion leaders); one or two agricultural advisors; one or two NGO representatives; 2–3 representatives of community organisations (e.g. women or youth groups) and agricultural cooperatives; as well as 2–3 representatives of other relevant society-rooted organizations such as WRUAs, CFAs, CPAs or faith-based groups. It is also important to include members from non-invaded neighbouring areas. The interactive engagement of the LIGs is achieved through a series of workshops, field days and training courses, as well as knowledge exchange and sharing events. The events are organized and facilitated by a proactive, innovative and motivated volunteer coordinator, who keeps LIG participants informed and involved through regular contacts and the distribution of information materials.

The bottom-up awareness creation and capacity development process of the LIG supports the participatory identification and implementation of promising SLM practices for *Prosopis* management. However, the scaling from individual SLM practices to *Prosopis* management at the landscape level requires different institutional structures that address critical planning and implementation requirements at technical and policy level while ensuring stakeholder participation at all levels and across sectors. This led to the idea, by the Woody Weeds + project, to create County Implementation Groups (CIGs) as the framework through which this objective could be achieved.

12.4.2 County Implementation Groups

The CIGs consist of 20–25 representatives from various stakeholder groups across all major sectors affected by *P. juliflora* and active from the county to the sublocation level, from policy makers to land users, and from planners to implementers. These include county government officials, administrators (down to the ward level) and representatives from the agriculture, environment, forestry, water and health sectors. County leadership, preferably a Member of the County Assembly, is also included. Additionally, the CIG consists of representatives of national government ministries, departments and agencies (down to location level), NGOs and community organizations (such as community and private conservancies, CPAs, WRUAs and CFAs). Locally active development partners and research scientists from national and international organizations serve as resource persons. The CIGs represent all ethnic groups and geographic regions in the county, as well as leaders of women and youth groups.

The CIGs coordinate the implementation of *P. juliflora* management. Their members develop a spatially explicit County *Prosopis* Management Plan and supervise its implementation at the subcounty and ward levels (Chapter 11, this volume). They also participate in awareness creation and capacity building events, serving as *Prosopis* management champions in both invaded areas and at the interface with non-invaded areas within the county and in neighbouring counties that are at risk of invasion. Additionally, CIG members engage with local stakeholders to implement *Prosopis* management activities and coordinate regular progress monitoring and periodic revision of management plans.

12.5 Selecting Control Methods to Enhance Multifunctionality

As a basis for the decision support process, we prepared an inventory of practices for the management of *Prosopis* tree species using the database of the World Overview of Conservation Approaches and Technologies (WOCAT) (www.wocat.net), which describes the objectives of SLM practices, the activities required to implement these practices and their costs, and the advantages and disadvantages of the practices. We included SLM practices that were not available in Kenya at the time the project was carried out but were applied in other parts of the global invaded range (e.g. Van Klinken and Campbell, 2001).

Using a deliberative multi-criteria decision (DMCD) support framework (Schwilch *et al.*, 2012) (Fig. 12.1), the LIG members reflected on the ecosystem services and disservices impacted by different land management practices, considering social, economic and ecological criteria. They then prioritized an area in their region and selected suitable SLM practices to implement the management of *Prosopis* trees in that area. The SLM practices were tested by LIG members or other land users at the identified location and their performance was assessed during joint field visits.

Valuing ecosystem services through stakeholder engagement remains methodologically challenging and novel in ecosystem multifunctionality research. However, it is critical since markets often fail to reflect the full value society places on or derives from these services. This failure to recognize the full value of ecosystem services contributes to ecosystem degradation. Developing SLM strategies for managing woody invasive alien plant species requires

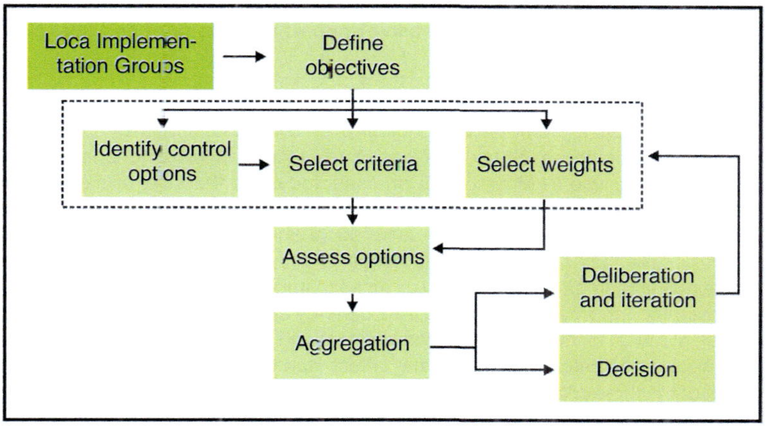

Fig. 12.1. Deliberative mult -criteria decision process used by Local Implementation Groups for the selection of locations prioritized for *Prosopis* management and of the management practice(s) to be tested in the case study areas of the Woody Weeds project (Afar region, Ethiopia; Baringo County, Kenya; Kahe area, Tanzania). Figure adapted from Schwilch *et al.* (2012).

a social-ecological systems approach. This approach must consider how ecosystem services are generated, how changes in these services influence human wellbeing and how changes in human wellbeing feed back into ecosystem service generation (Reyers *et al.*, 2013). The Woody Weeds project addressed ecosystem multifunctionality by collecting and sharing information on the effect of SLM practices on multiple ecosystem services and calculating ecosystem service multifunctionality indices for different stakeholder groups (Bardgett *et al.*, 2021; Linders *et al.*, 2021).

Besides the selection and testing of SLM practices, the LIG and CIG processes generated important additional outcomes. The structured decision support processes created an environment of trust and mutual understanding among the LIG and CIG members, including those benefiting from *Prosopis* trees, and changed their attitude towards environmental management. About 1 year after completion of the DMCD process, they were still more willing to sustainably manage *P. juliflora* and thus contribute to NRM than stakeholders who had not been involved (Box 12.1). Also, several LIG and CIG members served as champions and promoted the implementation of management practices that are aligned with the County *Prosopis* Management Plan.

12.6 Codesign and Uptake of County *Prosopis* Management Plans

While the LIGs made management decisions at the level of a particular location, a key objective of the CIGs was to develop and implement a spatially explicit County *Prosopis* Management Plan, which involved coordination and advocacy at various political and administrative levels. This process was crucial to achieving a locally adapted management plan and thus enhancing the chances of success of subsequent management efforts. Below, we briefly outline the steps covered through the series of codesign workshops.

The first CIG workshop addressed the constitution of the CIG and sought agreement, among its members, on its terms of reference. This required briefing of the members about the current extent of *Prosopis* invasion in the county, potential invasion trajectories and related current and future impacts. A draft version of the management plan was then codesigned based on a set of base maps, posters and factsheets describing various management practices

The identification and spatial arrangement of management objectives in the County *Prosopis* Management Plan depend on the current distribution of *P. juliflora*, the extent of areas that are

Box 12.1. Stakeholder perceptions about invasive species management and best practices: the role of participatory processes.

Stakeholder participation in land management initiatives is a key attribute of land degradation neutrality (UNCCD, 2019), as it promotes stakeholder ownership and implementation of sustainable land management (SLM) practices. Studies assessing the factors influencing land users' decisions to adopt and continuously use SLM practices primarily focus on the land user's socio-economic characteristics and on biophysical and institutional factors (e.g. de Graaff *et al.*, 2008; Banadda, 2010; Saguye, 2017). Other aspects related to land users' local knowledge, perceptions and values are often overlooked, despite their acknowledged potential to enhance the interest and commitment to adopt and continuous implementation of SLM practices (Jambo *et al.*, 2019). According to Schwilch *et al.* (2012), participatory approaches enhance transparency and ownership of decisions made, thus dispelling mistrust, a critical element in the implementation and scaling up and scaling out of selected SLM strategies. Also, participatory processes may affect stakeholders' weighing of decision criteria, e.g. in the context of the three pillars of sustainable development (socially acceptable, environmentally friendly and economically viable).

In the Woody Weeds project, we engaged members of Local Implementation Groups (LIGs) in a deliberative multi-criteria decision (DMCD) process to identify suitable *Prosopis* management practices for (i) prevention, (ii) early detection and rapid response, and (iii) control. DMCD is a structured process that combines the strengths of multi-criteria decision analysis with deliberations among citizens and scientists through knowledge sharing, negotiations and consensus-building (Proctor and Drechsler, 2006). One year after conducting the DMCD process with the LIGs, we compared perceptions and motivations of stakeholders who had participated in LIGs with a group of stakeholders representing the same stakeholder groups and originating from the same study area, but who were not involved in DMCD process. A total of 60 respondents were interviewed, 30 of whom were LIG members while the remaining 30 were not. Stakeholders from both groups were from the lowlands of Baringo County, Kenya, which are invaded by the non-native tree *Prosopis juliflora* (Mbaabu *et al.*, 2020).

The outcomes of the survey revealed a notable disparity ($p<0.05$) between stakeholders with experience and those without experience in DMCD processes regarding the necessity of managing *P. julilfora* (Box Fig. 12. 1). Most (89%) of the LIG members either agreed or strongly agreed that it is imperative to control the spread of *P. juliflora*, while only 11% expressed disagreement. None of the LIG members remained undecided, with all participants either agreeing or disagreeing on the necessity of managing *P. juliflora* invasion. In contrast, only 38% of the non-LIG members concurred or strongly concurred with the idea of managing *P. juliflora*, and 43% believed that it is not necessary to manage this invasive tree (Box Fig. 12.1). Equally noteworthy is that a considerable proportion (19%) of non-LIG participants remained uncertain about whether *P. juliflora* should be managed or not.

The respondents were then asked about the main criteria that would motivate them to implement the management of *P. juliflora*. Economic considerations were the main reason for motivating LIG (60%) and non-LIG (47%) members to implement SLM practices (Box Fig. 12.2). Concerns about the environmental impacts prompted 33% of the LIG members but only 19% of the non-LIG members to implement SLM practices. In contrast, the proportion of non-LIG members putting forward sociocultural criteria for implementing *P. juliflora* management was five times higher than that of LIG members.

At the beginning and end of the interview, the respondents were asked to rank different management practices according to their suitability for achieving specific management objectives. Two important results emerged from this exercise: first, non-LIG members preferred a management practice called 'management by utilization' which was promoted by governmental organizations but lacks scientific evidence that it indeed contributes to the management of *P. juliflora* (e.g. Mbaabu *et al.*, 2020). In contrast, none of the LIG members preferred this practice. Moreover, non-LIG members made more changes between the first and second ranking exercises than the LIG members, suggesting that LIG members felt more confident in taking decisions regarding the management of *P. juliflora*.

We conclude that the participatory integration of land users in the selection of SLM practices affected their perceptions about the need to manage *P. juliflora*, and thus contribute to natural resource

Continued

management (NRM). Stakeholders who had been engaged in a DMCD process were more committed to implement *P. juliflora* management than stakeholders who had not been involved. Participation in the DMCD process enhanced the knowledge of the environmental impacts of *P. juliflora*, which prompted LIG members to acknowledge and emphasize the urgency and need to manage the invasion, and it apparently made them more confident to make the right selection of SLM practices.

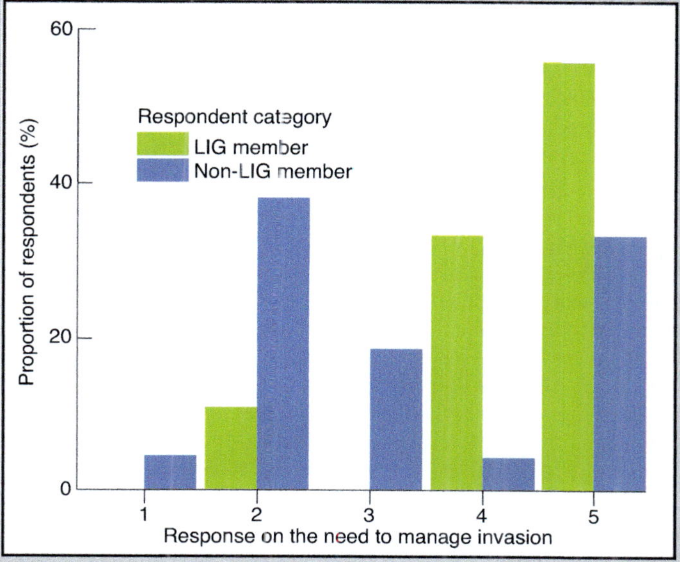

Box Fig. 12.1. A comparison of the perception about the need to manage the invasive tree *Prosopis juliflora* in Baringo County, Kenya, between Local Implementation Group (LIG) and non-LIG members. The response was recorded on a Likert scale ranging from 1 (strongly disagree) to 5 (strongly agree). Wilcoxon p = 0.023.

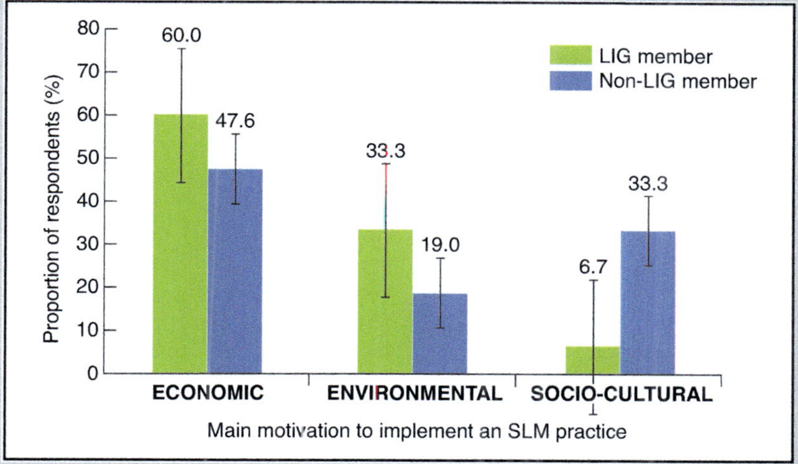

Box Fig. 12.2. A comparison of the motivations for Local Implementation Group (LIG) and non-LIG members for implementing a sustainable land management (SLM) practice. Numbers above the bars are means with standard deviations.

suitable for further invasion, the presence of important assets requiring protection (e.g. dry season grazing areas, cropland or water points) and the presence of *Prosopis* dispersal pathways (migration routes of wildlife and cattle, roads and rivers). These spatial criteria were integrated in a decision support tool presented in Chapter 11 (this volume).

During a community consultation (for feedback and validation) process, the draft plan was critically reviewed by community representatives in all subcounties of the targeted county, irrespective of whether those subcounties had already been invaded by *P. juliflora*. The feedback and recommendations by the community representatives were compiled in a report that was shared with the CIG.

In a second CIG workshop, an improved and validated version of the plans was drafted, based on the recommendations received from the community consultation process and considering new information gathered since the first workshop (e.g. on the current extent of *P. juliflora* invasion). Also, an estimation of the resources required for implementation of the plan was made. CIG members then adopted the revised plan and submitted it together with a memo to the County Executive and Assembly.

The spatially explicit County *Prosopis* Management Plan, the main output of the CIG codesign process, is a crucial element to guide management and restoration interventions and should be integrated into the official land use planning tools and documents of concerned counties. By doing so, it can help to preserve the supply of essential goods and ecosystem services in multifunctional landscapes that meet people's needs (Bardgett *et al.*, 2021). The County *Prosopis* Management Plans are based on the constitutional requirement that every county in Kenya must have a County Spatial Plan from which the CIDP is derived. Section 12.7 outlines how the management plan will be institutionalized through the CIDPs. The memo prepared by the CIG and submitted to the County Executive and Assembly was the first step towards this integration.

12.7 Institutionalizing the County Implementation Groups and the *Prosopis* Management Plan

Successful management of *P. juliflora* requires a long-term, perhaps a life-long, engagement in the form of collective actions aimed at reducing invasion densities and areas, and preventing spread to uninvaded areas. This requires institutional structures with sufficient means and legitimacy that are aligned with and embedded in existing governance structures at national and county levels, including the CIDPs. Experience from the Woody Weeds projects showed that initiatives which are not properly embedded in existing governance structures have limited chances of outliving the project period.

To achieve this, the first step was to identify the most suitable entry point within the implementation structures established by the NPS. The entry point had to guarantee seamless integration of CIGs into not only the national implementation structure but also the already established relevant county structures responsible for coordination or implementation of intervention programmes, e.g. sectors that include environment, health, livestock, water and spatial planning, among others. The NPS has an elaborate implementation framework that provides for the participation of all relevant state and non-state actors from national (Interministerial Steering Committee) through the county (County Implementation Committee) and subcounty (Subcounty Implementation Committee) to the ward level (Ward Implementation Committee) which links with community resource users, community-based natural resources management associations, extension providers and the national and county government representatives at the village level.

The County Implementation Committee, designated in the NPS as the responsible implementing organ, was identified as the most suitable entry point for the CIG. Yet, considering that the County Implementation Committee is a creation of the NPS and both operate under the county structures, the CIG was further modelled to become a permanent (technical) subcommittee of the County Implementation Committee or a Technical Working Group of the County Steering Group (or the County Environment Committee) which oversees the implementation of state and non-state supported intervention activities in a county. This way, the CIGs were embraced as part of the county implementation structures and therefore helped to streamline workflow and avoid conflicts in operation.

The second step at this level was to ensure that the ownership of the CIG's main product, the County *Prosopis* Management Plan, was escalated to the relevant executive and legislative institutions at county level, particularly the executive office of the governor and county assembly. To achieve this the CIG, with the support of relevant members of the County Executive Committee and the County Assembly, submitted a memorandum to the two institutions, giving an overview of the status of invasion in the county and the required management interventions to deal with the menace. The memorandum emphasized the need for the County Government to take ownership of the plan and support its implementation by ensuring that it was adopted by the executive and the County Assembly, and eventually integrated into the CIDPs and the subsequent Annual Development Plans for budgetary provision.

The CIG, with the support of the relevant member county departments, followed up to ensure that the County *Prosopis* Management Plan was integrated into the CIDP and that respective SLM practices proposed for priority intervention areas were mainstreamed into the annual workplans. The timing of the memorandum coincided with the preparation of the CIDPs in all the three counties covered by the Woody Weeds + project. This provided the opportunity to sensitize and lobby with especially the county finance and budget units, thus making the integration process faster. These measures, if continuously supported through adequate resourcing and political goodwill, should guarantee sustainability of the CIGs/Technical Working Groups of the County Steering Group and their activities in the affected counties. CIGs/Technical Working Groups of the County Steering Group can be duplicated in other counties should the invasion spread. To secure this in the long term, an effective policy dialogue should be initiated for including the CIGs in the relevant legislation. This was the case with the WRUAs when they operated informally in the initial stages (see Section 12.8). However, it may become necessary over time for the government to establish a special National *Prosopis* Fund and dedicate it to controlling the spread of this invasive tree species in the country. A dedicated fund will give the institution managing *P. juliflora* the degree of independence necessary to make rapid decisions for timely interventions to halt the rapid spread of the invasive weed.

12.8 Summary, Lessons Learned and Outlook

There is increasing recognition that global challenges need to be addressed in a holistic and integrated manner at the landscape level (Metternicht, 2017). Similarly, a multi-stakeholder approach is required to address biological invasions, as they are one of the main drivers of biodiversity loss and they substantially complicate NRM (IPBES, 2023). However, persistent sectoral silos, institutional fragmentation and jurisdictional boundaries, paired with inadequate ecological understanding and long-term funding (Reed *et al.*, 2021), are major hindrances to implementing effective coordinated management at the landscape level.

The management of landscapes requires institutional arrangements that foster horizontal (cross-sectoral) and vertical (across scales) collaboration among a diverse group of stakeholders with different needs, objectives and power positions. Such institutions are, however, largely non-existent in today's NRM governance structures in Kenya and other countries in eastern Africa, and thus the management of landscapes requires novel institutional arrangements that offer space for cross-sectoral and cross-scale communication, knowledge exchange, learning, negotiation and codecision processes.

In recent years, multi-stakeholder platforms (MSPs) have been created to strengthen the existing governance of sustainable NRM for safeguarding and improving the provision of ecosystem services and human wellbeing. MSPs can offer spaces for communication, considering both its informational and relational dimension (Hall and O'Rourke, 2014), deliberation and collaborative action. In a comparative analysis of MSPs either implemented top down (by government) or bottom up (by civil society), Ratner *et al.* (2022) found that all MSPs led to some improvements in communication, collaboration and trust between different stakeholders. In some cases, MSPs can play a coordination and advisory role in planning activities; in other cases, they were created for information exchange only, or meetings focused on a particular topic, with varying composition of stakeholders. Nevertheless, even in the latter case, MSPs provided a distinctive opportunity for community members to meet and engage and directly address their problems in a swift manner.

Sustainable management of invasive species, which is part of sustainable NRM, is a complex endeavour which must be coordinated across sectors and scales. It requires knowledge, skills, collaboration and influence from a diverse group of actors, including land and resource users, state actors from local administration to the national government, and from research, management and education, and non-state actors from community-based organizations to international NGOs. Thus, a multi-stakeholder institution which is supposed to design and support the implementation of a spatially explicit *Prosopis* Management Plan for a particular jurisdiction or landscape should have a well-balanced composition of resource users, decision makers, opinion leaders and resource persons and should operate according to jointly agreed terms of reference. If members of the multi-stakeholder institutions need to be replaced, they should ideally be replaced by persons with similar competences and influence.

In some ways, the LIGs are like the WRUAs and CFAs in that they are the lowest level governance structures for the management of *P. juliflora*. Similarly, the CIGs would resemble an NRM technical working group at either national or county level. As for the LIGs and CIGs, WRUAs were initiated in the frame of a programme, the larger Water Awareness Creation Campaign initiative of the mid-1990s, and were formed to address a commonly shared concern with a sense of urgency – water shortage and related disputes. Also, they were initially formed to address the need for a participatory, community-driven water management initiative. By 2002, Kenya had institutionalized the WRUAs, which now report to the Water Resources Authority whose main mandate is to regulate, monitor, assess, allocate water resources, manage and protect the water catchments and collaborate with other institutions for better water resource management in the country (the Water Act of 2002, which was revised to the Water Act 2016; Government of Kenya, 2016).

In the case of the CIGs, efforts are under way in Baringo and Isiolo County to assign them the status of a permanent Technical Working Group under the County Steering Group or the County Environment Committee. The incorporation of *P. julifora* management into the CIDPs in Baringo, Isiolo and Tana River counties allowed the mainstreaming of *P. juliflora* management into other spatially explicit planning activities and into the county's budgeting process. This increased the chances that the CIG will be institutionalized at the county level to a similar extent as the WRUAs and thus support the county's efforts to strengthen its NRM governance structure.

Multi-stakeholder platforms or institutions have wide-reaching aims for influencing the goals of governance of natural resources, which go further than the actual activities of the group. They offer space for social learning, which is most effective when embedded in structures and processes that enable joint actions (Schusler *et al.*, 2003). We support the notion that the stakeholders' motivation to participate in these processes – and thus the likelihood of social learning – is particularly high if the topic is a common concern, and when the various stakeholders realize that they have a say in the matter. A common sense of the urgency of the problem caused by the invasion of *Prosopis* species among the LIG and CIG members, including the representatives of the CPAs, was also evidenced by the fact that the same persons, including Members of the County Assembly and the County Executive Committee, participated in most activities.

Critics of MSPs argue that little power sharing takes place in such institutions or platforms (Ros-Tonen *et al.*, 2018). In a review of structured multi-stakeholder learning processes from SLM, Schwilch *et al.* (2012) concluded that the moderators of such processes have a critical role, also in dealing with power imbalances. As there is a risk that certain stakeholder groups might try to impose their ideology, the moderators of the LIG and CIG workshops should put emphasis on creating an environment of trust in which all participants feel that it is the content of their inputs that counts, not their position or perceived power. The outputs of the deliberation processes conducted with the LIGs and CIGs were approved by all members, including those benefiting from

P. juliflora; in the case of the outputs created by the CIG. This was evidenced by the fact that all members signed a memorandum that called for a spatially explicit *Prosopis* management plan in their home county which was shared with their governor. Also, in the end-of-project assessment, all the stakeholders interviewed stated that the Woody Weeds project facilitated collective knowledge creation and fostered an open and inclusive approach to considering different viewpoints.

In their review of structured multi-stakeholder learning events, Schwilch *et al.* (2012) reported that most stakeholders believed that what they had learned was useful, also for future work. Our evaluation of the LIG process in the Woody Weeds project points in the same direction, as the stakeholders participating in the LIG process continued to consider management of *P. juliflora* as part of NRM as more important than stakeholders who did not participate in the LIG process. Thus, our findings support the notion that participatory approaches do not only help increase mutual understanding and ownership over the challenges that are subject of the participatory approaches, such as invasive species management. They also foster trust and learning amongst stakeholders, who are then more likely to make deliberate decisions and commit themselves to implement them in the long term (Reed *et al.*, 2018).

It is too early to assess the sustainability of the novel institutional arrangements set up in the three pilot counties in Kenya and their impacts on the management of *P. juliflora* in the long term. A review of the Working for Water programme in South Africa revealed that during the first 23 years of the programme, the management of invasive plant species was primarily successful over relatively large areas if biological control was incorporated in the management toolbox (Henderson and Wilson, 2017). Thus, the success of the institutional arrangements set up in the three counties in Kenya will also depend on the management options available in Kenya, especially biological control. Nevertheless, we propose that the approach of stakeholder engagement and the creation of stakeholder platforms or institutions as described in this chapter can guide the implementation of the management of *Prosopis* species in other parts of the invaded range, and probably also the management of other invasive alien plants where a spatial management approach is warranted, e.g. for spatial prioritization of control activities at the landscape level (Meier *et al.*, 2014).

Acknowledgements

This work was supported by the Swiss Programme for Research on Global Issues for Development (r4d), funded by the Swiss National Science Foundation and the Swiss Agency for Development and Cooperation, through the project 'Woody invasive alien species in East Africa: Assessing and mitigating their negative impact on ecosystem services and rural livelihoods (Grant Number: 400440_152085), and by the Swiss Agency for Development and Cooperation through the project 'Woody Weeds +' (Project number 7F.10651.01). US was supported by CABI with core financial support from its member countries (see http://www.cabi.org/about-cabi/who-we-work-with/ key-donors/for full details). Partner institutions, Local Implementation Group and County Implementation Group members made in-kind contributions.

References

Banadda, N. (2010) Gaps, barriers and bottlenecks to sustainable land management (SLM) adoption in Uganda. *African Journal of Agricultural Research* 5, 3571–3580.

Bardgett, R.D., Bullock, J.M., Lavorel, S., Manning, P., Schaffner, U. *et al.* (2021) Combatting global grassland degradation. *Nature Reviews Earth and Environment* 2, 720–735.

Berkes, F. (2007) Community-based conservation in a globalized world. *Proceedings of the National Academy of Sciences of the United States of America* 104, 15188–15193.

Cumming, G.S., Cumming, D.H. and Redman, C.L. (2006) Scale mismatches in social-ecological systems: causes, consequences, and solutions. *Ecology and Society* 11, 14.

de Graaff, J., Amsalu, A., Bodnar, F., Kessler, A., Posthumus, H. *et al.* (2008) Factors influencing adoption and continued use of long-term soil and water conservation measures in five developing countries. *Applied Geography* 28, 271–280.

ElDidi, H., Rawat, S., Meinzen-Dick, R., Chaturvedi, R. and Sanil, R. (2024) Polycentric governance of commons through multi-stakeholder platforms: Insights from two case studies in India. *Environment, Development and Sustainability* 1, 1–27.

Food and Agriculture Organization (2022) *Voluntary Guidelines on the Responsible Governance of Tenure of Land, Fisheries and Forests in the Context of National Food Security.* United Nations Food and Agriculture Organization, Rome, Italy.

Government of Kenya (2016) *Water Act 2016*. Government of Kenya, Nairobi, Kenya.

Government of Kenya (2021) *Kenya State of Environment Report 2019–2021*. Government of Kenya, Nairobi, Kenya.

Hall, T.E. and O'Rourke, M. (2014) Responding to communication challenges in transdisciplinary sustainability science. In: Huutoniemi, K. and Tapio, P. (eds) *Transdisciplinary Sustainability Studies*. Routledge, Abingdon, UK, pp. 119–139.

Henderson, L. and Wilson, J.R. (2017) Changes in the composition and distribution of alien plants in South Africa: An update from the southern African plant invaders atlas. *Bothalia* 47, 1–26.

Intergovernmental Science-Policy Platform on Biodiversity and Ecosystem Services (IPBES) (2023) *Thematic Assessment Report on Invasive Alien Species and their Control of the Intergovernmental Science-Policy Platform on Biodiversity and Ecosystem Services*. Intergovernmental Science-Policy Platform on Biodiversity and Ecosystem Services, Bonn, Germany.

Jambo, I.J., Groot, J.C.J. and Descheemaeker, K. (2019) Motivations for the use of sustainable intensification practices among smallholder farmers in Tanzania and Malawi. *Journal of Life Sciences* 89, 100306.

Kameri-Mbote, P., Kibugi, R. and Kabira, N. (2023) *Environmental Governance in Kenya: Implementing the Constitutional Framework*. University of Nairobi, Nairobi, Kenya.

Kiteme, B.P. and Wiesmann, U. (2008) Sustainable river basin management in Kenya: Balancing needs and requirements. In: Hirsch Hadorn, G., Hoffmann-Riem, H., Biber-Klemm, S. *et al.* (eds) *Handbook of Transdisciplinary Research*. Springer, pp. 63–78.

Linders, T.E., Schaffner, U., Alamirew, T., Allan, E., Choge, S.K. *et al.* (2021) Stakeholder priorities determine the impact of an alien tree invasion on ecosystem multifunctionality. *People and Nature* 3, 658–672.

Lockwood, M., Davidson, J., Curtis, A., Stratford, E. and Griffith, R. (2010) Governance principles for natural resource management. *Society and Natural Resources* 23, 986–1001.

Mbaabu, P.R., Olago, D., Gichaba, M., Eckert, S., Eschen, R. *et al.* (2020) Restoration of degraded grasslands, but not invasion by *Prosopis juliflora*, avoids trade-offs between climate change mitigation and other ecosystem services. *Scientific Reports* 10, 1–13.

Mburugu, K.N. and Macharia, D. (2016) Resolving conflicts using indigenous institutions: A case study of Njuri-Ncheke of Ameru, Kenya. *International Journal of Science Arts and Commerce* 1, 18–33.

Meier, E.S., Dullinger, S., Zimmermann, N.E., Baumgartner, D., Gattringer, A. *et al.* (2014) Space matters when defining effective management for invasive plants. *Diversity and Distributions* 20, 1029–1043.

Metternicht, G. (2017) *Land Use Planning*. United Nations Convention to Combat Desertification, Bonn, Germany.

Muigua, K. (2018) *Devolution and Natural Resource Management in Kenya*. University of Nairobi, Nairobi, Kenya.

Proctor, W. and Drechsler, M. (2006) Deliberative multicriteria evaluation. *Environment and Planning C: Government and Policy* 24, 169–190.

Ratner, B.D., Larson, A.M., Barletti, J.P.S., ElDidi, H., Catacutan, D. *et al.* (2022) Multistakeholder platforms for natural resource governance: Lessons from eight landscape-level cases. *Ecology and Society* 27, 2.

Reed, J., Kusters, K., Barlow, J., Balinga, M., Borah, J.R. *et al.* (2021) Re-integrating ecology into integrated landscape approaches. *Landscape Ecology* 36, 2395–2407.
Reed, M.S., Vella, S., Challies, E., Vente, J., Frewer, L, *et al.* (2018) A theory of participation: What makes stakeholder and public engagement in environmental management work? *Restoration Ecology* 26, S7–S17.
Reyers, B., Biggs, R., Cumming, G.S., Elmqvist, T., Hejnowicz, A.P. *et al.* (2013) Getting the measure of ecosystem services: A social–ecological approach. *Frontiers in Ecology and the Environment* 11, 268–273.
Ros-Tonen, M.A.F., Reed, J. and Sunderland, T. (2018) From synergy to complexity: The trend toward integrated value chain and landscape governance. *Environmental Management* 62, 1–14.
Saguye, T. (2017) Determinants of adoption of sustainable land management (SLM) practices among smallholder farmers in Jeldu District, West Shewa Zone, Oromia Region, Ethiopia. *Journal of Resources Development and Management* 8, 96–116.
Schusler, T., Decker, D. and Pfeffer, M. (2003) Social learning for collaborative natural resource management. *Society and Natural Resources* 15, 309–326.
Schwilch, G., Bachmann, F., Valente, S., Coelho, C., Moreira, J. *et al.* (2012) A structured multi-stakeholder learning process for sustainable land management. *Journal of Environmental Management* 107, 52–63.
United Nations Convention to Combat Desertification (UNCCD) (2019) *Land Degradation Neutrality Transformative Projects and Programmes: Operational Guidance for Country Support*. United Nations Convention to Combat Desertification, Bonn, Germany.
van Klinken, R. and Campbell, S.D. (2001) The biology of Australian weeds 37. *Prosopis* L. species. *Plant Protection Quarterly* 16, 2–20.
World Bank (2005) In: *Integrating the Indigenous Knowledge of Borana Pastoralists into Rangeland Management Strategies in Southern Ethiopia*. Indigenous Knowledge (IK) Notes 81, World Bank, Washington, DC. Available at: http://documents.worldbank.org/curated/en/534131468023079129/Integrating-the-indigenous-knowledge-of-Borana-pastoralists-into-rangeland-management-strategies-in-Southern-Ethiopia (accessed 3 August 2024).

Part 4: Bridging Silos in Practice and Research

Environmental management is an issue that involves many people and institutions. The important groupings vary substantially with regard to their levels of understanding and expertise, interest and influence regarding particular aspects, their visions of desired outcomes, the degree to which they are able to influence important decisions, and their access to funding and resources. Understanding such complex environmental issues also requires collaboration across research disciplines through transdisciplinary research. Effectively dealing with the complex issues around managing invasive *Prosopis* trees requires holistic approaches that integrate insights from diverse fields such as ecology, economics and sociology. By encouraging integration between traditional silos, transdisciplinary research can foster innovative solutions, and it is essential for achieving the goals of sustainable development.

In this final part, we describe attempts to improve the chances of successful management of *Prosopis* tree invasions by encouraging the broad involvement of a wide range of parties, and by integrating research across disparate disciplines.

Chapter 13 by Ehrensperger *et al.* present the results of a stakeholder mapping exercise conducted in three study sites (Afar Region, Ethiopia; Baringo County, Kenya; and Amani Nature Reserve, Tanzania) affected by the invasion of *Prosopis julilfora* and other plant species. They derive relevant implications from these findings regarding the engagement of stakeholders in the management of invasive plant species across different jurisdictions and communities of practice. The chapter proposes a categorization of stakeholders that is determined by the level of interest in participating in addressing the problem, and their ability or capacity to influence the management of invasions. The chapter then proposes appropriate ways of engaging with people in each category to increase the overall effectiveness of management.

In Chapter 14, Eigenbrode *et al.* describe how research carried out by postgraduate students in the Woody Weeds project was conceived, designed and integrated across at least five scientific disciplines over ten years. It deals with the challenges of conducting research in social-ecological systems, including how to involve multiple stakeholders, adopt a collaborative culture across disciplines, integrate different types of data and deal with discipline-specific academic expectations. The lessons from this experience are relevant not only to the issue of invasive *Prosopis* trees in eastern Africa but to biological invasions research in general, as well as to the wider management of ecosystems and natural resources.

13 Stakeholders and their Involvement in the Management of *Prosopis juliflora* in Eastern Africa

Albrecht Ehrensperger[1]*, John R. Mbwambo[2], Simon Choge[3], Tena Alamirew[4], Brian W. van Wilgen[5] and Linus K. Munishi[6]

[1]Centre for Development and Environment (CDE), University of Bern, Switzerland; [2]Tanzanian Forestry Research Institute, Morogoro, Tanzania; [3]Kenya Forestry Research Institute, Marigat, Kenya; [4]Water and Land Resource Centre, Addis Ababa University, Ethiopia; [5]Centre for Invasion Biology, Department of Botany and Zoology, Stellenbosch University, South Africa; [6]The Nelson Mandela African Institute of Science and Technology, Arusha, Tanzania

Abstract

The management of invasive alien plant species, such as *Prosopis* trees, demands a collaborative approach involving stakeholders at various levels and from different sectors. Effective management hinges on integrating different perspectives and expertise, from individual land users to community groups and national authorities. However, stakeholders involved in managing invasive species often exhibit varying levels of awareness, recognition of the problem's severity and capacity to contribute to solutions. Challenges arise from differing perceptions of the benefits and impacts of invasive species, leading to divergent management strategies proposed by different sectors. This chapter examines stakeholders involved in *Prosopis* tree management and their levels of interest and influence. Actor mapping workshops conducted with stakeholders in three project sites identified a total of 125 stakeholders. National governmental agencies, local governments, international agencies and local communities were among the key stakeholders, each with varying levels of interest and influence. Stakeholders were categorized into groups such as influencers, regulators, extractors, beneficiaries and affectees, based on their roles and interests in invasive species management. The analysis revealed four idealized categories of engagement based on stakeholder levels of interest and influence. These categories can guide the engagement of (a) inconsequential stakeholders, (b) highly influential but disengaged stakeholders, (c) disempowered stakeholders, and (d) highly influential and engaged stakeholders. Recommendations are made to integrate these approaches into national strategies for invasive species management, ensuring the inclusion of all stakeholder categories and clear delineation of roles and responsibilities. However, the approach has limitations, including subjective categorizations and biases in stakeholder classification. The understanding of terms like 'influence' varied among participants, highlighting the need for careful consideration of language and cultural factors in future studies. Despite these limitations, the study provides valuable insights into stakeholder engagement strategies for effective invasive species management in the region.

*Corresponding author: albrecht.ehrensperger@unibe.ch

© CAB International 2025. *The Ecology and Management of Invasive Prosopis Trees in Eastern Africa* (eds U. Schaffner *et al.*)
DOI: 10.1079/9781800623644.0013

13.1 Introduction

The management of invasive alien plant species is a multifaceted and interconnected effort, involving multiple sectors and scales (as discussed in Chapters 11 and 12, this volume). Therefore, the success of such efforts depends on the integration of diverse perspectives and expertise into management strategies (Woodford *et al.*, 2016; Adoyo *et al.*, 2022). It hinges on the successful implementation of land management decisions by individual land users at the plot level, as well as by groups of land users within, for example, an irrigation scheme. In the case of grazing land and other communally owned resources, entire communities must collaborate to effectively address management. Managing invasive alien plant species in protected areas necessitates consensus building and collaborative problem solving among communities, conservation authorities and the tourism sector (Shackleton *et al.*, 2016). Finally, effectively addressing invasions across national borders requires a high level of coordination among national authorities of neighbouring countries. In short, effective management of invasive alien plant species requires the active involvement of diverse stakeholder groups throughout the planning, decision making and implementation stages (Novoa *et al.*, 2018).

However, the experiences gained from the Woody Weeds 'research for development' project in eastern Africa (www.woodyweeds.org; Section 14.3, Chapter 14, this volume), which was active from 2015 to 2022, and from other projects (e.g. Novoa *et al.*, 2016; Cordeiro *et al.*, 2020), have revealed that stakeholder groups influenced by or influencing invasive alien plant species and their management may exhibit significantly varied levels of awareness of the current or potential future challenges and impacts of the invasion. Moreover, these groups may differ in their recognition of the necessity to manage the issue, and might have distinct interests related to management objectives and priorities (Chapter 15, this volume). Additionally, stakeholders vary in their levels of capability, encompassing skills, financial and human resources, which can significantly influence their (perceived) ability to actively contribute to or participate in the required planning, decision-making and implementation processes (Shackleton *et al.*, 2016).

Consequently, mobilizing and engaging stakeholders in the management of invasive alien plant species can present challenges that range from limited participation when they are not perceived as a significant threat, to outright conflicts when stakeholders believe that the burden of management is unfairly distributed. Conflicts may also arise when certain stakeholder groups feel that their livelihoods are threatened, such as charcoal producers who are affected by the shortage of native wood and now depend on *Prosopis* trees for their trade. At the policy level, challenges may emerge as different sectors propose conflicting strategies for managing invasive alien plant species (Nuñez and Pauchard, 2010; Friedel *et al.*, 2011; Woodford *et al.*, 2016; Zengeya *et al.*, 2017).

Despite these challenges, involving all relevant stakeholder groups and acknowledging their diverse viewpoints in the design and subsequent implementation of a management strategy are unavoidable, primarily due to the normative nature of such a process. Inclusive engagement is essential to achieve an anticipatory rather than a reactive approach to conflict resolution (Crowley *et al.*, 2017). This implies that a thorough stakeholder analysis is required to identify the relevant stakeholder groups. Such an analysis can contribute to a responsive communication strategy which takes into consideration the current level of influence and motivation of the various stakeholder groups and how this affects invasive species management.

In this chapter, we describe the various stakeholders who are involved in, or who influence, the management of invasive alien plant species in eastern Africa and those who are influenced by activities and management decisions. Our focus is on the management of *Prosopis juliflora* in Ethiopia and Kenya, and of *Lantana camara* in Tanzania. The primary objective of this chapter is to provide insights that could lead to more effective stakeholder involvement in the context of invasive alien species management in the region. To achieve this, we (a) present a comprehensive stakeholder mapping, using stakeholders in Baringo (Kenya), Amani (Tanzania) and Awash (Ethiopia) as a representative sample; (b) propose a typology of these stakeholders, drawing from the approach by Newton and Elliott (2016), which facilitates targeted efforts and priority setting; and (c) derive relevant conclusions related to mobilizing and involving stakeholders in research, strategic development and implementation. We also discuss the national strategies that have been developed to deal with the problem and examine them in terms of the involvement, or exclusion, of stakeholders and what that may mean for their effectiveness. We conclude with the challenges that will face the wide range of stakeholders whose efforts will have to be effectively coordinated if the strategic goals of management are to be achieved.

13.2 Approach

We conducted a stakeholder mapping exercise to gather information on stakeholders, who are already actively participating in the management of *P. juliflora* and *L. camara*, who should be mobilized to participate in the same or who are, more generally, likely to become concerned about this challenge in the future.

Stakeholders were identified and classified during workshops with partners of the Woody Weeds project in three project sites: (1) the Amani Nature Reserve, Tanga Region, Tanzania (November 2015); (2) the Awash River Basin, Afar, Ethiopia (June 2016); and (3) Baringo County, Kenya (February 2017). The 25–30 participants in each workshop included land users, opinion leaders, chiefs, local to subnational government representatives, protected area managers and research institutes. During each workshop, participants identified stakeholders who are or might become concerned with the management of *Prosopis* trees (in Awash and Baringo) or *L. camara* (in Amani) and wrote their names on cards. They were then asked to position each card in a four-quadrant matrix in which the x-axis indicates (from left to right) increasing interest or motivation of the stakeholder to engage in the management of these species, and the y-axis indicates (from bottom to top) increasing influence of that stakeholder group, or ability to take decisions concerning the management of either *Prosopis* tree or *L. camara* invasions.

We used the typology introduced by Newton and Elliott (2016), which distinguishes between extractors, beneficiaries, affectees, regulators and influencers. The category of 'inputters' (those discharging or placing materials or infrastructure into the system) proposed by Newton and Elliott (2016) was excluded as it lacks relevance in the context of invasive alien species management. The remaining five categories (extractors, beneficiaries, affectees, regulators, influencers) provided a comprehensive understanding of the potential roles of stakeholder groups, facilitating the identification of suitable engagement approaches.

To assign stakeholders identified by workshop participants to these categories, a group of four experts conducted individual categorizations. They assigned one main category and one or several secondary categories to each of the 125 stakeholders identified by the workshop participants. If at least three out of four experts used the same main category, this category was used in the final list. Stakeholders receiving fewer than three identical main categories were discussed in a subsequent meeting and the final category assigned based on consensus.

13.3 Results

13.3.1 Stakeholder identification

Participants in the three workshops identified a total of 125 stakeholders or stakeholder groups who were involved in or influence the management of either *P. juliflora* (in Awash and Baringo) or *L. camara* (in Amani) or are influenced by activities and management decisions about these two species, and who should therefore be considered or even mobilized. There were around 40 different stakeholders or stakeholder groups in each project site (Fig. 13.1; see Appendix 13.1 for a full list).

National governmental agencies dominated, particularly in Baringo and Amani. Interestingly, participants in Afar identified more local than national governmental agencies. Local governments and international agencies and organizations (United Nations and Consultative Group on International Agricultural Research (CGIAR) agencies, CABI,

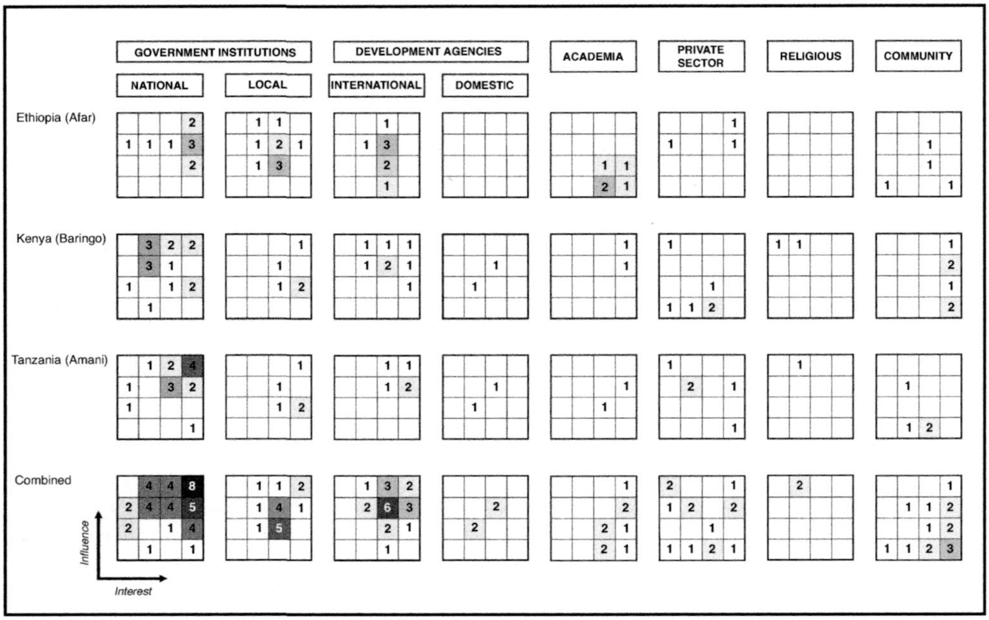

Fig. 13.1. Results of stakeholder mapping in Ethiopia (Afar), Kenya (Baringo) and Tanzania (Amani). The figure illustrates the number of institutional stakeholders identified, along with their interest in participating in invasive species management (*x*-axis of each 4 × 4 matrix) and their ability to influence that management (*y*-axis of each 4 × 4 matrix). The *x*- and *y*-axes utilize a four-point scale of interest or influence, ranging from very low, to low, high and very high, from left to right and bottom to top.

donors and others) were also well represented. They were followed by local communities and the private sector, then academia and finally domestic development agencies (local NGOs) and faith-based organizations. The level of interest and the ability of stakeholders to influence outcomes varied between stakeholders, with many having – in the opinion of the workshop participants – both a strong influence and a keen interest in the issue, especially at a national level (Fig. 13.1). These groupings are discussed briefly below.

13.3.2 Levels of interest

The stakeholders identified by the workshop participants seemed to have a fair level of interest in participating in the management of invasive species. A very low interest was only detected among some national institutions and private sector representatives (the Bureau of Agriculture and a sugar factory in Awash; the Ministry of Livestock, the national electricity supply company and the media in Amani; the national power supply company and a power plant in Baringo) as well as schoolteachers in Afar, who were categorized as community members. Academic institutions were rated as having a high interest, as well as local governmental institutions and communities, except for Amani (Tanzania), where the targeted invasive species (*L. camara*) is not perceived as being a serious threat as is the case for *P. juliflora*, the main invasive species in the two other project sites. Interest among private sector representatives was evenly distributed. The same applied to the national government, where some agencies have a very high interest and others none.

13.3.3 Levels of influence

There were several broad groupings that emerged from an examination of levels of interest (Fig. 13.1). These were (1) stakeholders who were perceived as mostly having a lot of influence in relation to invasive species management (national and local governments, international agencies, faith-based organizations); (2) others who have only medium influence (domestic development partners); (3) those who mostly do not have a lot of influence (communities); and (4) those where no clear pattern emerges (academia and the private sector). Each category also includes stakeholders who do not fit into this overall picture but the participants' reliance on or expectations of governmental and international agencies are clearly noticeable in this rating. Interestingly, the three private sector representatives with a very high level of influence included two media companies and one private investor. Another interesting aspect is that faith-based organizations were seen as having a high level of influence in Baringo and Amani (none was mentioned in Afar), but a rather low interest in putting that influence to use for the management of invasive species.

13.3.4 Typology of stakeholders

Out of the 125 stakeholders identified by the participants in the three workshops held in Ethiopia, Kenya and Tanzania, nearly half were placed into the category of *influencers* by the experts who conducted the categorization (see Table 13.1). This group encompassed a mix of international (and some local) governmental and non-governmental organizations, academic institutions, faith-based organizations, parastatal organizations and a few private sector entities. The second largest group consists of *regulators*, primarily composed of national and local institutions, along with a small number of parastatal organizations. Twelve stakeholders were classified as *extractors*, representing individuals or groups engaged in resource extraction from the region. They included honey producers, timber and tea plantations, a sugar factory, the tourism industry and large-scale farms. As mentioned in Section 13.2, one main category and one or several secondary categories were assigned to each stakeholder. For example, honey producers are in the main category 'extractors' but are at the same time 'affectees' and 'beneficiaries'. We refer here to the main category only. A limited number of stakeholders, such as charcoal producers and a power plant in Baringo, derived significant benefits from the presence of the invasive species (in this case, *P. juliflora*).

Finally, 12 stakeholder groups were classified as *affectees*, 10 of which belonged to communities (e.g. women, pastoralists or smallholders) and two represented protected areas. The affectees are mostly in the 'high interest and low influence' corner of the typology (Fig. 13.2). The three affectees with high influence were a farmer's association, a women's association and a national park.

The Cummins biomass power plant, listed among the *beneficiaries*, which was meant to benefit from *P. juliflora* processing in Baringo, was perceived as having low interest and influence, whereas the charcoal producers' association in the same area was seen as being influential. *Extractors* tended to have high

Table 13.1. Typology of stakeholders identified during the stakeholder mapping conducted in Baringo (Kenya), Afar (Ethiopia) and Amani (Tanzania). The typology used is adapted from Newton and Elliott (2016). Categories were assigned to stakeholders by seeking consensus from four senior and experienced lead scientists in the Woody Weeds research team.

	Afar	Baringo	Amani	Total
Affectees	6	4	2	12
Beneficiaries	0	2	0	2
Extractors	3	4	5	12
Influencers	20	20	19	59
Regulators	10	13	13	36
Total	39	43	39	121

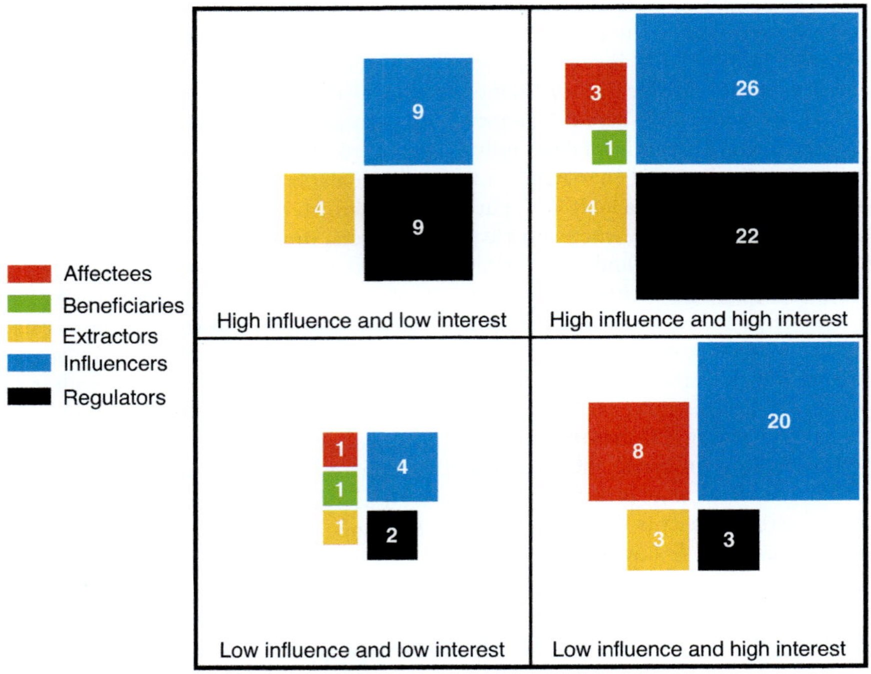

Fig. 13.2. Levels of interest and influence regarding the management of invasive plant species of five stakeholder categories in Afar (Ethiopia), Amani (Tanzania) and Baringo (Kenya). The size of squares represents the number of stakeholders in each category.

influence but with diverging levels of interest. Those with low interest included small companies, while those with high levels of interest included large-scale farmers and the tourism industry. *Influencers* were found in each quadrant of the matrix, particularly on the 'high interest' side. Those with low influence mainly included research and academic institutions, as well as extension services. Those with high influence were predominantly UN and CGIAR organizations, international NGOs and a few academic, parastatal and governmental institutions. Naturally, *regulators* were regarded as having high influence but several of them (such as various ministries and district administrations) had only low interest. Regulators with high interest and high influence included ministries, administrations, county assemblies and development authorities. Interestingly, this group also included customary institutions in Afar, chiefs in Baringo and forest rangers in Amani, i.e. stakeholders who are close to the community.

The distribution among affectees, beneficiaries, extractors, influencers and regulators is very similar in the three countries.

13.4 Discussion

13.4.1 Engagement types

The analysis presented above provides information that can be used to guide appropriate ways to engage various groups of stakeholders in invasive alien species management. The position of stakeholders in the four quadrants of the 'influence–interest' matrix (Fig. 13.2) is a useful basis for the identification of engagement approaches. Broadly speaking, these quadrants differentiate four idealized engagement approaches, outlined below.

13.4.1.1 Inconsequential stakeholders

When resources are limited, stakeholders in the lower left quadrant are typically those that would not be prioritized for engagement, as engaging them would necessitate both addressing their interest and endowing them with the required influence or ability to become active and effect change. However, in certain cases, seeking their participation might still be useful. For instance, during discussions in the Afar region, schoolteachers were placed in the lower left quadrant, with the argument that they do not exhibit significant concern about invasion by *P. juliflora* and possess limited influence on land-related issues. Nonetheless, with targeted awareness initiatives, they could potentially be used to raise awareness through their interactions with young people.

13.4.1.2 Highly influential but disengaged stakeholders

Stakeholders in the upper left quadrant represent an untapped potential. Often, their perceived lack of interest may arise from lack of awareness. Therefore, targeted awareness creation among the influential stakeholders in this group, which includes faith-based organizations, the media, parastatals, international NGOs and national government institutions, as well as regulators such as ministries, subnational administrations and resource management boards, could unlock significant opportunities A pertinent example is observed with faith-based organizations and the media, whose primary focus lies in community outreach, while their thematic interests may predominantly revolve around other issues.

13.4.1.3 Disempowered stakeholders

Stakeholders in the lower right quadrant are concerned supporters. While their interest in the management of invasive alien species is strong, they lack the agency and power to bring about significant changes. Empowering these stakeholders, which include numerous academic institutions and research-oriented governmental and parastatal organizations, would elevate their status from mere supporters to active contributors. This can be accomplished by facilitating networking among these stakeholders and fostering a community of like-minded individuals and institutions. Such collaboration would play a pivotal role in shaping the narrative concerning invasive alien species management in the region. For example, joint publications, collective participation in various events and policy-relevant declarations addressed to regulators would all contribute to amplifying their impact.

13.4.1.4 Highly influential and engaged stakeholders

Stakeholders in the upper right quadrant possess significant potential as drivers of change, given their substantial influence and interest in advancing the management of invasive species. Thus, any project or strategy seeking to effectively manage invasive alien species should anchor itself on strong partnerships and support from at least two or three stakeholders in this quadrant. To mobilize the influential and interested regulators and influencers, a possible strategy involves influencing their thematic agenda and priorities. Another approach would be to establish networks of counterparts, fostering connections between these stakeholders and affected groups within the local context. Such collaborative efforts would facilitate meaningful engagement and promote effective management practices for invasive species like *P. juliflora*.

13.4.2 Implications for the implementation of national strategies

Kenya and Ethiopia have developed strategies for the management of invasive *Prosopis* trees (Ministry of Livestock and Fisheries, 2017; Ndufa *et al.*, 2020). Tanzania has prepared a strategy to control and manage all invasive alien species, highlighting *Prosopis* trees as the major threat to ecosystems and livelihoods (Anonymous, 2019). Development of these strategies marks an important achievement for the region in adhering to several international obligations on protection of environment and livelihoods. Although there are differences between these strategies, they all aim to guide resource mobilization and coordination of effort across jurisdictions and affected stakeholders on invasive alien species control and management activities. The strategies also underline plans and actions by national governments, local communities, community-based organizations, non-governmental organizations, environmental agencies, industries and landowners.

For example, the Kenyan National *Prosopis* Strategy includes a stakeholder analysis listing 13 parastatals and government institutions, 14 NGOs and international agencies, two financial institutions, six entrepreneurs and an unspecified number of academic institutions. The annex of the strategy includes a table detailing the roles and responsibilities of various institutions for each outcome and activity. Around half of the 44 stakeholders that were identified during the workshop of the Woody Weeds project in Baringo are listed in either the stakeholder analysis or the annex of the national strategy. The former includes a mix of influencers and regulators, whereas the latter mainly lists regulators (ministries, national and county-level government offices, parastatals). Only one beneficiary (charcoal producers' associations) is mentioned explicitly. Affectees and extractors are mentioned in general terms or implicitly (e.g. as part of the term 'multi-stakeholder').

13.4.3 Recommendations

The above example of the Kenyan National *Prosopis* Strategy shows a tendency towards explicitly including influential and engaged stakeholders (upper right quadrant in our analysis) and a few influential but disengaged stakeholders (upper left quadrant), while listing the disempowered and inconsequential stakeholders (lower right and lower left quadrants) in generic terms only, without assigning clear roles and responsibilities to them. Further, there is a strong emphasis on regulators and influencers, while specific beneficiaries, affectees and extractors are rarely mentioned explicitly.

While these stakeholders might appear more distinctly in county-level management plans that are derived from the national strategy, we recommend integrating the four engagement types mentioned in Section 13.4 into the reflections leading to the development of national strategies. We further recommend integrating the stakeholder categorization into affectees, beneficiaries, extractors, influencers and regulators into stakeholder analyses used to develop such strategies, particularly in preparation of multi-stakeholder processes during which the consideration of these categories is of crucial importance.

13.4.4 Limitations of the approach

Our approach relies on subjective categorizations and rankings by workshop participants and experts. The outcome of such an approach thus strongly depends on the degree to which all participants and all experts share a common notion of variables such as influence and interest, or categories such as beneficiaries and extractors. During the workshops, participants insisted on getting a better understanding of 'influence' before positioning their cards in the matrix. For example, in Afar, land users and community representatives often assigned influence to economic power, while representatives from governmental organizations felt that influence was linked to actors directly involved in land management activities. This illustrated that understanding of the term might differ from one participant to another. A careful selection of terms used for the ranking exercise, accounting for language and cultural factors, is therefore crucial. Our stakeholder mapping might also be biased because institutional stakeholders are often easier for respondents to spot than individual stakeholders or stakeholder categories lacking a formal institutional identity.

Acknowledgements

This work was supported by the Swiss Programme for Research on Global Issues for Development (r4d), funded by the Swiss National Science Foundation and the Swiss Agency for Development and Cooperation, through the project 'Woody invasive alien species in East Africa: Assessing and mitigating their negative impact on ecosystem services and rural livelihood' (Grant Number: 400440_152085), and by the Swiss Agency for Development and Cooperation through the project 'Woody Weeds +' (Project number 7F.10651.01). BWvW thanks the National Research Foundation of South Africa for funding (Grant Numbers 150260 and 151653).

References

Adoyo, B., Schaffner, U., Mukhovi, S., Kiteme, B., Mbaabu, P.R. *et al.* (2022) Pathways towards the sustainable management of woody invasive species: Understanding what drives land users' decisions to adopt and use land management practices. *Land* 11, 550.

Anonymous (2019) *National Invasive Species Strategy and Action Plan (NISSAP) (2019-2029)*. Office of the Vice President, Dodoma, Tanzania.

Cordeiro, B., Marchante, H., Castro, P. and Marchante, E. (2020) Does public awareness about invasive plants pays off? An analysis of knowledge and perceptions of environmentally aware citizens in Portugal. *Biological Invasions* 22, 2267–2281.

Crowley, S.L., Hinchliffe, S. and McDonald, R.A. (2017) Conflict in invasive species management. *Frontiers in Ecology and the Environment* 15, 133–141.

Friedel, M.H., Grice, A.C., Marshall, N.A. and van Klinken, R.D. (2011) Reducing contention amongst organisations dealing with commercially valuable but invasive plants: The case of buffel grass. *Environmental Science & Policy* 14, 1205–1218.

Ministry of Livestock and Fisheries (2017) *National Strategy on Prosopis Juliflora Management*. Ministry of Livestock and Fisheries, Addis Ababa, Ethiopia.

Ndufa, J.K., Choge, S. and Chebciwo, J. (2020) *National Strategy for the Management and Control of the Invasive Prosopis Species*. Ministry of Environment and Forestry, Nairobi, Kenya.

Newton, A. and Elliott, M. (2016) A typology of stakeholders and guidelines for engagement in transdisciplinary, participatory processes. *Frontiers in Marine Science* 3, 230.

Novoa, A., Kaplan, H., Wilson, J.R.U. and Richardson, D.M. (2016) Resolving a prickly situation: Involving stakeholders in invasive cactus management in South Africa. *Environmental Management* 57, 998–1008.

Novoa, A., Shackleton, R., Canavan, S., Cybele, C., Davies, S.J. *et al.* (2018) A framework for engaging stakeholders on the management of alien species. *Journal of Environmental Management* 205, 286–297.

Nuñez, M.A. and Pauchard, A. (2010) Biological invasions in developing and developed countries: Does one model fit all? *Biological Invasions* 12, 707–714.

Shackleton, R.T., Le Maitre, D.C., van Wilgen, B.W. and Richardson, D.M. (2016) Identifying barriers to effective management of widespread invasive alien trees: *Prosopis* species (Mesquite) in South Africa as a case study. *Global Environmental Change* 38, 183–194.

Woodford, D.J., Richardson, D.M., MacIsaac, H.J., Mandrak, N.E., van Wilgen, B.W. *et al.* (2016) Confronting the wicked problem of managing biological invasions. *NeoBiota* 31, 63–86.

Zengeya, T., Ivey, P., Woodford, D.J., Weyl, O., Novoa, A. *et al.* (2017) Managing conflict-generating invasive species in South Africa: Challenges and trade-offs. *Bothalia* 47(2), a2160.

Appendix 13.1. Stakeholders identified as being involved in or affected by the management of invasive plant species or who might play an important role in this management in Amani Nature Reserve, Tanzania, Awash River Basin, Ethiopia, and Baringo County, Kenya. Levels of interest and levels of influence were scored on a four-point scale: 1 = very low; 2 = low; 3 = high; 4 = very high.

Actor	Type	Region	Type	Influence	Interest	Quadrant	Engagement type
Haramaya University	Academia	Afar	Influencer	0	3	Lower Right	Networking and empowerment
Addis Ababa University	Academia	Afar	Influencer	0	2	Lower Right	Networking and empowerment
Semera University	Academia	Afar	Influencer	1	2	Lower Right	Networking and empowerment
Ethiopian Institute of Agricultural Research	Academia	Afar	Influencer	1	3	Lower Right	Networking and empowerment
Ethiopian Institute of Water Resources	Academia	Afar	Influencer	0	2	Lower Right	Networking and empowerment
United Nations Environmental Programme (UNEP)	Consortium of International Agricultural Research Centres (CGIAR) and United Nations	Afar	Influencer	3	2	Upper Right	Mobilisation and networking
Schoolteachers	Community	Afar	Influencer	0	0	Lower Left	Awareness creation
Women	Community	Afar	Affectee	0	3	Lower Right	Networking and empowerment
Livestock keepers	Community	Afar	Affectee	1	3	Lower Right	Networking and empowerment
Community	Community	Afar	Affectee	1	2	Lower Right	Networking and empowerment
Peasant association	Community	Afar	Affectee	2	2	Upper Right	Mobilisation and networking
Swiss Agency for Development and Cooperation (SDC)	International NGO	Afar	Influencer	2	1	Upper Left	Awareness and mobilization
Centre for Agriculture and Bioscience International (CABI)	International IGO	Afar	Influencer	1	2	Lower Right	Networking and empowerment
Agricultural Technical Vocational Education and Training (ATVET) College	International NGO	Afar	Influencer	0	2	Lower Right	Networking and empowerment

Farm Africa	International NGO	Afar	Influencer	1	2	Lower Right	Networking and empowerment
Global Environmental Facility (GEF)	International NGO	Afar	Influencer	2	2	Upper Right	Mobilization and networking
German Agency for International Cooperation (GIZ)	International NGO	Afar	Influencer	2	2	Upper Right	Mobilization and networking
CARE Prime	International NGO	Afar	Influencer	2	2	Upper Right	Mobilization and networking
Environment and Land Administration	Local Government	Afar	Regulator	1	1	Lower Left	Awareness creation
Kebele Chairman	Local Government	Afar	Regulator	3	1	Upper Left	Awareness and mobilization
Amibara District Administration	Local Government	Afar	Regulator	2	1	Upper Left	Awareness and mobilization
Tulo Agricultural Extension Services	Local Government	Afar	Influencer	1	2	Lower Right	Networking and empowerment
Gewane Agricultural Extension Services	Local Government	Afar	Influencer	1	2	Lower Right	Networking and empowerment
Awash Fentale Extension Services	Local Government	Afar	Influencer	1	2	Lower Right	Networking and empowerment
Customary institutions	Local Government	Afar	Regulator	3	2	Upper Right	Mobilization and networking
Oromiya Pastoralist and Agriculture Bureau	Local Government	Afar	Regulator	2	2	Upper Right	Mobilization and networking
Gewane District Administration	Local Government	Afar	Regulator	2	3	Upper Right	Mobilization and networking
Tulo District Administration	Local Government	Afar	Regulator	2	2	Upper Right	Mobilization and networking
Pastoralist Standing Committee of the Parliament	National Government	Afar	Influencer	2	1	Upper Left	Awareness and mobilization

Continued

Appendix 13.1. continued

Actor	Type	Region	Type	Influence	Interest	Quadrant	Engagement type
Ministry of Livestock and Fisheries	National Government	Afar	Regulator	3	3	Upper Right	Mobilisation and networking
Awash Military Training Centre	National Government	Afar	Regulator	2	3	Upper Right	Mobilization and networking
Afar Agriculture and Pastoralism Bureau	Parastatal	Afar	Regulator	2	0	Upper Left	Awareness and mobilization
Awash National Park	Parastatal	Afar	Affectee	1	3	Lower Right	Networking and empowerment
Institute of Forestry Research	Parastatal	Afar	Influencer	1	3	Lower Right	Networking and empowerment
Alledeghi National Park	Parastatal	Afar	Affectee	2	3	Upper Right	Mobilization and networking
Ethiopian Institute of Biodiversity	Parastatal	Afar	Influencer	3	3	Upper Right	Mobilization and networking
Awash Basin Authority	Parastatal	Afar	Regulator	2	3	Upper Right	Mobilization and networking
Sugar factory	Private	Afar	Extractor	2	0	Upper Left	Awareness and mobilization
Tourism industry	Private	Afar	Extractor	2	3	Upper Right	Mobilisation and networking
Private investors	Private	Afar	Extractor	3	3	Upper Right	Mobilization and networking
Dar es Salaam University	Academia	Amani	Influencer	1	2	Lower Right	Networking and empowerment
Sokoine University	Academia	Amani	Influencer	2	3	Upper Right	Mobilization and networking

United Nations Development Programme (UNDP)	Consortium of International Agricultural Research Centres (CGIAR) and United Nations	Amani	Influencer	2	3	Upper Right	Mobilization and networking
Livestock keepers	Community	Amani	Affectee	0	1	Lower Left	Awareness creation
Honey producers	Community	Amani	Extractor	2	1	Upper Left	Awareness and mobilization
Farmers	Community	Amani	Affectee	0	2	Lower Right	Networking and empowerment
Schoolteachers	Community	Amani	Influencer	0	2	Lower Right	Networking and empowerment
Faith organizations	Faith Institutions	Amani	Influencer	3	1	Upper Left	Awareness and mobilization
World Wide Fund for Nature (WWF)	International NGO	Amani	Influencer	3	3	Upper Right	Mobilization and networking
Centre for Agriculture and Bioscience International (CABI)	International IGO	Amani	Influencer	2	3	Upper Right	Mobilization and networking
Eastern Arc Fund	International NGO	Amani	Influencer	2	3	Upper Right	Mobilization and networking
Aid agencies	International NGO	Amani	Influencer	2	2	Upper Right	Mobilization and networking
Tanga Water Office	Local Government	Amani	Regulator	1	2	Lower Right	Networking and empowerment
Village chairmen	Local Government	Amani	Regulator	1	3	Lower Right	Networking and empowerment
Village executive officer	Local Government	Amani	Regulator	1	3	Lower Right	Networking and empowerment
Local councillors	Local Government	Amani	Regulator	2	2	Upper Right	Mobilization and networking
Local government authority	Local Government	Amani	Regulator	3	3	Upper Right	Mobilization and networking

Continued

Appendix 13.1. continued

Actor	Type	Region	Type	Influence	Interest	Quadrant	Engagement type
Faida Mali NGO	Local NGO	Amani	Influencer	1	1	Lower Left	Awareness creation
Tanzania Forest Conservation Group	Local NGO	Amani	Influencer	2	2	Upper Right	Mobilization and networking
Ministry of Livestock	National Government	Amani	Regulator	2	0	Upper Left	Awareness and mobilization
Plant Health Service	National Government	Amani	Influencer	2	3	Upper Right	Mobilization and networking
National Assembly (Wabunge)	National Government	Amani	Regulator	2	2	Upper Right	Mobilization and networking
Ministry of Agriculture	National Government	Amani	Regulator	3	3	Upper Right	Mobilization and networking
Vice President's Office	National Government	Amani	Regulator	3	3	Upper Right	Mobilization and networking
Ministry of Natural Resources	National Government	Amani	Regulator	3	3	Upper Right	Mobilization and networking
Tanzania Electric Supply	Parastatal	Amani	Influencer	1	0	Lower Left	Awareness creation
Pangani Water Board	Parastatal	Amani	Regulator	3	1	Upper Left	Awareness and mobilization
Tanzania Pesticide Research	Parastatal	Amani	Influencer	0	3	Lower Right	Networking and empowerment
Tanzania Forest Research Institute	Parastatal	Amani	Influencer	3	2	Upper Right	Mobilization and networking
Amani Reserve	Parastatal	Amani	Influencer	3	3	Upper Right	Mobilization and networking
Tanzania Forest Fund	Parastatal	Amani	Influencer	2	2	Upper Right	Mobilization and networking
Tanzania Forest Service	Parastatal	Amani	Influencer	3	2	Upper Right	Mobilization and networking

National Biological Control Programme	Parastatal	Amani	Regulator	2	2	Upper Right	Mobilization and networking
Forest rangers	Parastatal	Amani	Regulator	2	3	Upper Right	Mobilization and networking
Tanzania Commission for Science and Technology (COSTECH)	Parastatal	Amani	Regulator	2	2	Upper Right	Mobilization and networking
Timber plantations	Private	Amani	Extractor	2	1	Upper Left	Awareness and mobilization
Tea company	Private	Amani	Extractor	2	1	Upper Left	Awareness and mobilization
Media	Private	Amani	Influencer	3	0	Upper Left	Awareness and mobilization
Butterfly farming	Private	Amani	Extractor	0	3	Lower Right	Networking and empowerment
Tourism industry	Private	Amani	Extractor	2	3	Upper Right	Mobilization and networking
Egerton University	Academia	Baringo	Influencer	2	3	Upper Right	Mobilization and networking
Nairobi University	Academia	Baringo	Influencer	3	3	Upper Right	Mobilization and networking
International Livestock Research Institute (ILRI)	Consortium of International Agricultural Research Centres (CGIAR) and United Nations	Baringo	Influencer	1	2	Lower Right	Networking and empowerment
Centre for International Forestry Research and World Agroforestry (CIFOR-ICRAF)	Consortium of International Agricultural Research Centres (CGIAR) and United Nations	Baringo	Influencer	2	2	Upper Right	Mobilization and networking
International Fund for Agricultural Development (IFAD)	Consortium of International Agricultural Research Centres (CGIAR) and United Nations	Baringo	Influencer	2	2	Upper Right	Mobilization and networking

Continued

Appendix 13.1. continued

Actor	Type	Region	Type	Influence	Interest	Quadrant	Engagement type
Food and Agricultural Organization (FAO)	Consortium of International Agricultural Research Centres (CGIAR) and United Nations	Baringo	Influencer	3	2	Upper Right	Mobilization and networking
Farmers' cooperative	Community	Baringo	Affectee	1	3	Lower Right	Networking and empowerment
Pastoralists	Community	Baringo	Affectee	0	3	Lower Right	Networking and empowerment
Small-scale farmers	Community	Baringo	Affectee	0	3	Lower Right	Networking and empowerment
Women	Community	Baringo	Affectee	2	3	Upper Right	Mobilization and networking
Charcoal producers	Community	Baringo	Beneficiary	2	3	Upper Right	Mobilization and networking
Large-scale farmers	Community	Baringo	Extractor	3	3	Upper Right	Mobilization and networking
Churches	Faith Institutions	Baringo	Influencer	3	1	Upper Left	Awareness and mobilization
Africa Forest Foundation	International NGO	Baringo	Influencer	3	1	Upper Left	Awareness and mobilization
World Vision	International NGO	Baringo	Influencer	1	3	Lower Right	Networking and empowerment
International Union for Conservation of Nature (IUCN)	International NGO	Baringo	Influencer	3	3	Upper Right	Mobilization and networking
Centre for Agriculture and Bioscience International (CABI)	International IGO	Baringo	Influencer	2	3	Upper Right	Mobilization and networking
County Assembly	Local Government	Baringo	Regulator	2	2	Upper Right	Mobilization and networking
Chiefs	Local Government	Baringo	Regulator	2	2	Upper Right	Mobilization and networking

Baringo County authorities	Local Government	Baringo	Regulator	3	3	Upper Right	Mobilization and networking
Subcounties	Local Government	Baringo	Regulator	2	2	Upper Right	Mobilization and networking
Rehabilitation of Arid Environments (RAE)	Local NGO	Baringo	Influencer	2	3	Upper Right	Mobilization and networking
Ministry of Labour	National Government	Baringo	Regulator	0	1	Lower Left	Awareness creation
Ministry of Water	National Government	Baringo	Regulator	2	1	Upper Left	Awareness and mobilization
Ministry of Finance	National Government	Baringo	Regulator	2	1	Upper Left	Awareness and mobilization
Ministry of Livestock	National Government	Baringo	Regulator	2	1	Upper Left	Awareness and mobilisation
Ministry of Agriculture	National Government	Baringo	Regulator	3	3	Upper Right	Mobilization and networking
Ministry of Environment	National Government	Baringo	Regulator	3	3	Upper Right	Mobilization and networking
Kenya Power	Parastatal	Baringo	Influencer	1	0	Lower Left	Awareness creation
Kenya Forest Service	Parastatal	Baringo	Influencer	3	1	Upper Left	Awareness and mobilization
Kenya Institute for Public Policy Research and Analysis	Parastatal	Baringo	Influencer	3	1	Upper Left	Awareness and mobilization
Kenya Wildlife Service	Parastatal	Baringo	Regulator	3	1	Upper Left	Awareness and mobilization
Kenya Agriculture and Livestock Research Organization	Parastatal	Baringo	Influencer	1	2	Lower Right	Networking and empowerment
National Environmental Management Agency	Parastatal	Baringo	Influencer	1	3	Lower Right	Networking and empowerment

Continued

Appendix 13.1. continued

Actor	Type	Region	Type	Influence	Interest	Quadrant	Engagement type
National Drought Management Agency	Parastatal	Baringo	Influencer	1	3	Lower Right	Networking and empowerment
Kenya Forestry Research Institute	Parastatal	Baringo	Influencer	3	2	Upper Right	Mobilization and networking
Kerio Valley Development Authority	Parastatal	Baringo	Regulator	2	2	Upper Right	Mobilization and networking
National Irrigation Board	Parastatal	Baringo	Regulator	3	2	Upper Right	Mobilization and networking
Cummins Power Plant	Private	Baringo	Beneficiary	0	0	Lower Left	Awareness creation
Bogoria Spa Resort	Private	Baringo	Extractor	0	1	Lower Left	Awareness creation
Media	Private	Baringo	Influencer	3	0	Upper Left	Awareness and mobilization
Baringo Island Camp	Private	Baringo	Extractor	0	2	Lower Right	Networking and empowerment
Honey Care	Private	Baringo	Extractor	1	2	Lower Right	Networking and empowerment
Tourists	Private	Baringo	Influencer	0	2	Lower Right	Networking and empowerment

14 Lessons for the Management of Cross-Disciplinary Research Teams

Sanford D. Eigenbrode[1]*, Sandra Eckert[2,3], René Eschen[4], Purity R. Mbaabu[5,6], Urs Schaffner[4] and Michael O'Rourke[7,8]

[1]Department of Entomology, Plant Pathology and Nematology, University of Idaho, Moscow, USA; [2]Centre for Development and Environment, University of Bern, Switzerland; [3]Institute of Geography, University of Bern, Switzerland, [4]CABI, Delémont, Switzerland; [5]Kenya Forestry Research Institute, Marigat, Kenya; [6]Institute for Climate Change and Adaptation, University of Nairobi, Kenya; [7]Toolbox Dialogue Initiative Center, Michigan State University, East Lansing, USA; [8]Department of Philosophy, Michigan State University, East Lansing, USA

Abstract

The spread of invasive woody weeds in Africa, the challenges it presents for human livelihoods and nature, and the difficulty of effectively managing them constitute a complex social-ecological problem. This chapter outlines the specific issue of invasive *Prosopis juliflora* in eastern Africa, then describes the conception, structure, implementation and legacy of a multi-year, cross-disciplinary, interinstitutional, international project to address it. The Woody Weeds project, which ran from 2015 to 2021, involved four universities and eight institutes in Ethiopia, Kenya, Tanzania, South Africa and Switzerland. It spanned at least five scientific disciplines, trained graduate students and engaged stakeholders from local to national levels. The chapter summarizes the perceived successes of the project and the difficulties it faced in achieving its aims. It concludes with a section of lessons learned to inform the design of other projects to address the problem of invasive weeds or similar complex social-ecological problems.

14.1 Introduction

Woody plant species are among the most important invasive alien species (IAS) worldwide (Binggeli, 1996; Richardson, 1998). The overall negative impact on ecosystem services and human wellbeing is often far more severe than the benefits that local actors can derive from these plants (Van Wilgen and Richardson, 2014). Many tree species have been introduced to Africa, often to provide benefits such as timber, firewood or erosion control. For example, more than 2000 tree species have been introduced into South Africa for these and other purposes (Glen and Van Wyk, 2016). Unfortunately, some of those species have become invasive and now cause negative ecological or livelihood impacts in addition to providing the intended benefits.

Various species of the genus *Prosopis*, including the evergreen *P. juliflora*, native to arid habitats in the Americas, were introduced into Africa in the 19th century to provide shade,

*Corresponding author: sanforde@uidaho.edu

© CAB International 2025. *The Ecology and Management of Invasive Prosopis Trees in Eastern Africa* (eds U. Schaffner *et al.*)
DOI: 10.1079/9781800623644.0014

fuelwood and wood for charcoal while fixing nitrogen in arid regions (Chapter 2, this volume). Several *Prosopis* species proved valuable for providing these services but eventually, negative effects became evident (Van Wilgen *et al.*, 2001; Shackleton *et al.*, 2014). They became invasive in southern and eastern Africa because they are highly competitive, spread rapidly and encroach on grasslands and other habitats where they deplete soil water and are difficult to remove and manage (Maundu *et al.*, 2009; Chapters 5–9, this volume). The beneficial and destructive effects of invasive *Prosopis* species affect various stakeholders to different degrees, resulting in a range of perceptions, motivations and policies concerning their management and use across the invaded range. Thus, they present a complex social-ecological dilemma that requires improved understanding of diverse aspects of the system to identify and implement appropriate responses.

The *Prosopis* invasion in eastern Africa exemplifies the multifaceted way in which IAS alter biodiversity and ecosystem functioning, thus changing the delivery of ecosystem services for human benefit and thereby affecting livelihoods (Fig. 14.1). For example, in the case of *P. juliflora*, people making charcoal may benefit from increasing abundance of the species, while agropastoralists suffer from the reduced availability and accessibility of pasture or rangeland (Chapter 7, this volume) and water resources (Chapter 8, this volume).

Many people and communities experience both beneficial and negative effects of *Prosopis* trees but when they become overabundant, the net effects to all are negative and apparent. Successful management of *Prosopis* trees, however, requires early implementation, before the negative effects become apparent to all (Chapter 1, this volume). Few projects have addressed the complexity of invasive woody plants comprehensively and this has impeded development of ecologically and economically sound and effective responses.

The Woody Weeds project (www.woodyweeds.org), which ran from 2015 to 2021 with funding from the Swiss National

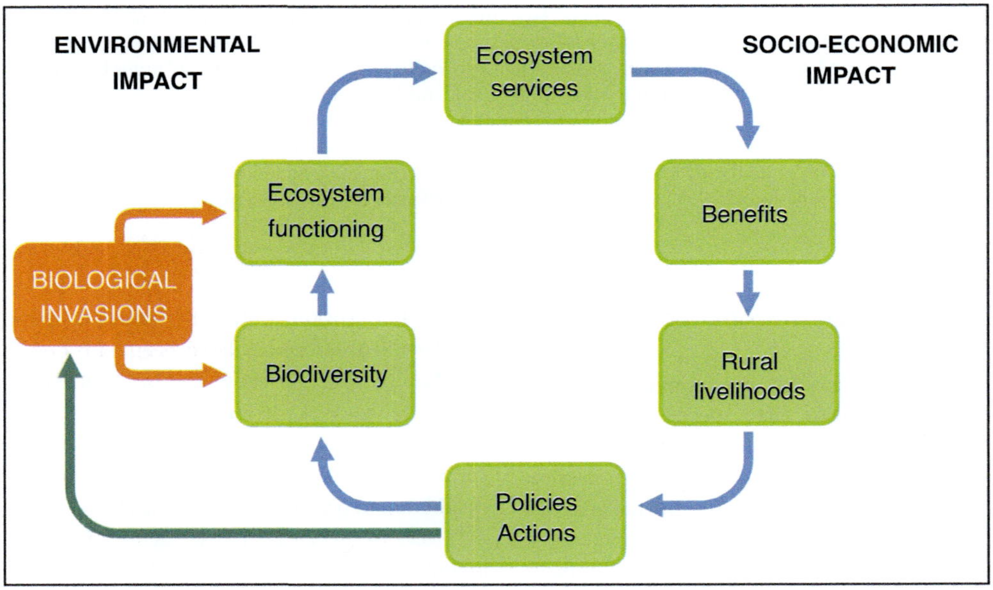

Fig. 14.1. Invasive alien species like *Prosopis juliflora* affect the environment and people in social-ecological systems, directly through changes in biodiversity or ecosystem functioning, or because such impacts affect the delivery of ecosystem services. Assessing impacts of these invasions on livelihoods and developing solutions therefore requires interdisciplinary and transdisciplinary approaches.

Science Foundation and the Swiss Agency for Development and Cooperation in the frame of the Swiss Programme for Research on Global Issues for Development, was designed to address the problem of invasive alien tree invasions in eastern Africa, particularly that of *P. juliflora*. It involved four universities and eight institutes in Ethiopia, Kenya, Tanzania, South Africa and Switzerland, enlisted at least five scientific disciplines, trained graduate students and was engaged with stakeholders from local to national levels. The project set out to assess the relationship between abundance or cover of *P. juliflora* and its impacts at local levels, map its current and potential future distribution, upscale impacts to the regional level and use this information to raise awareness of the problem and work with stakeholders to evaluate options and disseminate management strategies for implementation from local to national levels.

This chapter first outlines the challenges to conducting comprehensive transdisciplinary projects to address complex problems facing social-ecological systems like invasive alien trees. It then describes the specific elements of the Woody Weeds project. It closes with an assessment of the successes achieved and difficulties encountered and some recommendations for design and implementation of similar projects in the future.

14.2 Cross-Disciplinary Challenges to Conducting Research in Social-Ecological Systems

The *Prosopis* invasion in eastern Africa is a complex problem that operates across multiple dimensions and multiple scales simultaneously, affecting different communities in different ways. Transdisciplinary approaches are considered optimal for complex, social-ecological problems like these that have a range of overlapping, differential and incommensurable impacts across ecosystems and communities (Zscheischler *et al.*, 2017) (*transdisciplinary*$_b$, Box 14.1). These approaches involve experts from inside and outside the academy, including stakeholders such as community members, government and NGO representatives (Zscheischler *et al.*, 2017) (see Box 14.2), who commit to integrating their insights to respond to complex problems with the requisite and appropriate complexity (Lang *et al.*, 2012). In this section, we consider in turn the cross-disciplinary challenges to such projects organized into four sets important for the Woody Weeds project: problem conceptualization, developing a collaborative culture among project members, bridging disciplines and integrating diverse data types, incorporating different and sometimes conflicting contexts.

Social-ecological problems, such as sustainably managing a woody IAS, are wicked problems – complex problems exhibiting non-linear dynamics and emergent properties that are difficult to predict or understand. They have neither a 'definitive formulation' that all involved can agree upon nor a determinable set of possible solutions. Their solutions are not definitive and their success is relative. These characteristics pertain to the problem of invasive *Prosopis* trees in eastern Africa – their impacts differ on already degraded and relatively healthy land (Mbaabu *et al.*, 2020), there are ongoing debates about the merits of their utilization vs elimination, and the range of acceptable strategies differs between the affected countries. Approaches pursued in other countries (e.g. Australia, South Africa) are instructive but not definitive. Meanwhile, the invasion marches on, negatively affecting livelihoods.

A second set of challenges for transdisciplinary collaborations addressing complex problems within social-ecological systems concerns the motivation and capacity of the collaborators to work together effectively, which can be understood as their *collaborative culture*. Not only must the team members be complementary in terms of their research skills and perspectives (Pohl *et al.*, 2017), they must also possess other antecedent skills and attributes for successful collaboration (Stokols *et al.*, 2010). Once the team has formed, challenges related to power dynamics among collaborators can arise (Steger *et al.*, 2021), including turf battles over disciplinary rigour and superiority (Pellmar and Eisenberg, 2000), and different levels of influence and respect among academic researchers and collaborators from outside the academy (Connon, 2021). Since the participants will probably differ in personality, values, priorities and levels of investment, disagreement and

Box 14.1. A typology of cross-disciplinarity.

Projects that address social-ecological complexity typically depend on multiple disciplines to varying degrees to be successful. *Multidisciplinarity, interdisciplinarity* and *transdisciplinarity* are three common terms used to describe projects that depend on multiple disciplines, while *antidisciplinarity*, *pluridisciplinarity*, *postdisciplinarity* and *supradisciplinarity* are four less traditional examples that have also been discussed (see O'Rourke *et al*., 2019). Although these terms have been used for many decades – and in one case, *interdisciplinarity*, for almost a century (Klein, 2022) – there is little consensus surrounding their use, suggesting that the distinctions are meaningless (Rylance, 2015). We maintain, however, that the proliferation of these terms reflects the need for clarity in designing and facilitating collaboration involving multiple disciplines. In this chapter, we adopt the following definitions of four central terms, focusing specifically on research (Eigenbrode *et al*., 2007; O'Rourke *et al*., 2019).

- *Multidisciplinarity*. Research involving multiple disciplines that make contributions which are loosely integrated without the disciplinary perspectives themselves being integrated (cf. Mennes, 2020).
- *Interdisciplinarity*. Research involving multiple disciplines, where the disciplinary perspectives are integrated to produce research output that is more than the sum of the research inputs.
- *Transdisciplinarity*. There are two relevant senses, which we distinguish with the subscripts 'a' and 'b'.
  a. *Transdisciplinary*$_a$. Research involving multiple disciplines, where the disciplinary perspectives are integrated to produce research output that is more than the sum of the research inputs *and* the disciplinary inputs are themselves changed in the process (i.e. *transformative* interdisciplinarity) (cf. Klein, 2014).
  b. *Transdisciplinary*$_b$. Research involving the integration of perspectives from academic disciplines and knowledge systems outside the academy (e.g. communities, government, industry) (i.e. *transgressive* disciplinarity) (cf. Klein, 2014).
- *Cross-disciplinarity*. A generic term used to talk about integration involving multiple disciplines in its many manifestations, including multi-, inter- and transdisciplinarity.

These types can be understood as reflections of both a knowledge-related (or *epistemic*) dimension and source-related (or *social*) dimension (Box Fig. 14.1). When restricted to academic disciplines, multi-, inter- and transdisciplinarity mark increasing levels of integration among the constituent disciplines. The term *transdisciplinarity*$_b$ applies to projects where collaborators are drawn from outside the academy, regardless of the level of integration. The Woody Weeds project was a combination of interdisciplinary research, especially during Phase I, and *transdisciplinary*$_b$ research, given the involvement from the beginning of stakeholders from a range of communities.

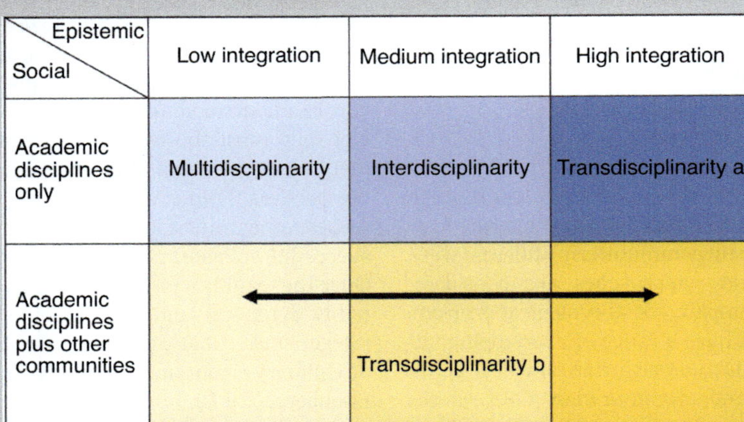

Box Fig. 14.1. Differences among terms used to describe complex, integrative research, analysed according to the level of integration of the knowledge contributed (i.e. epistemic) and two different combinations of knowledge sources (i.e. social). The arrow shows the positioning of work within the Woody Weeds project.

Box 14.2. The meanings of integration in social-ecological systems research.

The concept of *integration* is a key part of how complex, collaborative research is understood. Different modes of research practice can be distinguished according to how much integration there is among the contributed disciplinary inputs. It has become a 'mantra of our times' that we must meet complex problems with complex responses if we are to have any chance of generating helpful responses (Imbruce and Prazak, 2021), and this complexity involves not only a diversity of inputs but also their dynamic combination, i.e. their *integration*.

Complex phenomena like biological invasions are conceived as integrated wholes involving elements combined in ways that constitute them (cf. Mitchell, 2002). An adequate account of such a complex phenomenon will be *cross-disciplinary* (see Box 14.1) and it will represent the phenomenon in a way that ensures the causal connections underwriting its complexity are understood from the perspectives of the disciplines that understand them best. Because the phenomenon is an integrated whole, the account must be too if it is to accurately represent it.

But what is *integration*? There is no consensus on this but there have been various attempts to make sense of integration in the context of complex research. Most of these attempts begin with the idea of integration expressed in the vernacular sense, viz. 'combine or be combined to form a whole' (Soanes and Stevenson, 2008). In this chapter, we understand *integration* to be an input–output process that puts inputs into relations in the production of outputs that results in fewer outputs than inputs (O'Rourke *et al.*, 2016). Set in the context of a *transdisciplinary*$_b$ project like the Woody Weeds project (see Box 14.1), the inputs to be integrated include the people involved (i.e. the *social* dimension) and contributions that reflect their expertise (i.e. the *epistemic* dimension). See Bergmann *et al.* (2012) and especially Pohl *et al.* (2021) for related characterizations of transdisciplinary integration

conflict can lead to failure of the collaboration (Jakobsen *et al.*, 2004; Siebenhüner, 2018). Effective communication is essential for managing transdisciplinary collaboration to support the necessary relationships and the transfer of information among participants (Hall and O'Rourke, 2014). The Woody Weeds project participants came from diverse backgrounds and many had not previously collaborated. The requisite collaborative culture needed to be created and sustained.

Effectively addressing the issues facing social-ecological systems requires a *transdisciplinary*$_b$ approach with participants from both inside and outside the academy (see Box 14.1), which introduces the third set of challenges mentioned above. Transdisciplinary collaborators bring expertise rooted in different research worldviews and epistemologies (Tengö *et al.*, 2017; Elsawah *et al.*, 2020), supported by different training regimens (Eigenbrode *et al.*, 2007), and reflected in different technical languages (Bracken and Oughton, 2006). Despite appreciating the potential for conflict, collaborators can fail to negotiate differences in preferred methodologies (Jakobsen *et al.*, 2004) (e.g. quantitative and qualitative; Elsawah *et al.*, 2020), 'methodological standards' (Lang *et al.*, 2012), methods to address the social vs the ecological components of the social-ecological systems (MacMynowski, 2007; Elsawah *et al.*, 2020), and differences in temporal and spatial scale (Benda *et al.*, 2002). The diverse academic disciplines that contributed to the Woody Weeds project and the requisite engagement with stakeholders exemplify a *transdisciplinary*$_b$ approach.

The fourth set of challenges for projects addressing social-ecological systems arises because they extend across institutional, sectoral and geopolitical contexts (Dallimer and Strange, 2015). International teams face intercultural and linguistic challenges to communication (Wang *et al.*, 2019; Ward and Southwell, 2021). Multi-institutional teams confront differences in procedures, norms, professional expectations and timelines (Eigenbrode *et al.*, 2017; Fischke *et al.*, 2017). Institutions often differ in the valuing and incentives for cross-disciplinary work that can introduce professional risks for some collaborators (Roy *et al.*, 2013; Klein and Falk-Krzesinski, 2017). The Woody Weeds project participants came from multiple institutions and countries in Africa and Europe that introduced unique complexity. The participants also included PhD and MSc students and their mentors from these countries and institutions, adding an additional contextual dimension to the project.

14.3 Woody Weeds Project Design and Implementation

In this section, we describe the Woody Weeds project in detail, keeping in mind the challenges just articulated. After providing an overview of the project, we focus on its organization, how stakeholders were involved, how communication was managed and how students were trained. In the next section (14.4), with this description in place, we will revisit the challenges to highlight how the Woody Weeds project met them.

14.3.1 Overview of the Woody Weeds project

The overarching goal of the Woody Weeds project was to help mitigate the impacts of woody IAS on ecosystem services and human wellbeing in three eastern African countries (Ethiopia, Kenya and Tanzania). It was a partnership consisting of the Kenya Forestry Research Institute, Centre for Training and Integrated Research in Arid and Semi-arid Land Development (Kenya), the Tanzanian Forestry Research Institute, Sokoine University of Agriculture (Tanzania), Water and Land Resource Centre, Haramaya University (Ethiopia), Stellenbosch University (South Africa), Centre for Development and Environment of University of Bern, Switzerland, and CABI (Switzerland and Kenya). Within each of the three eastern Africa countries, the project focused on one or two regions affected by tree invasions; the case study regions affected by *P. juliflora* were the Awash River watershed in Afar Regional State, Ethiopia, Baringo County, Kenya, and Moshi Rural District in Kilimanjaro Region, Tanzania.

14.3.2 Project organization

The 7-year project was structured along temporal and spatial dimensions designed with stakeholder engagement from inception to implementation and development of scientific, social and practical knowledge required to address the problems caused by plant invasions. The project had an inception phase, a research phase (Phase 1) with two work packages, and an implementation phase (Phase 2), with one work package (Fig. 14.2). Within each phase, specific research tasks and their responsible entities were defined. The work was directed by a scientific team that included geneticists, ecologists, social scientists, economists and geographers.

In Phase 1, Work Package 1, guided by stakeholder interactions, the work delineated the distribution of invasive *Prosopis* trees and invasion processes, their ecological, hydrological and socio-economic impacts, stakeholder perceptions and management approaches in each of the case study areas. In Phase 1, Work Package 2, geoinformatics was used to map and project *Prosopis* tree infestations and scale up to make inferences about the problem at regional and national levels so that they would be more relevant for policy makers. During Phase 2, Work Package 3, guided by principles delineated during Phase 1, management practices were implemented by the team working closely with stakeholders. The Woody Weeds project scope also included other invasive alien woody species, but in this chapter we focus on the work done with *P. juliflora*. The research was primarily carried out by PhD and MSc students mentored by senior scientists from different universities, and two postdoctoral researchers from European institutions.

The organizational structure of the Woody Weeds project included the project leadership team and representatives from each of the nine institutions, many serving as supervisors of the PhD and MSc students. In-person, 3–4 day project meetings of all project personnel were held twice yearly from its inception in 2015 until COVID-19 restrictions in 2020, resuming for a final meeting in 2022. These meetings were held in locations that rotated among the three eastern African countries. During each of these in-person meetings, an entire day was allocated to interactions with stakeholders from the host country and from the focal area within that country, which was used for knowledge exchange between project scientists and stakeholders about project activities and results. Virtual meetings were also used more

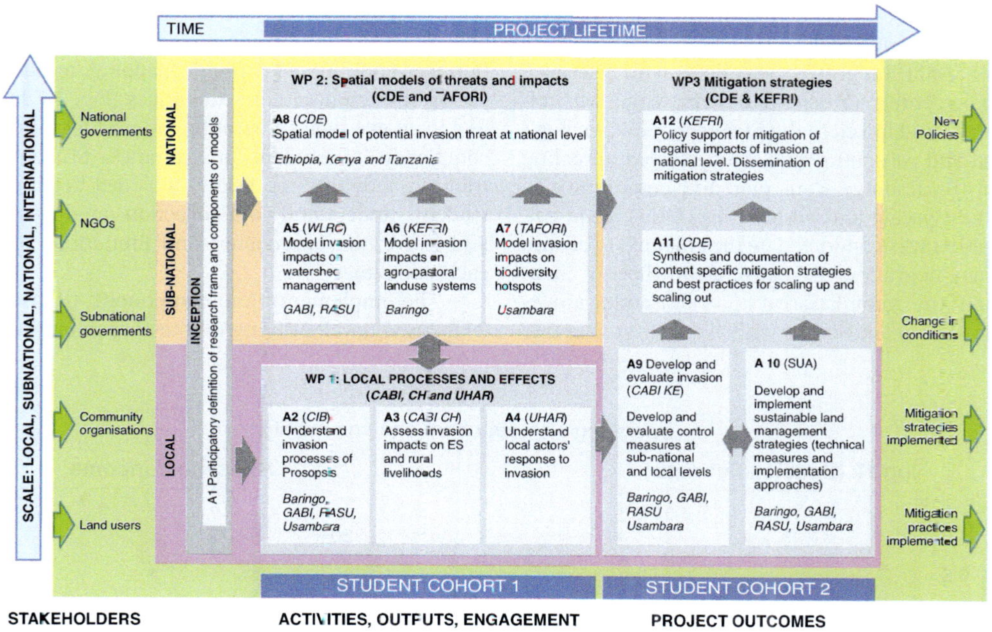

Fig. 14.2. The Woody Weeds Project structure. Objectives for the 12 project activities (A1–A12) and three work packages (WP 1 – WP 3) across three project phases: Inception, Phase 1 (WP 1 and WP 2) and Phase 2 (WP3). Institutional leaders of project activities are in parentheses: CABI CH, CABI Switzerland; CABI KE, CABI Kenya; CDE, Centre for Development and Environment (University of Bern); CIB, Centre for Invasion Biology (Stellenbosch University); SUA, Sokoine University of Agriculture; TAFORI, Tanzania Forest Research Institute; UHAR, University of Haramaya; WLRC, Water and Land Resource Centre. Student Cohorts 1 and 2 worked in Phase 1 and Phase 2.

frequently among participants throughout the project.

The research component (Phase 1) of the Woody Weeds project took place during its first 3 years and was designed to create a solid knowledge base. Six PhD students generated ecological, hydrological, genetic and socio-economic data, mapped the abundance of *Prosopis* species and created multidisciplinary outputs to provide a comprehensive overview of the impacts of *Prosopis* invasions in the three targeted regions. The research was interdisciplinary (as defined in Box 14.2) and sought to generate quantitative and spatially explicit results (e.g. geospatial data and findings, distribution and abundance maps, and areal statistics). This required compatibility of collected data across disciplines (i.e. measured in a predefined standard way, using the same physical and geospatial units and methods, including geolocation information). All data were collected with the potential for integration through remotely sensed or otherwise spatial data having a spatial pixel resolution typically of 30 × 30 m, although also at other resolutions. For example, ecological data were collected in 15 × 15 m plots within administrative communities and socio-economic data were collected through household interviews and a choice experiment.

While ecological field data were collected from known points in the landscape, sociological data were more difficult to integrate since they derived from surveys of agropastoralist households in regions where much of the grazing land is shared among users in a community or even larger area. To address this, the combined analysis of the relationship between *Prosopis* tree abundance and socio-economic and ecological parameters was carried out using the smallest administrative unit for sociological data as the basic experimental unit (e.g. sublocation in Kenya and kebele in Ethiopia).

Data integration was done with reference to percentage *Prosopis* tree cover, which was collected and linked to the socio-economic, ecological and geospatial data. This enabled upscaling of the findings from plot and community to county scale and beyond (Fig. 14.3). Data collection was conducted by the students to ensure this integration. To measure the effect of invasion, the administrative units chosen for analysis were in landscapes with average *Prosopis* tree cover ranging from close to 0% to the maximum values found in the study regions. Data collection was coordinated through mapping, anticipating the required data exchanges (Fig. 14.4) and to enable their integration for products and completion of student programmes. Student training activities (see below) included helping students understand data collection, analytical and inferential approaches and limitations of each discipline.

The implementation component (Phase 2) focused on identifying and testing potential

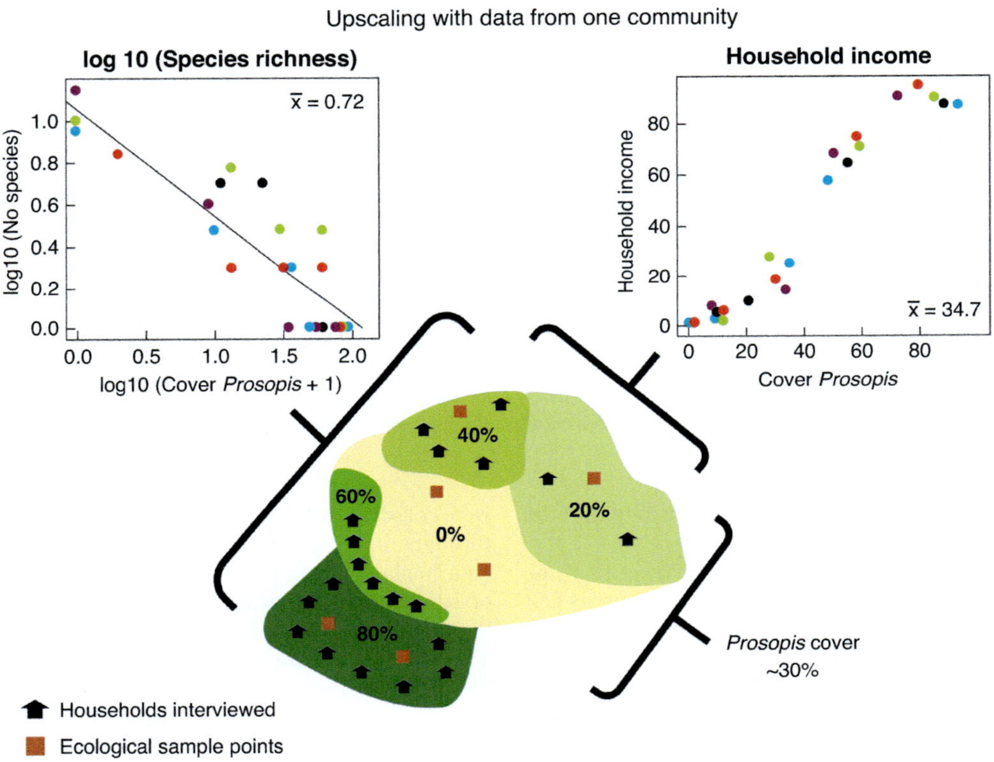

Fig. 14.3. Data from different disciplines are collected from different units, such as households or field plots, which is illustrated schematically in the figure. The map shows a hypothetical landscape with different average coverage with *Prosopis* from 0 to 80%. The locations where household interviews were conducted and ecological data were collected are shown on the lower map; polygons and average cover were based on remote sensing. Scatter plots show relationships to other factors within the socio-economic and ecological datasets but share the comparable *x*-axis (*Prosopis* cover). Colours on points for scatter plots indicate data representing different parts of the hypothetical community. Combined analysis of multidisciplinary datasets on a landscape or national scale therefore requires scaling to a common spatial unit. The Woody Weeds project achieved this by aggregating data collected in administrative units, which were characterized based on remotely sensed LULC maps and *Prosopis* fractional cover maps.

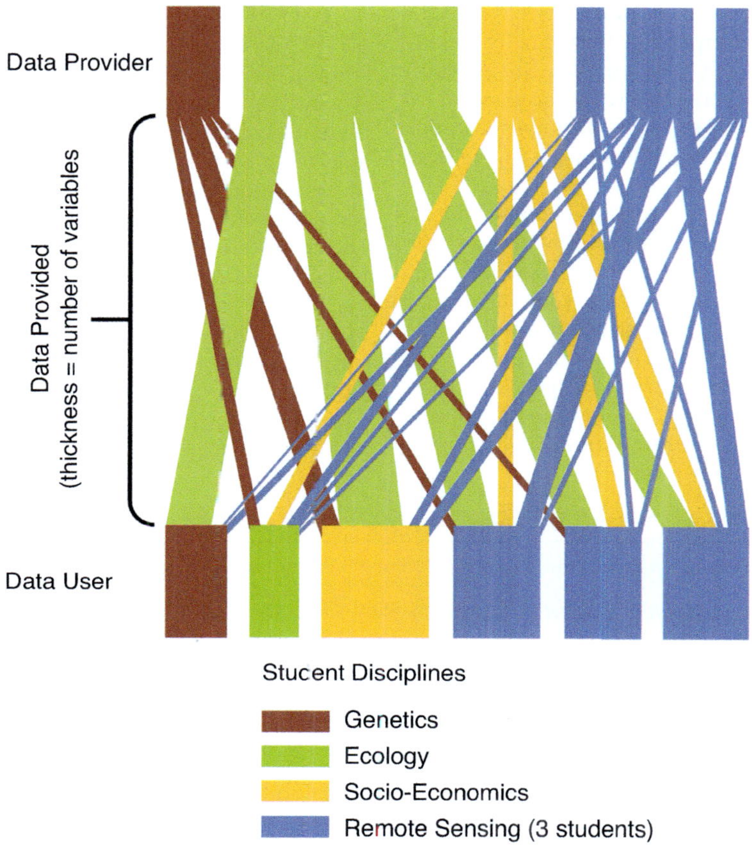

Fig. 14.4. From the outset of the Woody Weeds project, it was recognized that data from different project activities would need to be analysed jointly to produce the desired outputs. Six PhD students working on these activities therefore identified what data would be shared and designed their data collection to enable timely delivery. Thickness of the lines indicates the approximate number of variables to be exchanged and colours of the boxes indicate disciplines of the students.

sustainable land management (SLM) practices by diverse groups of stakeholders. The stakeholders were organized in Local Implementation Groups (LIGs) (Schwilch *et al.*, 2012a; b) that comprised people representing local and subnational interests from agriculture, pastoralism, charcoal makers, women's groups, administrations, NGOs, scientists from the Woody Weeds project and others. Management practices for the control of *Prosopis* trees that were evaluated included biological, physical, chemical and cultural measures. Each participant came with her/his own knowledge and experience and together, the group went through a deliberate decision-making process to identify potential SLM practices.

Practices were identified based on local knowledge and supplemented by knowledge from elsewhere or from scientific literature. Identified practices were scored based on criteria that relate to ecological, economic and sociocultural sustainability aspects of the practice, followed by a ranking that resulted in a weighting of the criteria. The ranking was, again, informed by local knowledge and experience, as well as by research outputs from the Woody Weeds project. Relevant criteria were then prioritized by LIG participants, which involved negotiation and joint decision making (Schwilch *et al.*, 2012b).

At least one practice each was selected by LIGs in Ethiopia, Kenya and Tanzania for testing during the Woody Weeds project. The practices were tested experimentally on farmland or on rangeland, and data about the implementation as well as effectiveness for managing *Prosopis* trees were collected.

Project findings were disseminated during in-person interactions with relevant stakeholders, in scientific publications, policy briefs and extension materials. Written summaries were used as part of the project's involvement on national committees that informed the development of national strategies for management of invasive *Prosopis* trees in Kenya and of invasive species more generally in Tanzania.

14.3.3 Involving stakeholders

Stakeholder involvement was essential for the Woody Weeds project in achieving its aims. Stakeholders included institutions, groups or individuals who could either affect or be affected by the research and stand to gain if the project was successful or lose if it was not (Box 14.3). Their involvement was supported from framing to implementation as recommended for transdisciplinary projects (Lang *et al.*, 2012) and included coconceptualization of the problem, codevelopment of knowledge and solutions, and validation, dissemination and uptake of management strategies to mitigate the negative effects of woody IAS.

Box 14.3. Who are stakeholders in social-ecological systems research?

The concept *stakeholder* is widely used across many literatures, e.g. business, tourism, education, environmental management and transdisciplinary research. Most directly, *X* is a *stakeholder* if *X* holds a *stake* in *Y*, i.e. how things fare with *Y* will have an impact on *X*. It is a relational concept relating the entity that holds the stake and the entity in which the stake is held. What this relation amounts to, however, varies depending on the context (cf. Donaldson and Preston, 1995, p. 69).

As a concept, *stakeholder* has received its most considerable development in the business and management literatures, where it is the focus of *stakeholder theory*. In this context, a stakeholder stands to profit if the firm or corporation profits (Friedman and Miles, 2002). Freeman (1984), widely regarded as the first canonical contribution to stakeholder theory, offers a much discussed definition of 'stakeholder': 'any group or individual who can affect or is affected by the achievement of the organization's objectives' (p. 46). This is a very broad definition where the relation between the stakeholder and the organization in which the stake is held concerns the causal influence of one on the other. Although the direct definition in the first paragraph looks asymmetrical, this approach highlights how *X* can hold a stake in *Y* by influencing *Y* and the pursuit of its objectives, thus profiting from *Y*'s success.

We can work toward a definition of 'stakeholder' in the social-ecological systems context by beginning with Freeman's (1984) definition. Although Freeman was focusing on stakeholders in relation to corporations, the generality of this definition makes it useful as a framework for identifying research stakeholders, i.e. groups or individuals who can affect or be affected by the achievement of a research project's objectives. Here the nature of the stake is tied to the research context: one is a stakeholder if one can affect the research pursued by a project or is affected by that research should it prove successful.

We narrow this broad definition down in two ways. First, we follow Donaldson and Preston (1995) in taking a stake to be a 'potential benefit' (p. 86), which supplies the concept *stakeholder* with a positive valence: if things fare well for *Y* – e.g. if it is a research, implementation and development project like the Woody Weeds project and it achieves its objectives – then the stakeholder *X* will benefit; if things fare badly for *Y*, things will not go so well for *X*. This highlights the second way in which we narrow down the Freeman definition, by following Clarkson (1994) in associating the stakeholder concept with *risk*. By virtue of the stake *X* has in research project *Y*, they assume some measure of risk if the project doesn't work out, i.e. they stand to lose (or at least not gain) as a consequence.

This broad definition of *stakeholder* allows us to identify project participants as stakeholders, since they have taken on risks by becoming involved in the project and they stand to gain if the project is successful. But it also broadens the stakeholder category to include many organizations, including governmental and non-governmental, as well as communities, universities and individuals.

From the start, the Woody Weeds project interacted closely with stakeholders across different sectors (e.g. agriculture, environment, water, tourism and health) and scales (local to national), assessing their needs (e.g. missing knowledge, access to existing information, access to management tools) and understanding their main criteria for adopting *Prosopis* tree management and environmental management in general. At the international scale, NGOs and research institutes participated. Nationally, ministries dealing with invasive species, livestock, forestry and wildlife were represented, along with institutions such as forestry services or agriculture. More locally, participating stakeholders included subnational governments and NGOs, community-based organizations, natural resource management organizations, commercial institutions such as charcoal producer associations, communities (e.g. chiefs, elders) and community conservancies (e.g. chairs of the board). This was facilitated through a number of activities within the project as outlined below.

Interacting social systems include institutions at different jurisdictional scales from local to international (Berkes, 2007) across which perspectives of stakeholders will vary. Thus, a challenge in sustainable management of natural resource issues like IAS is to involve stakeholders not only from different sectors but acting at different scales. Stakeholder interactions were a principal focus of the main inception workshop of the Woody Weeds project, to which primarily national, regional and county government representatives, research institutions and NGOs from Ethiopia, Kenya and Tanzania were invited, and the three regional inception workshops, one in each case study area, to which representatives of administrative level 2 (e.g. county) and level 3 (e.g. subcounty), local communities and NGOs were invited.

Presentation of project objectives and structure to stakeholders during these inception workshops had multiple purposes. First, stakeholder feedback reframed the study object and refined or complemented the original research questions or hypotheses. Second, discussions among stakeholders identified topics where there was uncertainty or disagreement among stakeholders. These aspects were incorporated into scenario building and facilitated decision processes at later stages of the project. Third, in regional workshops, a stakeholder mapping exercise was conducted that included identifying whether all stakeholder views were represented within each region and how different stakeholders were positioned on a continuum of power vs motivation to address specific issues.

Regional workshops also revealed that stakeholder knowledge, perception and willingness to manage invasions by *Prosopis* trees differed among the regions, especially between those who still had large areas of healthy ecosystems and those with large areas degraded by *Prosopis* trees and between those experiencing an early stage of invasion compared with those experiencing a later stage. For example, interviews with local stakeholders in the Afar region in Ethiopia revealed that 79% of the respondents thought positively about *Prosopis* trees when they were introduced in the 1980s, while today only 10% think that the benefits still outweigh the negative effects of this tree (Seid *et al.*, 2020).

A key output of the regional workshops was the creation of the LIGs, which put researchers in partnership with representatives of affected communities, extension services, NGOs active in the case study regions and representatives of local, subnational and national administrations. The LIGs also included 'negative' or 'hidden' stakeholder groups which benefited from *Prosopis* trees, such as charcoal producer associations (Pirozzi, 2019). Using a structured deliberative multicriteria decision process (Schwilch *et al.*, 2012b), the LIGs rated ecosystem services affected by *Prosopis* tree management practices (including doing nothing) and reflected on trade-offs and synergies while integrating multiple social, economic and ecological dimensions. This is transdisciplinary research and cocreation of knowledge with joint responsibility for implementation of solutions.

Each LIG jointly agreed on the *Prosopis* tree management objectives and SLM practices to be tested in its region. Discussions within LIGs helped resolve conflicts between those viewing *Prosopis* trees as an asset and those seeking to limit their spread. Even stakeholders benefiting from *Prosopis* trees came to understand that continued spread threatened grasslands and croplands at the landscape scale. A compromise included management to enable continued utilization while other sources of income were developed, including planting non-invasive trees. LIG members found they generally agreed on the ecological impacts of invasive *Prosopis* trees, while agreement and certainty varied around the economic implications. The three lessons learned from the LIGs were that:

- diverse stakeholders agreed on several aspects related to *Prosopis* trees
- diverse stakeholder interests can only be approached via spatially explicit management scenarios
- more robust economic and social data were needed regarding the different management scenarios.

The LIGs also provided a platform for training trainers and identifying champions who could disseminate knowledge about woody IAS and management options among communities and local institutions beyond the lifetime of the Woody Weeds project.

The Woody Weeds project also worked from the outset with decision makers at subnational and national levels in the three target countries about upcoming project events and new project results. In Kenya, this led to an invitation from the Kenyan Ministry for Environment and Forestry to provide a summary of the key findings from Phase 1. Based on that summary, the Kenyan authorities began drafting a National *Prosopis* Strategy which differed considerably from what the authorities had been promoting over the previous 10–20 years. In Tanzania, three project members were invited to join a taskforce to develop a National Invasive Species Strategy and Action Plan for that country, with one project member serving as the chair of the taskforce and reporting directly to the Vice-President's Office.

Outputs were also generated for the wider public and the scientific community (https://woodyweeds.org/?page_id=2). The project has published more than 50 popular articles in print and online media. National and regional TV and radio channels reported on *Prosopis* and on the work of the Woody Weeds project in all target countries. Research results were presented at scientific conferences across the world and published in more than 30 peer-reviewed scientific publications, many of them reflecting the project's transdisciplinary approach (see Box 14.1) and offering solutions to manage *Prosopis* trees as an exemplary 'conflict-of-interest' genus.

14.3.4 Promoting effective communication within the project

Successful large social-ecological system projects require effective communication among collaborators (Eigenbrode *et al.*, 2007; Hall and O'Rourke, 2014; Morton *et al.*, 2015; Hoffmann *et al.*, 2017a, b; Pischke *et al.*, 2017; Wang *et al.*, 2019; Steger *et al.*, 2021; Urton and Murray, 2021). In a survey of 139 'Science with Society' transdisciplinary projects, ineffective communication ranked fourth of nine barriers to success (Steger *et al.*, 2021). Virtually all the sets of challenges facing these projects, however, can be mitigated through deliberate strategies to improve communication. The diverse backgrounds, physical locations, institutional structures and expectations of project participants within the Woody Weeds project introduced significant communication challenges. To address these, the project employed several measures to enhance communication among scientists, between senior scientist mentors and students, and between scientific participants and diverse stakeholders (Table 14.1).

14.3.5 Training project participants

Most of the research in the Woody Weeds project was led and conducted by students. Because of this, student training was inherently interdisciplinary. In fulfilling their responsibilities for designing their research and data collection, sharing data (Fig. 14.4), and preparing a required first-authored interdisciplinary paper for publication, students gained exceptionally thorough experience with the challenges and means for successful integration from project conception through data gathering, analysis and writing. Because the project was

Table 14.1. Activities within the Woody Weeds project designed to promote communication among participants and with stakeholders.

Parties involved	Nature of communication	Description of activity
Project leadership, participants and stakeholders	National inception workshop	A one-day event at the beginning of the project with national and regional representatives from Ethiopia, Kenya and Tanzania; presentation of general project outline; listen to feedback and input from stakeholders; reflection on whether the project has identified key general questions
	Regional inception workshops	A one-day event at the beginning of the project with local and regional stakeholders; listen to feedback and input from stakeholders on the project (selection of species, regions and research questions)
	Local Implementation Group meetings	Engagement with representatives of key stakeholder groups to include them in decision making about the implementation of invasive species management in their localities
	Meetings with policymakers	Ad hoc meetings to discuss relevance of project findings for the development of appropriate policies
	Final project meeting	A meeting held in the final year of the project with participation from all project scientists, many of the students, including those who had graduated, and representative stakeholders at the local, regional and national levels
Scientific collaborators	Project meetings	Physical meetings with the whole project team approximately every 9 months and including progress reports, particularly by students; planning and coordination of next activities
	Postmeeting tours	One-day field trips by scientists and students to local areas near meeting venues
	Toolbox workshops	Workshops designed to facilitate cross-disciplinary understanding and communication
	Communication surveys	Annual online surveys of project participants to assess effectiveness of communication efforts in the project
	Ad hoc face-to-face project interactions	Gatherings of project participants to focus on specific issues
	Electronic or virtual communications	Ongoing communication using email, phone, WhatsApp, Skype or Zoom
	Monthly reporting by PhD students to work package leaders	Preparation of written reports
	Quarterly reporting by work package leaders to the project director	Preparation of written reports

Continued

Table 14.1. Continued

Parties involved	Nature of communication	Description of activity
Academic supervisors and students	Student training days	Two training days attended by all postgraduate students before each of the project meetings; training provided by senior project members; students drafted interdisciplinary research questions and received training in data collection, data entry and analysis
	PhD student dinners	Joint dinners of all PhD students, without supervisors, on one of the evenings of the project meetings
	Student–supervisor interactions	Meetings between students and supervisors, as part of typical student mentoring but also sometimes with project leadership to help to align the mentoring with project needs
	Codesigning data collection	As required to meet the cross-disciplinary goals of the project, students were guided in coordinating the types of data collected and the timing of data collection to enable data sharing and integration
	Joint field work	Students typically worked independently but planned joint data collection activities to help with mutual understanding of methods, approaches and constraints to improve data integration
	Student cohort knowledge transfer events	Meetings of PhD students from Phase 1 (primarily research) and Phase 2 (primarily implementation)
	Interdisciplinary research papers	Publication of papers integrating data from at least two students from different disciplines
	Coauthored perspective manuscript	Jointly authored white paper by PhD students on the components of social-ecological research and implementation and their experience in training
	PhD paper writing workshop	Physical meeting of PhD students of Phase one in Switzerland

thoroughly integrated with stakeholders, all students worked directly with stakeholders and were privy to efforts to incorporate stakeholder perspectives into the project's research and implementation.

To support their cross-disciplinary development, the Woody Weeds project provided training to students. This included training on data integration and analysis, and on the skills required for working across disciplines. Much of this took place during the project's biannual meetings. One or 2 days immediately prior to these meetings were set aside as 'student training days' during which students received practical training in study design, data collection, documentation and sharing, remote sensing, geospatial and statistical data analysis, cross-disciplinary communication, manuscript writing and dissemination of scientific findings to different stakeholders. In addition to including a compulsory interdisciplinary chapter in their theses, the students worked together on a manuscript describing their training within the Woody Weeds project addressing an issue within a complex social-ecological system. Many of the measures taken to enhance communication within the Woody Weeds project (Table 14.1) were specifically designed and implemented for students.

Student training confronts many of the challenges to transdisciplinary research, either directly since students contribute to the diversity of the research community or indirectly because students in such projects are exposed to and learn many of the skills required for transdisciplinary efforts in their careers.

14.4 Lessons for Other Complex Social-Ecological Problems

As summarized in Section 14.3, the Woody Weeds project employed a multifaceted project design to facilitate interdisciplinary and transdisciplinary integration, and this implicitly entailed mitigating the four sets of challenges outlined in Section 14.2. In this section, we revisit those challenges and highlight how the Woody Weeds met them and its successes and difficulties in doing so.

14.4.1 Problem conception and stakeholder involvement

Identifying and involving stakeholders in project conception, execution and implementation is a well-recognized idea (Lang *et al.*, 2012). One of the key components arising from such involvement would be a shared conception of the problem at hand. The Woody Weeds partners preidentified stakeholders at the national and regional levels and invited representatives to participate in the project's main inception workshop. The preselection of stakeholders was reviewed by conducting stakeholder mapping exercises at the regional inception workshops. These early meetings were critical for project design. For example, based on stakeholder input, the need for the project to focus on *Prosopis* trees rather than on other invasive taxa in all three countries was established. Also, the project framing and emerging research questions differed among case study regions, even if they had been invaded by the same species at the same time. Stakeholders participated in each of the project's annual meetings and in participatory processes during the project. All activities were designed to support two-way communication involving them. The input of the LIGs (Schwilch *et al.*, 2012a; b) was especially important during Phase 2, i.e. when codefining management objectives for jointly selected priority areas and test implementing SLM practices.

Participating scientists in Kenya, Tanzania and Ethiopia had existing relationships with stakeholders and these, together with regular interactions during the project, were essential in building the trust and dialogue the project required. Stakeholders at various levels (Box 14.3) helped to realize implementation at the local level, uptake of the practices at the regional scale and changes to regional and national policies informed by the findings and recommendations of the Woody Weeds project. An evaluation of the effect of the participatory LIG process on shaping stakeholders' perceptions and actions revealed that those who had undergone the LIG process considered more criteria in IAS management decisions and considered the need and urgency to manage IAS more than persons from the same region and the same stakeholder groups who had not been involved in the participatory process (Adoyo *et al.*, 2022).

There were some difficulties, however. The Woody Weeds project team experienced significant challenges in maintaining the level of stakeholder interactions established during the first part of the project, particularly during the peak of the COVID pandemics. It was only towards the end of the project that stakeholder involvement could be at least partially restored.

In feedback provided near the end of the project, some stakeholders reported that their substantive involvement with the Woody Weeds project came after the project concept was largely developed. Although the focus on *Prosopis* trees had broad support among stakeholders, views on what to prioritize differed. Some stakeholders from local to national levels were not convinced of the need for the aggressive management that became the main thrust of the project and favoured a strategy focused on utilization, primarily for charcoal, especially among stakeholders in Kenya and Tanzania. Some proponents of utilization felt sidelined, at least at the outset. Eventually, this was addressed. Representatives of charcoal producer associations contributed to all activities, including the participatory LIG process. The different views were also incorporated into policy that acknowledges an income-generating role for utilization of *Prosopis* biomass in heavily invaded areas, but that management requires an integrated approach combining physical, chemical, cultural and biological control.

The challenge of acquiring project coownership by stakeholders could be addressed by funding agencies investing more resources to involve stakeholders very early in project concept and design during proposal development, prior to awarding full funding. To our knowledge, such an approach is rarely implemented. In addition, more effort could be devoted to communicating project activities to stakeholder groups outside those actively engaged throughout the life of the project.

14.4.2 Developing a collaborative culture

The Woody Weeds project leadership also aimed to develop the collaborative culture among project partners necessary for success in addressing problems in complex social-ecological systems (Goring *et al.*, 2014; Eigenbrode *et al.*, 2018). This was done through opportunities for community building that were part of the project's in-person meetings, involving scientists, students and stakeholders, and by encouraging an environment of productivity, cocreation and celebration of shared successes. The project achieved success in forming a collaborative culture, but that required diligence from leadership. While student training days and other joint activities created personal relationships among those who collected the data, the role of university supervisors of the students must be carefully discussed to make sure that they are fully integrated in the project team. The extra burden or transactional costs of large projects like the Woody Weeds project can be managed by recognizing and attending to them early and consistently during project execution (e.g. Eigenbrode *et al.*, 2017).

Many of the activities that were part of project face-to-face meetings appeared to be effective in promoting a collaborative culture. These included student training days, the PhD dinners during annual meetings and joint field trips. A sense of collaborative community was apparent during these meetings. However, the sense of community tended to weaken between all-project meetings, depending on the level of commitment from participants from different institutions and backgrounds, and the powerful traditional incentives for individual effort that can hold sway outside the project. This required leadership to remind participants of the importance of collaboration and renew the collaborative culture at each meeting. For all involved, the extra effort required to succeed with broadly cross-disciplinary projects as compared with those with a disciplinary focus can seem daunting and even discouraging if not anticipated. This should be addressed early and frankly but it is often difficult even for leadership to appreciate this. Students at times felt this extra burden in the Woody Weeds project.

Effective collaboration also depends on communication at multiple levels and among all participants. The Woody Weeds project design included several measures intended to improve communication. Throughout, attention was paid to both the informational and relational dimensions of communication (Keyton, 1999). The experience of the Woody Weeds project with respect to communication was consistent with the widely recognized importance of in-person interactions as critical for successful collaborations (Hampton and Parker, 2011). The most effective occurred during in-person project meetings or meetings among subsets of participants. On the other hand, a multiyear, international project must employ a mix of in-person and electronic communication tools that ideally are complementary. Achieving balance between face-to-face meetings and virtual tools may be project specific. Based on feedback from students, stakeholders and scientists during the project and in an end-of-project survey, the most effective strategies were the biannual multiday project meetings held in one of the countries near or within different implementation sites; student training days prior to these meetings; postmeeting tours; annual Toolbox workshops (Box 14.4); LIG workshops; the annual, anonymous online communication survey; the planned knowledge transfer activities between the first and second cohort of students at roughly the midpoint of the project; and student data integration workshops that were part of student training days. Most included a mixture of structured and social time and all of them took place as components of in-person project-wide annual meetings. The annual communication survey was effective in providing midyear assessments of the effectiveness of communication within the project that was used to guide planning.

Box 14.4. TDI and the Toolbox workshop.

The Toolbox Dialogue Initiative (TDI) is a US-based research and capacity-building consortium that has been a partner with the Woody Weeds project since its inception. TDI was brought on board to facilitate internal communications within the project, focusing specifically on enhancing the ability of project participants to communicate across their differences (e.g. differences in discipline, institutions, culture).

Formerly known as the 'Toolbox Project', TDI emerged in 2005 out of a US National Science Foundation-sponsored Integrative Graduate Education and Research Traineeship (IGERT) project at the University of Idaho that required graduate students to produce publishable research in interdisciplinary teams (Bosque-Pérez *et al.*, 2016). The Toolbox dialogue method was designed by a group of faculty and graduate students associated with the IGERT project to help enhance mutual understanding across disciplinary differences and enable the students to take each other's perspective on their common project. By enhancing mutual understanding and perspective taking ability, the Toolbox dialogue method builds collaborative capacity in cross-disciplinary teams (Rinkus and O'Rourke, 2020).

The Toolbox dialogue method builds collaborative capacity through structured dialogue in a workshop setting. The first step in applying the method involves examining the differences in research and practice worldviews among research teammates using philosophical concepts and methods (Eigenbrode *et al.*, 2007). These differences are then articulated as statements associated with a disagree–agree rating response scale. These statements are written to get at fundamental differences in worldview for each team; for Woody Weeds, they included 'Curiosity-driven research has more academic value than applied research' and Scientific research (applied or basic) must be hypothesis driven'. What results is a survey-like instrument – the *Toolbox* – that is distributed at the beginning of a Toolbox workshop. Participants in the workshop complete the instrument and then have a roughly 90 min conversation about the issues raised by the prompts, after which they complete the instrument again. Data are collected during the workshop (e.g. rating response scores, audio recording of the dialogue) and are subsequently analysed to help identify and reinforce gains made during the dialogue (Hubbs *et al.*, 2020).

Less effective strategies to facilitate communication in the Woody Weeds project were online communication mechanisms, including email and WhatsApp or Zoom meetings, which despite facilitating more dimensions of human communication than other electronic methods were not viewed as especially successful. The annual communication survey received an approximate 50% response rate, which was adequate, but respondents could have been non-random with respect to their views about collaboration.

In summary, the Woody Weeds project was successful in mentoring students through cross-disciplinary training that generated high-quality cross-disciplinary publications. The experience was appreciated and valued by these students, who are continuing in productive careers. The success required a project structure that was designed to support this integration and extra effort from some mentors and project leaders. This additional effort is probably inevitable in projects like the Woody Weeds project and should be anticipated.

14.4.3 Bridging disciplines and integrating data types

The project design recognized the challenges inherent in bridging disciplines, with their often disparate epistemologies and expectations for scholarship. Project leadership engaged the services of the Toolbox Dialogue Initiative (TDI) (https://tdi.msu.edu), who worked closely with project leadership, conducting a Toolbox workshop in the Woody Weeds project launch meeting and workshops at subsequent annual meetings throughout the project to improve communication needed to bridge epistemological divides and ways of understanding (Hubbs *et al.*, 2020). The workshops involved students, scientists and stakeholders at specific points in the project life cycle to allow them to engage with potentially contentious differences in perspectives on issues such as the approach to integrating scientific and stakeholder knowledge. Although some participants, including students, did not understand the Toolbox workshops and their purpose at the start of the project, this improved with experience. Reports from these workshops were shared with leadership and project members.

In addition, senior scientists from different disciplines and national and subnational administrators were invited to project meetings to present current practices in their disciplines relevant to addressing the social-ecological challenges caused by *Prosopis* tree invasions.

In summary, project-long commitment to improving communication across diverse channels of a project like Woody Weeds has benefits for project cohesion, understanding a common purpose and helping participants to bridge interdisciplinary divides while collaborating. Success in this aspect was enabled by strong commitment from leadership to the work required. On the other hand, because the importance of communication and broad collaboration is valued unequally, leaders should be prepared for the need to be strong advocates despite possible resistance. Interventions to enhance communication, like Toolbox workshops, can be effective but their design and delivery should reflect the diversity of the project participants and their preparedness to participate in such interventions.

The integration of data collected by students from different disciplines is another feature of cross-disciplinary projects. Diverse data types, e.g. ecological data from project sites, social data on human perceptions of issues and challenges, remotely sensed and modelled geospatial data, were successfully integrated at the regional and subcontinental level relevant for policy makers (e.g. Fig. 14.3). This resulted in many student-led cross-disciplinary publications. For example, Linders *et al.* (2019) used environmental and economic data collected in the same administrative units in two eastern African regions to assess how the relationships between ecosystem service supply and use changed with increasing levels of invasion by *P. juliflora*, and whether this relationship differed between regions. Bekele *et al.* (2022) compared local stakeholders' perceptions of the effects of these invasions on the environment with detailed ecological data collected in the same areas. These two and other student-led outputs provide insights into the social-ecological aspects of invasions that can inform effective policies for its management and may inform other IAS management strategies.

Integration across disparate data types is nonetheless inherently challenging because of how data are collected and the differing spatial and temporal resolutions that must be reconciled analytically. In the early stages, much effort was required to identify variables that could be measured for meaningful cross-disciplinary analyses. The work was performed by students, so their training had to include disciplinary and cross-disciplinary considerations. Successful integration depends on the timing of data acquisition, but data acquisition was constrained by differing institutional requirements for completion of coursework prior to conducting research.

In summary, project development and planning should acknowledge difficulties created by differences in data types and methodologies across disciplines, institutions and countries and allocate resources to resolving such issues and facilitating collaboration. For projects like Woody Weeds that rely on postgraduate students, this must be incorporated into their training and differences in disciplinary and institutional expectations of student programmes should be anticipated.

14.4.4 Incorporating different and sometimes conflicting contexts

Collaboration also had to be nurtured among students from differing cultural and disciplinary backgrounds. The research in Woody Weeds was conducted by PhD and MSc students in training, and the project sought to provide first-rate training in specific disciplines while at the same time requiring a component of cross-disciplinary research leading to publications.

The students were productive, and had produced 20 peer-reviewed publications by May 2023, most of which had elements of interdisciplinary or transdisciplinary work (https://

woodyweeds.org/?page_id=2). Students also contributed to the development of policy briefs, practice briefs and insight briefs. Training included student training days emphasizing data integration, cocollection of data and Toolbox workshops designed to improve cross-disciplinary communication by students. Students were required to publish at least one interdisciplinary paper as part of their training, and all met this requirement.

Overall, the student experience was valued by students for its effectiveness based on a postproject formal survey and interviews. They learned that invasive species management, and natural resource management in general, is often context-dependent, and one way to incorporate conflicting views and needs is to take a spatial approach, which also allows integration of management with other land use planning activities. Nonetheless, the project experienced the common tension between expectations of disciplinary mentors and institutions and integrating goals of the project (Morse *et al.*, 2007; Bosque-Pérez *et al.*, 2016; Eigenbrode *et al.*, 2017). For example, research institutions in eastern Africa often require PhD students to complete course work and defend their PhD proposal, which may delay data collection by more than a year. This necessitated additional effort from project leadership to guide students in their cross-disciplinary work.

14.5 Implications for Management

The problem of invasion by *Prosopis* trees is complex, and finding sustainable solutions to management will require leadership from people who are not just domain specialists but who are able to see the bigger picture and to integrate information and views from diverse disciplines. The Woody Weeds project has trained students who have obtained postgraduate qualifications specifically tailored to address these needs, but more will be needed. This is urgent, because the problem of invasion by *Prosopis* species, and many other invasive alien species, is growing rapidly. The approach described in the chapter provides a framework in which researchers and managers can collaborate in ways that hopefully will increase the relevance of research outputs and facilitate their uptake.

Acknowledgements

This work was supported by the Swiss Programme for Research on Global Issues for Development (r4d), funded by the Swiss National Science Foundation and the Swiss Agency for Development and Cooperation, through the project 'Woody invasive alien species in East Africa: Assessing and mitigating their negative impact on ecsystem services and rural livelihood' (Grant Number: 400440_152085). RE and US were supported by CABI with core financial support from its member countries (see http://www.cabi.org/about-cabi/who-we-work-with/ key-donors/for full details).

References

Adoyo, B., Schaffner, U., Mukhovi, S., Kiteme, B., Mbaabu, P.R. *et al.* (2022) Pathways towards the sustainable management of woody invasive species: Understanding what drives land users' decisions to adopt and use land management practices. *Land* 11, 550.

Bekele, K., Linders, T.E., Eschen, R., Shiferaw, H., Haji, J. *et al.* (2022) How well do local stakeholders' perceptions of environmental impacts of an invasive alien plant species relate to ecological data. *Ecological Indicators* 137, 108748.

Benda, L.E., Poff, L.N., Tague, C., Palmer, M.A., Pizzuto, J. *et al.* (2002) How to avoid train wrecks when using science in environmental problem solving. *BioScience* 52, 1127.

Bergmann, M., Jahn, T., Knoblock, T., Krohn, W., Pohl, C. *et al.* (2012) *Methods for Transdisciplinary Research: A Primer for Practice*. Campus, Berlin, Germany.

Berkes, F. (2007) Community-based conservation in a globalized world. *Proceedings of the National Academy of Sciences of the United States of America* 104, 15188–15193.

Binggeli, P. (1996) A taxonomic, biogeographical and ecological overview of invasive woody plants. *Journal of Vegetation Science* 7, 121–124.

Bosque-Pérez, N.A., Klos, P.Z., Force, J.E., Waits, L.P., Cleary, K. *et al.* (2016) A pedagogical model for team-based, problem-focused interdisciplinary doctoral education. *BioScience* 66, 477–488.

Bracken, L.J. and Oughton, E.A. (2006) What do you mean?' The importance of language in developing interdisciplinary research. *Transactions of the Institute of British Geographers* 31, 371–382.

Clarkson, M.E. (1994) *Risk-based Model of Stakeholder Theory*. Centre for Corporate Social Performance and Ethics, Toronto, Canada.

Connon, I.L.C. (2021) Failure to consider local political processes and power relations in the development of a transdisciplinary research project plan: Learning lessons from a stormy start. In: Fam, D. and O'Rourke, M. (eds) *Interdisciplinary and Transdisciplinary Failures: Lessons Learned from Cautionary Tales*. Routledge, New York, USA, pp. 131–146.

Dallimer, M. and Strange, N. (2015) Why socio-political borders and boundaries matter in conservation. *Trends in Ecology & Evolution* 30, 132–139.

Donaldson, T. and Preston, L.E. (1995) The stakeholder theory of the corporation: Concepts, evidence, and implications. *Academy of Management Review* 20, 65–91.

Eigenbrode, S.D., O'Rourke, M., Wulfhorst, J.D., Althoff, D.M., Goldberg, C.S. *et al.* (2007) Employing philosophical dialogue in collaborative science. *BioScience* 57, 55–64.

Eigenbrode, S.D., Martin, T., Morton, L.W., Colletti, J., Goodwin, P. *et al.* (2017) Leading large transdisciplinary projects addressing social-ecological systems: A primer for project directors. Available at: https://nifa.usda.gov/leading-transdisciplinary-projects (accessed 22 July 2024).

Eigenbrode, S.D., Binns, W.P. and Huggins, D.R. (2018) Confronting climate change challenges to dryland cereal production: A call for collaborative, transdisciplinary research, and producer engagement. *Frontiers in Ecology and Evolution* 5, 164.

Elsawah, S., Filatova, T., Jakeman, A.J., Kettner, A.J., Zellner, M.L. *et al.* (2020) Eight grand challenges in socio-environmental systems modeling. *Socio-Environmental Systems Modelling* 2, 16226.

Freeman, R.E. (1984) *Strategic Management: A Stakeholder Approach*. Pitman Publishing, Boston, USA.

Friedman, A.L. and Miles, S. (2002) Developing stakeholder theory. *Journal of Management Studies* 39, 1–21.

Glen, H. and Van Wyk, B. (2016) *Guide to Trees Introduced into South Africa*. Penguin Random House, Cape Town, South Africa.

Goring, S.J., Weathers, K.C., Dodds, W.K., Soranno, P.A., Sweet, L.C. *et al.* (2014) Improving the culture of interdisciplinary collaboration in ecology by expanding measures of success. *Frontiers in Ecology and the Environment* 12, 39–47.

Hall, T. and O'Rourke, M. (2014) Responding to communication challenges in transdisciplinary sustainability science. In: Huutoniemi, K. and Tapio, P. (eds) *Transdisciplinary Sustainability Studies: A Heuristic Approach*. Routledge, Oxford, UK, pp. 119–139.

Hampton, S.E. and Parker, J.N. (2011) Collaboration and productivity in scientific synthesis. *BioScience* 61, 900–910.

Hoffmann, S., Pohl, C. and Hering, J.G. (2017a) Exploring transdisciplinary integration within a large research program: Empirical lessons from four thematic synthesis processes. *Research Policy* 46, 678–692.

Hoffmann, S., Pohl, C. and Hering, J.G. (2017b) Methods and procedures of transdisciplinary knowledge integration: Empirical insights from four thematic synthesis processes. *Ecology and Society* 22, 27.

Hubbs, G., O'Rourke, M. and Orzack, S.H. (2020) *The Toolbox Dialogue Initiative: The Power of Cross-Disciplinary Practice*. CRC Press, Boca Raton, USA.

Imbruce, V. and Prazak, M. (2021) The challenges of studying place: Learning from the failures of an experimental interdisciplinary and community-engaged environmental studies course. In: Fam, D. and O'Rourke, M.R. (eds) *Interdisciplinary and Transdisciplinary Failures: Lessons Learned from Cautionary Tales*. Routledge, New York, USA, pp. 181–197.

Jakobsen, C.H., Hels, T. and McLaughlin, W.J. (2004) Barriers and facilitators to integration among scientists in transdisciplinary landscape analyses: A cross-country comparison. *Forest Policy and Economics* 6, 15–31.

Keyton, J. (1999) Relational communication in groups. In: Gouran, L. and Poole, M.S. (eds) *The Handbook of Group Communication Theory and Research*. Sage, Thousand Oaks, USA, pp. 192–222.

Klein, J.T. (2014) Discourses of transdisciplinarity: Looking back to the future. *Futures* 63, 68–74.

Klein, J.T. (2022) *Beyond Interdisciplinarity: Boundary Work, Communication, and Collaboration*. Oxford University Press, Oxford, UK.

Klein, J.T. and Falk-Krzesinski, H.J. (2017) Interdisciplinary and collaborative work: Framing promotion and tenure practices and policies. *Research Policy* 46, 1055–1061.

Lang, D.J., Wiek, A., Bergmann, M., Stauffacher, M., Martens, P. *et al.* (2012) Transdisciplinary research in sustainability science: Practice, principles, and challenges. *Sustainability Science* 7, 25–43.

Linders, T.E.W., Schaffner, U., Eschen, R., Abebe, A., Choge, S.K. *et al.* (2019) Direct and indirect effects of invasive species: Biodiversity loss is a major mechanism by which an invasive tree affects ecosystem functioning. *Journal of Ecology* 107, 2660–2672.

MacMynowski, D.P. (2007) Pausing at the brink of interdisciplinarity: Power and knowledge at the meeting of social and biophysical science. *Ecology and Society* 12, 20–34.

Maundu, P., Kibet, S., Morimoto, Y., Imbumi, M. and Adeka, R. (2009) Impact of *Prosopis juliflora* on Kenya's semi-arid and arid ecosystems and local livelihoods. *Biodiversity* 10, 33–50.

Mbaabu, P.R., Olago, D., Gichaba, M., Eckert, S., Eschen, R. *et al.* (2020) Restoration of degraded grasslands, but not invasion by *Prosopis juliflora*, avoids trade-offs between climate change mitigation and other ecosystem services. *Scientific Reports* 10, 20391.

Mennes, J. (2020) Putting multidisciplinarity (back) on the map. *European Journal for Philosophy of Science* 10(2), 18.

Mitchell, S.D. (2002) Integrative pluralism. *Biology & Philosophy* 17, 55–70.

Morse, W.C., Nielsen-Pincus, M., Force, J.E. and Wulfhorst, J.D. (2007) Bridges and barriers to developing and conducting interdisciplinary graduate-student team research. *Ecology and Society* 12(2), 8.

Morton, L.W., Eigenbrode, S.D. and Martin, T.A. (2015) Architectures of adaptive integration in large collaborative projects. *Ecology and Society* 20, 5.

O'Rourke, M., Crowley, S. and Gonnerman, C. (2016) On the nature of cross-disciplinary integration: A philosophical framework. *Studies in History and Philosophy of Biological and Biomedical Sciences* 56, 62–70.

O'Rourke, M., Crowley, S., Laursen, B., Robinson, B. and Vasko, S.E. (2019) Disciplinary diversity in teams, integrative approaches from unidisciplinarity to transdisciplinarity. In: Hall, K.L., Vogel, A.L. and Croyle, R.T. (eds) *Strategies for Team Science Success: Handbook of Evidence-Based Principles for Cross-Disciplinary Science and Practical Lessons*. Springer Nature, Cham, Switzerland, pp. 21–46.

Pellmar, T.C. and Eisenberg L. (2000) *Bridging Disciplines in the Brain, Behavioral, and Clinical Sciences*. National Academy Press, Washington, DC, USA.

Pirozzi, M. (2019) Stakeholders, who are they? *PM World Journal* 8, 1–10.

Pischke, E.C., Knowlton, J.L., Phifer, C.C., Gutierrez Lopez, J., Propato, T.S. *et al.* (2017) Barriers and solutions to conducting large international, interdisciplinary research projects. *Environmental Management* 60, 1011–1021.

Pohl, C., Krütli, P. and Stauffacher, M. (2017) Ten reflective steps for rendering research societally relevant. *GAIA – Ecological Perspectives for Science and Society* 26, 43–51.

Pohl, C., Klein, J.T., Hoffmann, S., Mitchell, C. and Fam, D. (2021) Conceptualising transdisciplinary integration as a multidimensional interactive process. *Environmental Science & Policy* 118, 18–26.

Richardson, D.M. (1998) Forestry trees as invasive aliens. *Conservation Biology* 12, 18–26.

Rinkus, M.A. and O'Rourke, M. (2020) Qualitative analyses of the effectiveness of toolbox dialogues. In: Hubbs, G., O'Rourke, M. and Orzack, S.H. (eds) *The Toolbox Dialogue Initiative: The Power of Cross-Disciplinary Practice*. CRC Press, Boca Raton, USA, pp. 142–161.

Roy, E.D., Morzillo, A.T. and Seijo, F. (2013) The elusive pursuit of interdisciplinarity at the human–environment interface. *BioScience* 63, 745–753.

Rylance, R. (2015) Grant giving: Global funders to focus on interdisciplinarity. *Nature* 525(7569), 313–315.

Schwilch, G., Bachmann, F. and Graaff, J. (2012a) Decision support for selecting SLM technologies with stakeholders. *Applied Geography* 34, 86–98.

Schwilch, G., Bachmann, F., Valente, S., Coelho, C., Moreira, J. *et al.* (2012b) A structured multi-stakeholder learning process for sustainable land management. *Journal of Environmental Management* 107, 52–63.

Seid, O., Haji, J. and Legesse, B. (2020) Rural households' perception on the effects of *Prosopis juliflora* invasion: The case of Amibara District of Afar National Regional State, Ethiopia. *Pastoralism* 10, 1–9.

Shackleton, R.T., Le Maitre, D.C., Pasiecznik, N.M. and Richardson, D.M. (2014) *Prosopis*: A global assessment of the biogeography, benefits, impacts and management of one of the world's worst woody invasive plant taxa. *AoB PLANTS* 6, lu027.

Siebenhüner, B. (2018) Conflicts in transdisciplinary research: Reviewing literature and analysing a case of climate adaptation in northwestern Germany. *Ecological Economics* 154, 117–127.

Soanes, C. and Stevenson, A. (2008) Integrate. In: *Concise Oxford English Dictionary*, 11th rev. edn. Oxford University Press, Oxford, UK.

Steger, C., Klein, J.A., Reid, R.S., Lavorel, S., Tucker, C. *et al.* (2021) Science with society: Evidence-based guidance for best practices in environmental transdisciplinary work. *Global Environmental Change* 68, 102240.

Stokols, D., Hall, K.L., Moser, R.P., Feng, A., Misra, S. *et al.* (2010) Cross-disciplinary team science initiatives: Research, training, and translation. In: Frodeman, R., Klein, J.T. and Mitcham, C. (eds) *The Oxford Handbook of Interdisciplinarity*. Oxford University Press, Oxford, UK, pp. 471–493.

Tengö, M., Hill, R., Malmer, P., Raymond, C.M., Spierenburg, M. *et al.* (2017) Weaving knowledge systems in IPBES, CBD and beyond – lessons learned for sustainability. *Current Opinion in Environmental Sustainability* 26, 17–25.

Urton, D. and Murray, D. (2021) Project manager's perspectives on enhancing collaboration in multidisciplinary environmental management projects. *Project Leadership and Society* 2, 100008.

van Wilgen, B.W. and Richardson, D.M. (2014) Challenges and trade-offs in the management of invasive alien trees. *Biological Invasions* 16, 721–734.

van Wilgen, B.W., Richardson, D.M., Le Maitre, D.C., Marais, C. and Magadlela, D. (2001) The economic consequences of alien plant invasions: Examples of impacts and approaches to sustainable management in South Africa. *Environment, Development and Sustainability* 3, 145–168.

Wang, J., Aenis, T. and Siew, T.F. (2019) Communication processes in intercultural transdisciplinary research: Framework from a group perspective. *Sustainability Science* 14, 1673–1684.

Ward, W.S. and Southwell, A.F. (2021) Intercultural relations in international agricultural research teams: Perceived issues influencing communication effectiveness. *Journal of Intercultural Communication Research* 50, 122–145.

Zscheischler, J., Rogga, S. and Busse, M. (2017) The adoption and implementation of transdisciplinary research in the field of land-use science – A comparative case study. *Sustainability* 9, 1926.

15 Implications of Recent Research Findings for the Management of *Prosopis* Species in Eastern Africa

Brian W. van Wilgen[1]*, Albrecht Ehrensperger[2], Ketema Bekele[3] and Urs Schaffner[4]

[1]Centre for Invasion Biology, Department of Botany and Zoology, Stellenbosch University, South Africa; [2]Centre for Development and Environment (CDE), University of Bern, Switzerland; [3]School of Agricultural Economics, Haramaya University, Dire Dawa, Ethiopia; [4]CABI, Delémont, Switzerland

Abstract

This chapter provides a review of the key findings of recent research on the ecology and management of invasive *Prosopis* trees in eastern Africa and discusses their management implications. The *Prosopis juliflora* invasion is relatively young in eastern Africa, and large areas of suitable habitat remain uninvaded. This underscores the importance of prioritizing the prevention of further spread to areas suitable for invasion but that are currently uninvaded. Invasive *Prosopis* trees have substantial negative impacts on biodiversity, water resources, rangeland productivity and human and animal health, yet they do have some positive benefits. Handling this tension can be challenging, and management goals should be based on considerations of the net value (benefits minus adverse impacts) and not on single aspects or one-dimensional solutions. While management techniques are available and can be effectively applied at relatively small scales, this is not the case for invasions over larger areas where cost-efficient and sustainable options hardly exist. In addition, many stakeholders are involved, and they differ considerably in terms of their levels of awareness, how affected they are by the invasion, how empowered they are to deal with it and which outcomes they would prefer. This is a complex socio-ecological problem that demands transdisciplinary solutions codeveloped with communities and other actors at subnational and national scales. Effective solutions will have to be based on knowledge of (1) the ecosystem, (2) the target outcomes that would be both acceptable and possible, and (3) how the system can be transformed to achieve the desired outcomes. It is clear, though, that the use of biological control will probably offer the only way in which these invasions can be effectively and affordably managed. The utilization of biomass from the invading trees may bring some economic benefits but has demonstrably not been effective as a control method. Great care will therefore need to be taken to ensure that any programme of utilization does not interfere with the goal of bringing invasions under control.

15.1 Introduction

Research into the ecology and management of invasive alien *Prosopis* trees in eastern Africa over the past decade has increased our understanding and highlighted the complex interactions between the invading trees, the recipient ecosystems and the people who live

*Corresponding author: bvanwilgen@sun.ac.za

© CAB International 2025. *The Ecology and Management of Invasive Prosopis Trees in Eastern Africa* (eds U. Schaffner *et al.*)
DOI: 10.1079/9781800623644.0015

in them. Much of the recent work has deliberately followed a transdisciplinary approach. In this chapter, we discuss the implications arising from these recent research findings. We examine how this understanding may help us to manage *Prosopis* trees better in future and what approaches could be used to guide management that will maximize benefits and minimize harm. We focus on the ecology and management of *P. juliflora* in eastern Africa. However, we believe that our findings can also inform research, policy and practice on *P. juliflora* or other *Prosopis* species in other parts of the world, or on invasive plants that are causes for conflicts in general.

15.2 Findings and Implications

15.2.1 History, evolution and spread

15.2.1.1 Key findings relating to introduction and spread

The invasion by *Prosopis* trees in eastern Africa is still young. While the first *Prosopis* species were introduced into Africa in the 19th century (Shackleton *et al.*, 2014), the widespread introductions in eastern Africa were only made between the 1970s and 1990s (Chapter 2, this volume). The invasion by *P. juliflora* started soon after these introductions and was rapid. To date, this tree has invaded probably close to 10 million ha of land in eastern Africa (Chapter 4, this volume). In the Afar region of Ethiopia alone, *P. juliflora* has invaded ~1.2 million ha within 35 years, and the invasions continue to spread (Shiferaw *et al.*, 2019). Shiferaw *et al.* (2019) found that features of the landscape, such as distance to rivers, roads and villages, were more significant predictors of *Prosopis* tree distribution in the Afar region than climatic variables. This suggests that *Prosopis* trees are likely to continue spreading and increasing in abundance due to human action if they are not effectively regulated and managed.

Eckert *et al.* (2020) showed that large parts of the presently uninvaded lowland areas in eastern Africa are at risk of being invaded soon. Their climatic suitability models suggest that 77% of Kenya's lowlands, 51% of Tanzania's lowlands and 45% of Ethiopian lowlands are threatened by *Prosopis* invasion (Chapter 4, this volume). The level of invasion within the climatically suitable regions may vary depending on non-climatic factors, such as soil type, but it is clear that a very substantial area is still at risk of being invaded.

From a scientific point of view, the relatively recent invasion by *Prosopis* species in eastern Africa provided an opportunity to study eco-evolutionary processes that underlie successful invasion, and to assess differences in invasiveness between closely related species with the same history of introduction. Most of the original plantations of the *Prosopis* species introduced to Kenya in the early 1980s are still present today (Mbaabu *et al.*, 2019; Choge *et al.*, 2022), which provided a unique opportunity to compare the originally introduced genotypes of a non-native plant species with those at the invasion front. Using a combination of field and laboratory experiments, Castillo *et al.* (2021a) provided evidence of contemporary evolution in *P. juliflora*, despite the relatively short time that the species had been present (Chapter 3, this volume). Their study is in line with a growing body of evidence for rapid evolution in populations of invasive alien plant species post introduction, which may increase their adaptation to the new environment and enhance their invasiveness.

Hybridization is one of the factors that can enhance plant performance (Schierenbeck and Ellstrand, 2009). There is evidence that hybridization enhances the invasiveness of *Prosopis* in countries such as Australia and South Africa. Yet, this does not appear to be the case in eastern Africa, and Castillo *et al.* (2021b) found that only the tetraploid *P. juliflora* has become invasive in this region. Despite similar introduction histories of *P. juliflora* and *P. pallida*, the diploid *P. pallida* has not become invasive, and the few hybrids between *P. juliflora* and *P. pallida* found in Kenya were only detected at sites where these two species were originally planted together, probably reflecting that *P. juliflora* × *P. pallida* hybrids are triploid and therefore sterile. As discussed in Chapter 3 (this volume), the different mechanisms underlying *Prosopis* invasions in different parts of the world serve as a powerful example of the context-dependency of biological invasions and illustrate the need for further research to evaluate the contributions of different processes that allow alien species to become invasive in different environments.

15.2.1.2 Implications for policy and practice

The invasion of *P. juliflora* is still relatively recent and large areas in eastern Africa suitable for *Prosopis* trees have not yet been invaded. Management strategies aimed at reducing the negative impacts of this invasive tree species should therefore emphasize measures that help to slow down or prevent its further spread (Chapter 4, this volume). Shiferaw *et al.*'s (2019) finding that people and their livestock are predominantly responsible for promoting invasions (see Section 15.2.1.1 above) suggests that human-assisted transfer of propagules to uninvaded locations and planting of *Prosopis* trees where they do not yet occur should be strongly discouraged. Also, care should be taken to prevent the establishment of new infestations by long-distance transfer of livestock from invaded areas affected by drought to uninvaded areas with better forage availability, or from water resources to dry-season grazing areas. This can be achieved by quarantining livestock from areas invaded by *Prosopis* trees for 7–10 days in livestock holding areas to allow them to excrete ingested *Prosopis* seeds before the animals are moved to uninvaded areas. Moreover, *Prosopis* seeds collected for animal feed should be processed (e.g. milled) where they are collected before being transported to areas without *Prosopis* trees.

The spread of invasive *Prosopis* trees is accelerated by establishing new plantations in areas where they do not yet occur, and if the plantations are extensive and/or sourced from multiple areas, the risk of rapid invasion is amplified due to increased propagule pressure and greater genetic diversity. For example, Le Roux (2023) noted that propagule pressure was a common feature that characterized invasions by Australian *Acacia* species globally. There is a global drive, in present policies and public debates on environment and climate, to plant more trees. While forest protection and reforestation are important objectives in many places, greater care is required to avoid unwanted negative impacts (Bond *et al.*, 2019; Parr *et al.*, 2024).

First, it is crucial to acknowledge that not all tree species are desirable at any location and that it is important to exclude invasive trees such as *P. juliflora* from these programmes (Brundu *et al.*, 2020). Second, it is equally important to exclude such programmes from pristine biomes and ecosystems with no or low natural tree cover. Regrettably, current land degradation monitoring frequently categorizes extensive grasslands and savannas as degraded solely based on their low tree cover, disregarding the fact that they are natural open ecosystems (Veldman *et al.*, 2015; Bond, 2019) that are rich in unique biodiversity and provide multiple ecosystem services that deliver substantial benefits to people. Planting trees or allowing *P. juliflora* to invade such areas could lead to significant unintended consequences, including biodiversity loss, reduced water resources and decreased carrying capacity of rangelands (see also Section 15.2.2 below; Parr *et al.*, 2024).

Measures to slow down or stop the invasion process should also target the leading edge of *Prosopis* tree invasions, where the densities are still very low and commercial utilization therefore is of little interest. Areas bordering existing infestations should be regularly surveyed for new incursions and any *Prosopis* trees detected should be rapidly removed (early detection and rapid response, Chapter 10, this volume). The regular use of remote sensing to monitor the fractional cover of *Prosopis* tree invasions would be useful in this regard (Chapter 4, this volume), and physical or chemical control measures may be particularly suited for managing *P. juliflora* at the invasion front.

The size of the problem and the potential for its growth mean that the resources to effectively manage invasions by *Prosopis* trees will always be a limiting factor. Experience in Australia and South Africa has revealed that control efforts based on mechanical and chemical approaches alone cannot be effective over larger areas unless they are supplemented by biological control (Kleinjan *et al.*, 2021; Sun *et al.*, 2022; see also Section 15.3.2 below, and Chapter 10, this volume).

15.2.2 Impacts and benefits

15.2.2.1 Key findings relating to impacts and benefits

Although *Prosopis* trees were introduced to eastern Africa for the benefits that they could provide, these benefits have subsequently been cancelled out and are now considerably exceeded by negative impacts, and the net negative outcome is set to grow. The spread of these trees results in fundamental changes to the structure and functioning of communal rangelands, croplands, wetlands and protected areas. The diversity of native plants and animals is markedly reduced and open savannas become dense, impenetrable stands of trees with little or no grass below them (Chapter 5, this volume). *Prosopis* trees reduce water resources, especially groundwater reserves. In densely invaded areas, they can consume ~50% of the annual precipitation of the region, thereby exacerbating the effects of a changing climate (Chapter 6, this volume). Moreover, there have been significant declines in livestock and wildlife populations in invaded areas due to displacement of the native herbaceous and woody species that provide the natural nutrition necessary for them to survive (Chapter 7, this volume). The sustainability of traditional rural livelihoods has thus become threatened across the region, and many rural pastoralists feel that they are no longer in control of their lives (Chapter 8, this volume). Invasions by *Prosopis* trees have brought about a multidimensional sustainability challenge requiring a multisectoral response, as they impact widely on the health of plants, animals, people and the environment (Chapter 9, this volume).

The problem is complicated because there are also benefits that can be derived from utilization of invasive *Prosopis* trees, primarily through the production and sale of charcoal and feed supplements. For that reason, most people perceive *Prosopis* trees as beneficial during the early stages of an invasion. Their perceptions and attitudes change as negative impacts become apparent, but there is a lag phase between when the impacts manifest themselves and when people's attitudes start to change (Seid *et al.*, 2020; Bekele *et al.*, 2022) (Fig. 15.1). In addition, views diverge in

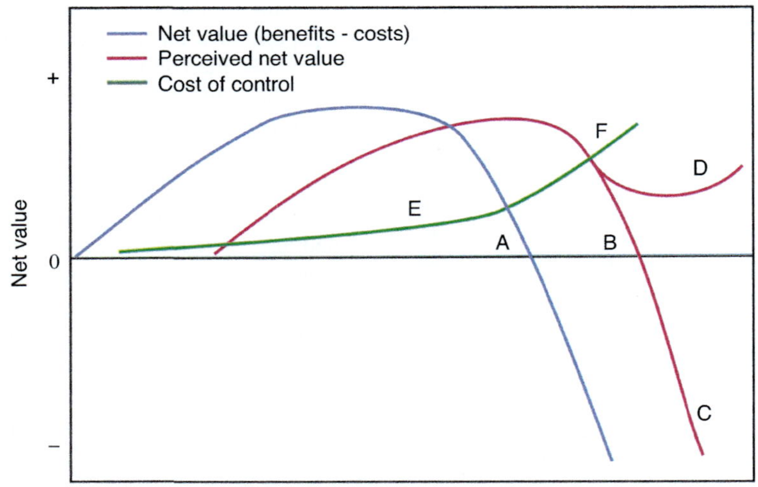

Fig. 15.1. Trends in the net value associated with an invasion of *Prosopis* trees over time, showing how people's perceptions and the costs of control change in response to increasing *Prosopis* tree cover. A: Point beyond which the value of benefits is exceeded by the cost of impacts. B: Point beyond which the perceived value of benefits is exceeded by the perceived cost of impacts. C: Costs perceived by people who suffer the consequences of impact. D: Costs perceived by people who gain from benefits. E: Cost of control at the point at which negative impacts begin to reduce the value of benefits. F: Cost of control at the point at which people perceive that the net benefits become negative.

the later stages of invasion. The few people who benefit from *Prosopis* trees tend to retain their positive views, while the majority who carry the costs hold increasingly negative views (Linders *et al.*, 2021; Bekele *et al.*, 2022) (Fig 15.1).

The consequences of these trends are twofold. First, by the time that there is sufficient negative sentiment to trigger attempts to control the spread of *Prosopis* trees, they would already have invaded substantial areas, escalated the negative impacts and reduced the feasibility of gaining control. Second, discussions on potential management strategies to deal with the problem become polarized, leading to ineffective compromises (such as 'control through utilization') or even conflicts (Box 2.1, Chapter 2, this volume).

Further confusion is introduced to the debates on the value of *Prosopis* trees by the global drive to increase tree cover in degraded areas to sequester carbon and reduce the rate of climate change (Box 2.2, Chapter 2, this volume). However, the extent to which afforestation or tree invasions can lead to increased soil carbon stocks depends on prior land use or land cover, climate and the tree species, so assuming that carbon stored by woody plant invasions can offset emissions may be incorrect for eastern Africa (Jackson *et al.*, 2002; Mbaabu *et al.*, 2020; Hasler *et al.*, 2024). In fact, the expansion of trees or shrubs into grasslands can lead to a depletion of valuable soil carbon stocks (Conant and Paustian, 2002; Jackson *et al.*, 2017), and the proposals for widespread tree planting are undesirable for a range of other reasons (Bond *et al.*, 2019; Parr *et al.*, 2024). Instead, restoration of grasslands in tropical semi-arid regions can both serve as a measure for climate change mitigation and help restore key ecosystem services important for pastoralist and agropastoralist communities.

When considered in monetary or economic terms, the impacts of *Prosopis* tree invasions are significant. The current combined economic effects of impacts of invasion can be expressed as an aggregate economic welfare loss. This is a measure of the degree to which people are worse off because of certain circumstances (e.g. the presence of invasive *Prosopis* trees) compared to where they would have been had the situation not existed. For the people living in two study areas (Baringo County in Kenya and the Afar region in Ethiopia), these losses have been estimated to be USD 4.2 and 6 million per year respectively (Chapter 8, this volume). The annual value for these losses across three countries (Ethiopia, Kenya and Tanzania) was estimated to be USD 357 million annually. *Prosopis* trees are spreading rapidly (Chapter 4, this volume) and if these invasions are not effectively contained, the projected net present value of their impacts (i.e. the value of all future costs and benefits in monetary terms discounted to the present) for Ethiopia, Kenya and Tanzania is estimated to be in the region of USD 7 billion, assuming a discount rate of 3% (Chapter 8, this volume).

15.2.2.2 Implications for policy and practice

The ways in which the impacts of invasion by *Prosopis* trees and other conflict-of-interest species are generated are multifaceted and interconnected, and they cannot be effectively addressed without an integrated and holistic approach. Proposed solutions based on single benefits (e.g. charcoal production or carbon credits) or negative impacts (e.g. reduced water resources or rangeland carrying capacity) would not necessarily deliver optimal outcomes. Rather, a course of action should be determined based on a joint consideration of net outcomes (all positive benefits minus all negative impacts). Examples of approaches that could be used to inform policy include estimations of total economic value (Chapter 8, this volume) or holistic, multifactoral impact assessments (Bardgett *et al.*, 2021; Chapter 9, this volume).

The estimations of aggregated economic welfare loss give an indication of the level of spending that would be justifiable if that spending was able to eliminate the cause of the loss. In the case of Baringo County in Kenya and the Afar region in Ethiopia, for example, it would be economically justifiable to spend between USD 4.2 and 6 million per year to bring *Prosopis* invasions under control (Bekele *et al.*, 2018). A one-off budget of USD 7 billion for Ethiopia, Kenya and Tanzania today would thus be justifiable if it was sufficient to prevent these ongoing welfare losses (Chapter 8, this volume).

15.2.3 Management

15.2.3.1 Key findings relating to management

As with other invasive plant species, objectives for the management of *P. juliflora* differ depending on the level of invasion. If this invasive tree has not yet established in an area, the objective would be to prevent its introduction; if *P. juliflora* has established but the infestations are still sparse, the objectives would be to remove all the trees and prevent reinvasion (i.e. early detection and rapid response); and in areas where *P. juliflora* has established widely and complete removal is unlikely or impossible, the objective would be to contain the species, reduce its densities to mitigate negative impacts and/or to protect important assets. Methods are available to deal effectively with invasions at relatively small scales (Chapter 10, this volume). If these methods are used, the return on investment for the removal of *Prosopis* trees on cropland becomes positive after 2–3 cropping seasons (Fig. 10.4, Chapter 10, this volume), and if the collected wood can be used for charcoal or biochar production, the return on investment can be realized more quickly.

However, pastoralists and other land users confronted with extensive invasions over large areas are often left without cost-effective and sustainable management options. This is due to several factors, including the scale of the problem (Chapter 4, this volume), the attributes that make the species an aggressive invader (Box 1.2, Chapter 1, this volume), lack of resources to deal with the invasions (Chapter 8, this volume), lack of coordination across all stakeholders (Chapter 12, this volume), and proposals for simple, one-dimensional solutions (Box 2.1, Chapter 2, this volume). The introduction of *P. juliflora* to eastern Africa cannot be undone and eradication cannot be achieved as the species is now too widespread. Thus, the goals of management can never be to eradicate *Prosopis* trees from eastern Africa; rather, management must aim to keep the negative impacts of *Prosopis* species to a minimum and preferably to attain an overall net positive benefit when all aspects are considered jointly (Chapter 9, this volume). To do this, multiple stakeholders need to be engaged, and resources need to be pooled to effectively achieve this goal over relatively extensive areas.

Planning is also important. Spatial planning can strengthen land governance, improve economic opportunities based on the sustainable management of land resources, and develop land use options that reconcile conservation and development objectives (Metternicht, 2017). In the context of invasive species management, it also offers the opportunity to spatially integrate the three main management objectives of preventing introduction to new areas, detecting new incursions and dealing with them rapidly and effectively containing established invasions and reducing their impact (Chapter 11, this volume). Using a spatially explicit management plan is also valuable because invasive species like *P. juliflora* are spread over a mosaic of various land use and habitat types that are managed by different stakeholders with different management objectives, and because limited funding usually requires a prioritization of management interventions. Spatial plans, ideally jointly developed and approved by stakeholders suffering negative consequences and those benefiting from the invasive tree species, can identify priorities and integrate the management of *P. juliflora* – and other invasive plant species – into existing planning and budgetary processes (Chapter 11, this volume).

Experience in Australia and South Africa has shown that biological control is an essential component for managing *Prosopis* tree invasions over large areas (Chapter 12, this volume). In Australia, some success has been achieved using biological control (van Klinken, 2012). Although the use of biological control for *Prosopis* trees in South Africa was originally restricted to seed-feeding insects to slow their spread but not to damage the trees themselves, further damaging biological control agents have now been released. To quote Kleinjan *et al.* (2021), these agents were released 'in response to recognition in South Africa that there is no other route to successful management of *Prosopis*'. The big advantage of classical biological control is that it is self-sustaining and does not require continuous action by land users (Sun *et al.*, 2022). In a review of the 'Working for Water' programme in South Africa, Henderson and Wilson (2017) found that during the first 23 years of the programme, the increase in invasive plants only slowed down, or the invaded area contracted, when classical biological control was part of the integrated weed management approach. In all other cases, invasions continued to spread despite considerable efforts to contain them using mechanical and chemical methods (van Wilgen *et al.*, 2020a).

The governments of Ethiopia, Kenya and Tanzania have all recently adopted strategies for dealing with invasive *Prosopis* trees, in response to growing concerns regarding the impacts of these invasions and the predictions that they are set to grow. The strategies all demonstrate a clear intention to deal with the problem, and although they agree on several approaches, each still has shortcomings that will need to be addressed if these strategies are to be effective (Chapter 2, this volume).

15.2.3.2 Implications for policy and practice

Recent research on the management of invasive alien *Prosopis* trees in eastern Africa has covered a broad range of issues. First, it is clear that this is a complex problem, that clear policies on how to deal with it need to be adopted, and that broad strategic approaches need to be developed at national and subnational levels to give effect to these policies. Legislation needs to be promulgated so that governments can effectively regulate relevant practices. Given the relatively recent but rapidly growing nature of *Prosopis* tree invasions, it will also be very important to prioritize the prevention of further spread. Management plans with explicit and achievable goals will need to be developed, with clear guidance on issues such as monitoring and adaptive management, the use of biological control and the utilization of products such as charcoal and feed supplements from invasive *Prosopis* trees. Finally, it will be vitally important to engage with all potentially affected stakeholders, and to adopt approaches that cater for their diverse needs and expectations. Each of these issues is discussed briefly below.

LAWS AND REGULATIONS. It would be useful to review the various laws relevant to the management of biological invasions in eastern Africa countries. Laws are formal, binding rules that govern the behaviour of individuals, organizations and government itself. The primary purpose of laws is to provide a structured legal framework that defines acceptable conduct and consequences for violations. Regulations are rules and directives issued by government agencies to implement and enforce laws. The primary purpose of regulations is to ensure the consistent and effective application of laws, translating legal provisions into practical requirements for individuals and businesses.

One finding that is relevant to regulating further introduction of *Prosopis* species is that diploid species that are allopatric in their native ranges will hybridize if cointroduced into new environments, and hybrids can become invasive. It would therefore be logical to treat all diploid *Prosopis* species as a single taxon for the purposes of regulation (Chapter 3, this volume).

POLICIES AND STRATEGIES. Ideally, nations should have clear policies on how they intend to deal with biological invasions, and they should develop strategies to guide the implementation of those policies. Government policy refers to a broad set of principles, objectives and guidelines that a government adopts to address specific issues or achieve overarching goals. The primary purpose of government policy is to articulate the government's vision and intentions, serving as a framework for decision making and action. Government strategy is a detailed plan of action designed to achieve the specific objectives outlined in government policies. The primary purpose of government strategy is to outline the steps and resources required to realize the objectives set out in policies.

First, it is not clear whether any of the eastern African countries have adopted policy positions on the issue of biological invasions and it would arguably be important for that to be done (Lukey and Hall, 2020). Second, some revisions to existing strategies will be needed to make them more effective. For example, goals need to be framed in terms of outcomes rather than inputs, goals need to be made more specific and/or realistic, and more guidance is needed on key issues such as the use of biological control, whether or under what circumstances to allow the utilization of products from invasive *Prosopis* trees, or the role that *Prosopis* trees should play in increasing tree cover (Chapter 2, this volume).

There is also a need for regional (international) strategies that will coordinate activities across the borders of neighbouring countries.

PLANNING. The adage says that failing to plan is planning to fail. Management plans that cover larger areas (for example, for protected areas, large farms or communally owned rangelands) are therefore an essential input to management (Chapter 10, this volume). Management plans also need to be spatially explicit because the interventions will differ depending on the stage of invasion in different sections of any area covered by a plan (Chapter 11, this volume; Fig. 10.1, Chapter 10, this volume). Spatial planning can also help to resolve different and sometimes conflicting stakeholder interests regarding the desired outcomes of management. Given that perceptions of the value of invasive *Prosopis* trees are slow to change (Fig. 15.1), it has proven useful to have plans drawn up jointly by opinion leaders from both invaded and uninvaded or lightly invaded areas. In such cases different courses of action can be debated among peers who have experienced first hand the consequences of not taking early action and those who have not (Chapter 12, this volume).

PREVENTION AND EARLY DETECTION. There is ample evidence that early and rapid interventions to manage invasive alien plants when populations are small and sparsely distributed will lower the costs and increase the likelihood of success (Higgins *et al.*, 2000; van Wilgen *et al.*, 2001). Therefore, preventing further spread has become a primary goal that is recognized in the national *Prosopis* strategies of eastern African nations (Chapter 2, this volume). Regulations can be developed to reduce the risk of spread, for example by not allowing planting in new areas and regulating the movement of seeds or livestock, but regular and effective monitoring (discussed below) will be needed and responses to new incursions will need to be rapid to increase the chances of success.

MONITORING AND ADAPTIVE MANAGEMENT. Effective and regular monitoring will be needed for two reasons. First, monitoring will be needed to detect new incursions or to locate the leading edges of existing invasions, so that appropriate responses can be implemented in the right places. These responses could either be rapid action taken to eliminate new incursions or focusing efforts on the leading edges of invasions to contain spread. This monitoring could either utilize remote sensing (Chapter 4, this volume) or be based on regular ground surveys along high-risk pathways of dispersal (Shiferaw *et al.*, 2019).

Second, regular monitoring is necessary to assess progress towards the goals of management as set out in management plans, and as an input to adaptive management (Holling, 1978). Adaptive management is defined as an intentional approach to making decisions and adjustments in response to new information and changes in context. Adaptive management implies that changes will often be necessary as understanding improves or as environmental conditions or societal values change. Ecosystem management interventions, such as controlling invasive alien plants, must proceed in the absence of a full understanding of the complexity of ecological interactions, so progress towards goals needs to be monitored. In cases where progress is not satisfactory, it may be necessary to adapt the management approach or even change the goals to more realistic ones.

BIOLOGICAL CONTROL. Current evidence suggests that biological control will have to be used to its full potential to assist in slowing, halting or reversing the spread of *Prosopis* trees and helping to avoid substantial increases in their negative impacts (Chapter 10, this volume). Not only is biological control safe, effective and sustainable (van Wilgen *et al.*, 2013), but it also delivers very attractive returns on investment compared to other forms of management (van Wilgen *et al.*, 2020b). Without effective biological control, it will remain very difficult, if not impossible, to effectively manage *Prosopis* tree invasions that cover large areas.

UTILIZATION. Utilization of the biomass provided by invasive *Prosopis* trees (for example, in the form of charcoal or feed supplements) is frequently put forward as a means to gain control of these invasions (Chapter 2, this volume). However, the effectiveness of utilization in achieving control has yet to be demonstrated. On the other hand, where such projects have formally been put in place, they were wholly ineffective as spread simply continued despite the utilization (Ng *et al.*, 2017; Mbaabu *et al.*, 2019).

This is not to say that all forms of utilization should be discontinued; in heavily invaded areas it may be one of the few remaining possibilities for rural communities to buffer their losses in income (e.g. from livestock production) due to the invasion of *Prosopis* trees (Linders *et al.*, 2020). However, it needs to be made very clear that such utilization will not be effective as a form of control, and can merely offset losses in areas where it has become dominant (Chapter 10, this volume). For example, when developing spatial management plans, it would be acceptable to focus control efforts on the leading edges of the invasion and to utilize the biomass in areas away from the leading edge where *Prosopis* trees occur at high densities. However, great care must be taken to ensure that any programme of utilization does not interfere with the goal of bringing invasions under control, because well-controlled invasions will ultimately deliver much higher net benefits to most people (Chapter 9, this volume).

ADOPTING A BROAD SYSTEMS PERSPECTIVE. There is a risk that ecologists may regard technical solutions as adequate for achieving the goals of management, but this overlooks the complexity of the situation and the need for considering the different types of knowledge required for the successful implementation of effective management (Box 15.1). A particular challenge associated with the implementation of management approaches is that it implies the acceptance, adoption and continued use of management practices by different groups of land users and stakeholders. Studies on the adoption of sustainable land management practices (De Graaff *et al.*, 2008; Wiesmann *et al.*, 2011) have highlighted the role of psychological thresholds, aspirations and beliefs that strongly influence the potential acceptance of such practices, the roles of anticipated benefits and the capability

Box 15.1. The importance for successful management of considering different types of knowledge.

Prosopis tree invasions in eastern Africa are an example of a complex social-ecological problem. Problems like these can only be effectively managed if three basic and interrelated questions are all considered (Hirsch *et al.*, 2006).

- The first is about system knowledge, where the characteristics of invaded ecosystems, the interactions between invasive species and native organisms, and the diverse impacts of invasive species need to be understood.
- The second is target knowledge, which addresses the question of what the desired outcomes of management are, and whether it will be possible to achieve them. Target knowledge plays a crucial role in reconciling the diverse management priorities and objectives that exist, with an intention to establish a sustainable and desirable future state that aligns with the interests of all stakeholders. This requires the crafting of clear and shared objectives for invasion management.
- The third is transformation knowledge, which considers how the system (understood through system knowledge) needs to be transformed to achieve the desired outcome (determined by target knowledge). Transformation knowledge can be used to identify and implement transformative policies and actions that address ways to mitigate the impacts of invasions, guiding the path towards the jointly defined target state.

To ensure the effectiveness of future management strategies, a robust foundation must be built on these three interconnected knowledge types. It is only by integrating these knowledge types that researchers and practitioners can develop comprehensive approaches to combat the challenges posed by biological invasions.

(financial, human resources and skills) of stakeholders to facilitate their adoption, as well as the roles of motivation and perceived benefits in securing a continued use of such practices. Hence, management strategies that rely solely on promoting technical solutions and practices (e.g. many of those outlined in Chapter 10, this volume) could well be destined to fail, as they overlook the perceptions of concerned actors and the underlying drivers that influence their decision making.

Instead, these management strategies should adopt a systems perspective and amalgamate technical solutions with governance, economy and finance, individual and collective action to achieve effective systemic change (United Nations, 2019). In the context of managing invasive *Prosopis* trees, a systems perspective might entail prioritizing the preservation of environmental commons, such as grazing areas and conservancies, or fostering the development of sustainable food systems. By taking a holistic approach and considering the broader landscape and socio-economic factors, such strategies can address the root causes of the invasion problem and enhance the likelihood of successful and sustainable outcomes.

INVOLVING STAKEHOLDERS. Community engagement will be essential for the management of *Prosopis* invasions to succeed. There are many stakeholder groups that can affect the outcomes of management interventions. Some of these groups are more influential than others, and some are directly affected by the problem while others may only be affected marginally or not at all. These stakeholders exhibit varying levels of awareness, recognition of the problem's severity and capacity to contribute to solutions. For management to be successful, management plans should be developed in close collaboration with all stakeholders. Multi-stakeholder platforms are increasingly being incorporated into the governance and management of natural resources by governments and civil society. They help to implement inclusive processes that address power inequities, support adaptive learning and yield more equitable patterns of stakeholder interaction, with the aim of promoting outcomes that favour social-ecological resilience (Chapters 12 and 13, this volume).

15.3 Future Challenges and Opportunities

The introduction of *P. juliflora* to eastern Africa is a classic example of a well-intentioned action gone wrong. This scenario has often been observed with alien trees, initially introduced to serve as sources of timber or firewood or for other benefits, only to lead to invasion and ensuing negative consequences (Richardson, 1998; Rouget *et al.*, 2002). The likelihood of introduced trees becoming problematic invaders has historically been substantially increased by the careful selection of species that possess ecological adaptations to make them eminently suitable for the new environment (see, for example, Poynton, 1977) and by increased propagule pressure arising from widespread plantings (Chapter 2, this volume).

In this book, we have attempted to highlight the intricate and multilayered challenges of managing *Prosopis* tree invasions and achieving sustainable outcomes. These issues lack easy, one-size-fits-all solutions. Instead, collaboration across scales and sectors will be needed if agreed objectives are to be met. Of course, it would be very important to derive broader lessons from the experience with *Prosopis* trees in eastern Africa, and to take steps to avoid repeating the mistakes of the past in other parts of the world or with different species. One option would be to promote the use of native species rather than introducing potentially invasive alien species. Another would be to require rigorous risk assessments and analyses for any proposals to introduce new alien species to the region. Supplying a burgeoning human population with food and fibre depends heavily on alien plants, and the quest to find and establish further useful species will continue.

Not all alien species become problematic invaders, but imports of new species need to be carefully screened and regulated to minimize as far as possible the risk of introducing another problem species. Several models have been developed to support such assessments. For example, Reichard and Hamilton (1997) used several attributes of woody plants to derive a model for predicting potential invasiveness. Their model allows users to place species into one of three categories (allow admission, deny admission or delay admission pending further analyses). These authors also recommended that species that are known to be invasive elsewhere should not be admitted. Invasion scientists have also recently developed a robust framework for objectively assessing the risk of a new import becoming invasive (Kumschick *et al.*, 2020), and this should be widely adopted by countries when developing regulations to reduce the risk of introducing invasive alien species.

Acknowledgements

This work was supported by the Swiss Programme for Research on Global Issues for Development (r4d), funded by the Swiss National Science Foundation and the Swiss Agency for Development and Cooperation, through the project 'Woody invasive alien species in East Africa: Assessing and mitigating their negative impact on ecosystem services and rural livelihoods' (Grant Number: 400440_152085), and by the Swiss Agency for Development and Cooperation through the project 'Woody Weeds +' (Project number 7F.10651.01). BWvW thanks the National Research Foundation of South Africa for funding (Grant Numbers 150260 and 151653). US was supported by CABI with core financial support from its member countries.

References

Bardgett, R.D., Bullock, J.M., Lavorel, S., Manning, P., Schaffner, U. *et al.* (2021) Combatting global grassland degradation. *Nature Reviews Earth & Environment* 2, 720–735.

Bekele, K., Haji, J., Legesse, B. and Schaffner, U. (2018) Economic impacts of *Prosopis* spp. invasions on dryland ecosystem services in Ethiopia and Kenya: Evidence from choice experimental data. *Journal of Arid Environments* 158, 9–18.

Bekele, K., Linders, T.E., Eschen, R., Shiferaw, H., Haji, J, *et al.* (2022) How well do local stakeholders' perceptions of environmental impacts of an invasive alien plant species relate to ecological data. *Ecological Indicators* 237, 108748.

Bond, W.J. (2019) *Open Ecosystems: Ecology and Evolution Beyond the Forest Edge*. Oxford University Press, Oxford, UK.

Bond, W.J., Stevens, N., Midgley, G.F. and Lehmann, C.E.R. (2019) The trouble with trees: Afforestation plans for Africa. *Trends in Ecology & Evolution* 34, 963–965.

Brundu, G., Pauchard, A., Pyšek, P., Pergl, J., Bindewald, A.M. *et al.* (2020) Global guidelines for the sustainable use of non-native trees to prevent tree invasions and mitigate their negative impacts. *NeoBiota* 61, 65–116.

Castillo, M.L., Schaffner, U., van Wilgen, B.W. and Le Roux, J.J. (2021a) The contribution of phenotypic traits, their plasticity and rapid evolution to invasion success: Insights from an extraordinary natural experiment. *Ecography* 44, 1035–1050.

Castillo, M.L., Schaffner, U., van Wilgen, B.W., Montaño, N.M., Bustamante, R.O. *et al.* (2021b) Genetic insights into the globally invasive and taxonomically problematic tree genus *Prosopis*. *AOB PLANTS* 13(1), laa069.

Choge, S., Mbaabu, P.R. and Muturi, G.M. (2022) Management and control of the invasive *Prosopis juliflora* tree species in Africa with a focus on Kenya. In: Puppo, M.C. and Felker, P. (eds) *Prosopis as a Heat Tolerant Nitrogen Fixing Desert Food Legume*. Academic Press, London, UK, pp. 67–81.

Conant, R.T. and Paustian, K. (2002) Potential soil carbon sequestration in overgrazed grassland ecosystems. *Global Biogeochemical Cycles* 16, 1143.

De Graaff, J., Amsalu, A., Bodnar, F., Kessler, A., Posthumus, H. *et al.* (2008) Factors influencing adoption and continued use of long-term soil and water conservation measures in five developing countries. *Applied Geography* 28, 271–280.

Eckert, S., Hamad, A., Kilawe, C.J., Linders, T.E., Ng, W.T. *et al.* (2020) Niche change analysis as a tool to inform management of two invasive species in Eastern Africa. *Ecosphere* 11, e02987.

Hasler, N., Williams, C.A., Denney, V.C., Ellis, P.W., Shrestha, S. *et al.* (2024) Accounting for albedo change to identify climate-positive tree cover restoration. *Nature Communications* 15, 2275.
Henderson, L. and Wilson, J.R. (2017) Changes in the composition and distribution of alien plants in South Africa: An update from the Southern African plant invaders Atlas. *Bothalia* 47, 1–26.
Higgins, S.I., Richardson, D.M. and Cowling, R.M. (2000) Using a dynamic landscape model for planning the management of alien plant invasions. *Ecological Applications* 10, 1833–1848.
Hirsch, G.H., Bradley, D. and Pohl, C. (2006) Implications of transdisciplinarity for sustainability research. *Ecological Economics* 60, 119–128.
Holling, C.S. (1978) *Adaptive Environmental Assessment and Management*. Wiley, New York, USA.
Jackson, R.B., Banner, J.L. and Jobbágy, E.G. (2002) Ecosystem carbon loss with woody plant invasion of grasslands. *Nature* 418, 623–626.
Jackson, R.B., Lajtha, K., Crow, S.E., Hugelius, G., Kramer, M.G. *et al.* (2017) The ecology of soil carbon: Pools, vulnerabilities, and biotic and abiotic controls. *Annual Review of Ecology, Evolution, and Systematics* 48, 419–445.
Kleinjan, C.A., Hoffmann, J.H., Heystek, F., Ivey, P. and Kistensamy, Y. (2021) Developments and prospects for biological control of prosopis (Leguminosae) in South Africa. *African Entomology* 29, 859–874.
Kumschick, S., Wilson, J.R.U. and Foxcroft, L.C. (2020) A framework to support alien species regulation. *NeoBiota* 62, 213–239.
Le Roux, J.J. (2023) Invasion genetics of Australian *Acacia* species. In: Richardson, D.M., Le Roux, J.J. and Marchante, E. (eds) *Wattles: Australian Acacia Species Around the World*. CABI, Wallingford, UK, pp. 93–102.
Linders, T.E., Bekele, K., Schaffner, U., Allan, E., Alamirew, T. *et al.* (2020) The impact of invasive species on social-ecological systems: Relating supply and use of selected provisioning ecosystem services. *Ecosystem Services* 41, 101055.
Linders, T.E., Schaffner, U., Alamirew, T., Allan, E., Choge, S.K. *et al.* (2021) Stakeholder priorities determine the impact of an alien tree invasion on ecosystem multifunctionality. *People and Nature* 3, 658–672.
Lukey, P. and Hall, J. (2020) Biological invasion policy and legislation development and implementation in South Africa. In: van Wilgen, B.W., Measey, J., Richardson, D.M., Wilson, J.R. and Zengeya, T.A. (eds) *Biological Invasions in South Africa*. Springer, Cham, Switzerland, pp. 515–551.
Mbaabu, P.R., Ng, W.T., Schaffner, U., Gichaba, M., Olago, D. *et al.* (2019) Spatial evolution of *Prosopis* invasion and its effects on LULC and livelihoods in Baringo, Kenya. *Remote Sensing* 11, 1217.
Mbaabu, P.R., Olago, D., Gichaba, M., Eckert, S., Eschen, R. *et al.* (2020) Restoration of degraded grasslands, but not invasion by Prosopis juliflora, avoids trade-offs between climate change mitigation and other ecosystem services. *Scientific Reports* 10, 20391.
Metternicht, G. (2017) Land use planning. *Global Land Outlook (Working Paper)* 2, 25–31.
Ng, W.T., Rima, P., Einzmann, K., Immitzer, M., Atzberger, C. *et al.* (2017) Assessing the potential of Sentinel-2 and Pléiades data for the detection of *Prosopis* and *Vachellia* spp. in Kenya. *Remote Sensing* 9, 74.
Parr, C.L., Te Beest, M. and Stevens, N. (2024) Conflation of reforestation with restoration is widespread. *Science* 383(6684), 698–701.
Poynton, R.J. (1977) *Tree Planting in Southern Africa: The Pines*. Department of Forestry, Pretoria, South Africa.
Reichard, S.H. and Hamilton, C.W. (1997) Predicting invasions of woody plants introduced into North America. *Conservation Biology* 11, 193–203.
Richardson, D.M. (1998) Forestry trees as invasive aliens. *Conservation Biology* 12, 18–26.
Rouget, M., Richardson, D.M., Nel, J.A. and Wilgen, B.W. (2002) Commercially important trees as invasive aliens – towards spatially explicit risk assessment at a National scale. *Biological Invasions* 4, 397–412.
Schierenbeck, K.A. and Ellstrand, N.C. (2009) Hybridization and the evolution of invasiveness in plants and other organisms. *Biological Invasions* 11, 1093–1105.
Seid, O., Haji, J. and Legesse, B. (2020) Rural households' perception on the effects of *Prosopis juliflora* invasion: The case of Amibara district of Afar National Regional State, Ethiopia. *Pastoralism* 10, 21.
Shackleton, R.T., Le Maitre, D.C., Pasiecznik, N.M. and Richardson, D.M. (2014) Prosopis: A global assessment of the biogeography, benefits, impacts and management of one of the world's worst woody invasive plant taxa. *AoB PLANTS* 6, plu027.

Shiferaw, H., Schaffner, U., Bewket, W., Alamirew, T., Zeleke, G. *et al.* (2019) Modelling the current fractional cover of an invasive alien plant and drivers of its invasion in a dryland ecosystem. *Scientific Reports* 9, 1576.

Sun, Y., Müller-Schärer, H. and Schaffner, U. (2022) Fighting neobiota with neobiota: Consider it more often and do it more rigorously. *Biological Conservation* 268, 109506.

United Nations (2019) *Global Sustainable Development Report 2019: The Future is Now - Science for Achieving Sustainable Development*. United Nations, New York,USA.

van Klinken, R.D. (2012) *Prosopis* spp. mesquite. In: Julien, M. and McFadyen, R.E. (eds) *Biological Control of Weeds in Australia*. CSIRO Publishing, Canberra, Australia, pp. 477–485.

van Wilgen, B.W., Richardson, D.M. and Higgins, S.I. (2001) Integrated control of invasive alien plants in terrestrial ecosystems. *Land Use and Water Resources Research* 1, 1732–2016.

van Wilgen, B.W., Moran, V.C. and Hoffmann, J.H. (2013) Some perspectives on the risks and benefits of biological control of invasive alien plants in the management of natural ecosystems. *Environmental Management* 52, 531–540.

van Wilgen, B.W., Wilson, J.R., Wannenburgh, A. and Foxcroft, L.C. (2020a) The extent and effectiveness of alien plant control projects in South Africa. In: Wilgen, B.W., Measey, G.J. and Richardson, D.M. (eds) *Biological Invasions in South Africa*. Springer, Cham, Switzerland, pp. 581–613.

van Wilgen, B.W., Raghu, S., Sheppard, A.W. and Schaffner, U. (2020b) Quantifying the social and economic benefits of the biological control of invasive alien plants in natural ecosystems. *Current Opinion in Insect Science* 38, 1–5.

Veldman, J.W., Overbeck, G.E., Negreiros, D., Mahy, G., Le Stradic, S. *et al.* (2015) Where tree planting and forest expansion are bad for biodiversity and ecosystem services. *BioScience* 65, 1011–1018.

Wiesmann, U., Ott, C., Ifejika Speranza, S., Kiteme, B.P., Müller-Böker, U. *et al.* (2011) A human actor model as a conceptual orientation in interdisciplinary research for sustainable development. In: Wiesmann, U. and Hurni, H. (eds) *Research for Sustainable Development: Foundations, Experiences, and Perspectives*. University of Bern, Bern, Switzerland, pp. 231–256.

Index